W0107479

Photochemical Probes in Biochemistry

NATO ASI Series

Advanced Science Institutes Series

*A Series presenting the results of activities sponsored by the NATO Science Committee,
which aims at the dissemination of advanced scientific and technological knowledge,
with a view to strengthening links between scientific communities.*

The Series is published by an international board of publishers in conjunction with the
NATO Scientific Affairs Division

A Life Sciences	Plenum Publishing Corporation
B Physics	London and New York
C Mathematical	Kluwer Academic Publishers
and Physical Sciences	Dordrecht, Boston and London
D Behavioural and Social Sciences	
E Applied Sciences	
F Computer and Systems Sciences	Springer-Verlag
G Ecological Sciences	Berlin, Heidelberg, New York, London,
H Cell Biology	Paris and Tokyo

Series C: Mathematical and Physical Sciences - Vol. 272

Photochemical Probes in Biochemistry

edited by

Peter E. Nielsen
Department of Biochemistry B, The Panum Institute,
University of Copenhagen, Copenhagen, Denmark

Kluwer Academic Publishers

Dordrecht / Boston / London

Published in cooperation with NATO Scientific Affairs Division

Proceedings of the NATO Advanced Research Workshop on
Photochemical Probes in Biochemistry
Copenhagen, Denmark
14–19 August 1988

Library of Congress Cataloging in Publication Data

NATO Advanced Research Workshop on Photochemical Probes in
  Biochemistry (1988 : Copenhagen, Denmark)
    Photochemical probes in biochemistry : proceedings of the NATO
  Advanced Research Workshop on Photochemical Probes in Biochemistry,
  Copenhagen, Denmark, August 14-19, 1988 / edited by Peter E.
  Nielsen.
      p.   cm. -- (NATO ASI series. Series C, Mathematical and
  physical sciences ; vol. 272)
    Includes indexes.
    ISBN-13:978-94-010-6905-2   e-ISBN-13:978-94-009-0925-0
    DOI: 10.1007/978-94-009-0925-0

    1. Photoaffinity labeling--Congresses. 2. Biochemistry-
  -Congresses.   I. Nielsen, Peter E., 1951-   . II. Title.
  III. Series: NATO ASI series. Series C, Mathematical and physical
  sciences ; no. 272.
  QP519.9.P48N37 1988
  574.19'285--dc19   89-31038
                                                        CIP

ISBN-13:978-94-010-6905-2

Published by Kluwer Academic Publishers,
P.O. Box 17, 3300 AA Dordrecht, The Netherlands.

Kluwer Academic Publishers incorporates the publishing programmes of
D. Reidel, Martinus Nijhoff, Dr W. Junk and MTP Press.

Sold and distributed in the U.S.A. and Canada
by Kluwer Academic Publishers,
101 Philip Drive, Norwell, MA 02061, U.S.A.

In all other countries, sold and distributed
by Kluwer Academic Publishers Group,
P.O. Box 322, 3300 AH Dordrecht, The Netherlands.

printed on acid free paper

All Rights Reserved
© 1989 by Kluwer Academic Publishers
Softcover reprint of the hardcover 1st edition 1989

No part of the material protected by this copyright notice may be reproduced or
utilized in any form or by any means, electronic or mechanical, including photo-
copying, recording or by any information storage and retrieval system, without written
permission from the copyright owner.

TABLE OF CONTENTS

PREFACE

The concept of using photochemical probes in the study of biological systems was developed by Westheimer who published the first photoaffinity labeling experiments more than twenty years ago (J.Biol. Chem. 237, (1962) 3006). Since then the concept has been used successfully in various areas of biochemistry and recently several new interesting and exciting aspects of the concept have been developed. It is the general opinion by scientists in the "field" that the full potential of photochemical probes in biochemical studies has far from been exploited yet. This is mostly due to the interdisciplinary character of the concept involving photochemistry, synthetic chemistry as well as biochemistry/molecular biology.

The perspective of the NATO advanced workshop on "Photochemical Probes in Biochemistry", held in Holte (Copenhagen) Denmark 14-19, August, 1988, was several fold. The workshop was to give an account of the "state of the art" of using photochemical probes in biochemistry as well as to bring together specialists in photochemistry, synthetic chemistry and molecular biology in order to analyze advantages as well as the inherent problems and pitfalls of the concept and provide suggestions and guidelines for future research.

Furthermore, it is the hope of the editor that the present publication which gives an account of the lectures presented at the workshop, will provide an introduction to scientists who are not familiar with photochemical probes, but to whom these could help answer central and pertinent questions.

I hope this book will serve as a source of general information about photochemical probes in biochemistry, and an entrance to the relevant original literature as well as a source of inspiration for novel applications and experimental designs.

This workshop and the present publication was generously supported by NATO Division of Scientific Affairs. I also want to thank the participants of the workshop and the contributors to this publication for their enthusiasm and cooperation. Support from Boehringer Mannheim and Amersham, Denmark, is also acknowledged.

Peter E. Nielsen
August, 1988

PHOTOLABELING REAGENT DESIGN

Ole Buchardt, Ulla Henriksen, Troels Koch, and
Peter E. Nielsen (a)
Research Center for Medical Biotechnology, The
H. C. Ørsted Institute, University of Copenha-
gen, Universitetsparken 5, DK-2100 Copenhagen
Ø, Denmark; (a) Department of Biochemistry B,
The Panum Institute, Blegdamsvej 3, DK-2200
Copenhagen N, Denmark

ABSTRACT. Photolabeling reagents are composed of some or
all of the following ligands: Photoprobes, thermal probes,
labels, linkers, and cleavable sites. The design and function of such ligands are discussed on the basis of our
own attempts to design and test novel ligands and reagents. Novel ligands include diazocyclopentadienylcar-
bonyloxy photoprobes, biotinyl labels and selenium containing linkers which can be cleaved by very mild oxidation.

INTRODUCTION

The use of photochemically activated reagents in biology
and medicine has steadily gained importance, and considering their wealth of possibilities we can expect to see
further increase in the range and application of photochemistry as a tool in these areas. They can be divided
into (i) photolabeling, (ii) photoaffinity labeling, and
(iii) photocrosslinking reagents, and are composed of some
or all of the following ligands (Scheme 1): Photoprobes
(P), thermal probes (T), label (L), and a linker which may
contain a cleavable ligand (X).
    (i) Photolabeling reagents are used to attach a label
(a tag) on biological (macro)molecules and are usually
composed of a photoprobe and a label, often connected via
a linker, but the label may also be in the form of radio-
isotope substitution on the photoprobe.
    (ii) Photoaffinity labeling reagents, which are used
to identify and analyze receptors on biological macromolecules by photochemical attachment of a label to the target. They are as a rule modified receptor substrates, and

1

P. E. Nielsen (ed.), Photochemical Probes in Biochemistry, 1–9.
© *1989 by Kluwer Academic Publishers.*

for low molecular substrates, their design and syn-thesis are too varied to allow any generalization and shall not be dealt with here. When the substrate is large or macro-molecular, reagents of a more general nature can be used to modify them. Such reagents contain as a rule one (or photoprobe, a thermal probe (or a second photoprobe), a linker and a label.

(iii) Photocrosslinking reagents are used to study interaction between two or more (macro)molecules. They usually contain one thermal probe and one photoprobe (or two photoprobes) bound together by a linker, which may be cleavable. Furthermore, they often are required to be supplied with a label.

The present paper contains a general discussion of the design and utility of reagents for photolabeling, photoaffinity labeling, and photocrosslinking, with par-ticular emphasis on the authors's own work in this area, *i.e.*, with protein-DNA photoaffinity labeling and photo-crosslinking, protein-protein photocrosslinking, and DNA photolabeling reagents.

1. Reagent design

A priori, only the shortcomings of the synthetic chemist limit the design of reagents, and we will undoubtedly see many elegant and ingenious designs in the future.

In principle, the synthetic chemist is asked to com-bine a certain number of ligands in such a way that a certain number of chemical events can be timed in the resulting reagent (*e.g.*, Scheme 1). Furthermore, the re-

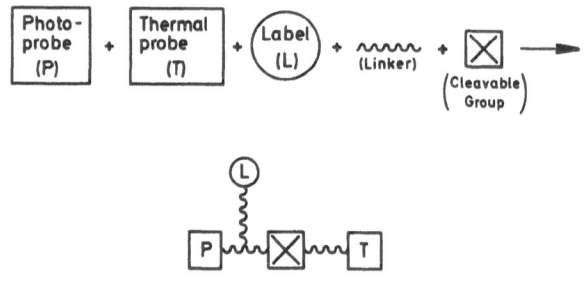

Scheme 1

agent must be sufficiently soluble in water, and the pho-toprobe must still be activable with light of a suitable wavelength ($\lambda \gg 300$ nm is preferable, but it is not necessarily prohibitive to use light of a shorter wave-length in some cases). Most important, the efficiency of the reagents must be good; but what is efficiency?

For the photochemist, efficiency will as a rule be understood to be quantum yield. However, due to the complicated system which often is irradiated in biological experiments, even quite low quantum yields do not necessarily militate against using a probe. Efficiency in the present context is amount of probe attached to an otherwise unharmed or little harmed target.

2. The photoprobe

A number of photoprobes have been described (Scheme 2), some of which are commercially available in the form of reactive derivatives, like, e.g., active esters or active halides.

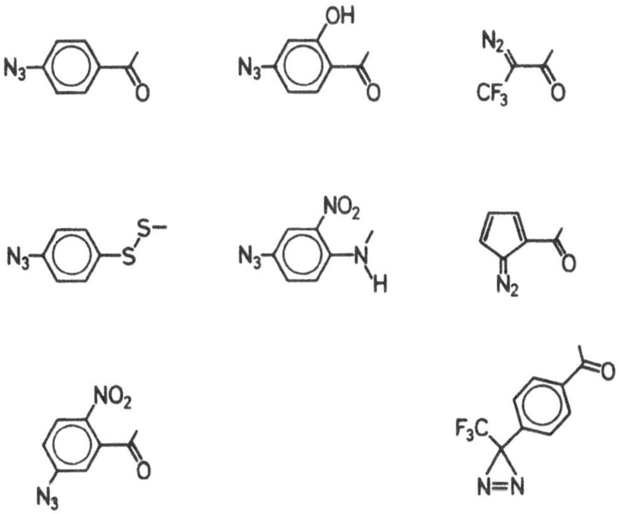

Scheme 2.   Aryl azido, diazo and diazirino photoprobes.

In practice, by far the most work has been based on aryl azides with diazo compounds as the second most used type. In a few cases, ketones (Chapter 10) or nitromethoxyphenyl derivatives (chapter 2) have been used.
    It is frequently stated that both azido and diazo photoprobes react with the target via nitrene or carbene species. While this makes good sense in the latter case, it has been shown to be fallaceous in the former, where all evidence now indicates that the labeling species are azacycloheptatetraene derivatives, which react with nucleophilic functions like -NH$_2$, =NH, -SH or -OH.[4] Furthermore, mounting evidence also strongly indicates that

only singlet excited azidoaryl photoprobes lead to covalent binding to any large degree [1,2] (cf. Chapter 3
(G. Schuster)).

Thus, we have in the azidoaryl probes a rather select
reactivity, whereas the carbenes generated from diazo or
diazirino probes must be assumed to react in situ by
either insertion or addition [3]. However, no experimental
evidence on biological systems or reliable models has been
forthcoming. Furthermore, it is uncomfortable for the
designer of photoreagents that virtually all experiments
on the photochemistry of aryl azides and diazo compounds
have been performed in aprotic solvents or in a few cases
alcohols. Thus photochemical studies of the behavior of
aryl azides and diazo compounds in buffered aqueous solutions are highly warranted.

We have tested a number (Table 1) of photoprobes for
their covalent binding to protein, DNA and RNA in a model
system in which the photoprobe was attached to a 9-amino-
acridinyl ligand [4]. The acridinyl ligand served both as
an "equalizer" of physicochemical properties and as a
fluorescence label.

Table 1.

Probe	Labeling Spec	Labeling efficiency, %, $\lambda > 300$ nm			
		BSA	HISTONES	RNA	DNA
$RN(H)$–⟨⟩–N_3, NO_2	?	9 ± 3	17	8	10 ± 1
$RN(H)$–C(O)–⟨⟩–N_3	(pyridyl ketone)	20 ± 11	35	40	18 ± 5
$RN(H)$–C(O)–C(N_2)–CO_2Et	(carbene CO_2Et)	0.8 ± 0.7	0.4	0.25	–
RO–C(O)–(cyclopentadiene N_2)	(cyclopentadienylidene)	9 ± 2	13	25	19 ± 2
RO–C(O)–(cyclopentadiene N_2)	(cyclopentadienylidene)	5	2	8	–

$R =$ (hexyl chain)–NH–(9-aminoacridinyl) $\xrightarrow{H_2O/OH^-}$ (acridone)

These results show that the 4-azidobenzoyl ligand is a good choice for most purposes and although this ligand does not absorb strongly above 300 nm, photoactivation is easily accomplished using Pyrex filtered UV light (λ > 300 nm) [4]. Furthermore, we have observed that photoactivation by light absorbed by the acridinyl chromophore ($\lambda \sim$ 400 nm) takes place, probably via an electron transfer mechanism [5].

3. The thermal probe

So far, N-hydroxysuccinimide (NHS) esters have been used most in practice when amino groups were available in the target. Other reactive esters have had a more limited use, and thiol specific probes are also known (Scheme 3).

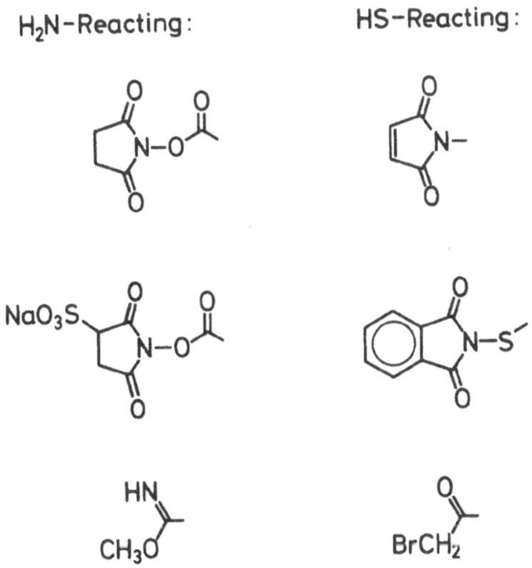

Scheme 3.   Thermal probes.

However, there are good reasons to search for new, preferentially selective thermal probes.

4. The label

For very high sensitivity, radioactivity is still to be preferred as the label, and it will undoubtedly be the choice for many future applications. The ideal radioactive label is only inserted synthetically at the very last moment, e.g., immediately before the final analytical step (cf. chapter 11 (M. A. Schwartz)). Furthermore, UV absorbance and better yet, fluorescence, can be used, and for many applications fluorescence with a sufficiently high quantum yield is excellent. We have very often placed a 9-acridinylamino ligand in our reagents not only in order to secure DNA-binding by intercalation, but also because it can be hydrolyzed in almost quantitative yield to acridone, cf. Table 1, which has a quantum yield of almost 1 [4].

We are presently attempting to insert biotin [6] into various reagents, the results of which so far look promising (see also below).

5. The linker

The linker is required to join the various components of the reagent, and is used to obtain the optimal distances (and configuration in the case of non-flexible linkers) between the ligands. The composition of the linker can also greatly influence the physical properties of the reagent. Thus hydrophilic linkers may confer water solubility to a reagent.

Introducing cleavability to a linker has mainly been done using disulphides which are cleaved by mild reduction (which unfortunately also cleaves disulfide bridges holding protein subunits together), 1,2-diols which are cleaved oxidatively with periodate or an ester linkage, which can be cleaved by hydrolysis. The oxidative or hydrolytic treatment may, however, be deteriorating to both carbohydrates of glycosylated proteins as well as to the peptides themselves.

We have tried to utilize the well known selenide-oxidation-elimination principle in crosslinking with some success [7].

Our initial reagent (1, Scheme 4) is excellent for crosslinking and it is cleaved under extremely mild oxidative conditions, thereby making it the reagent of choice for problems where less mild oxidation, reduction or hydrolysis is useless or unpractical. Having had nice results with the first example, we proceeded to make an extended series of compounds, most of which unfortunately were not very efficiently oxidatively cleaved.

However, most recently better results have been obtained, and we now have in hand an improved thermal crosslinking reagent (2) with excellent cleaving abilities by mild oxidation as well as a photoprobe-thermal probe analog (3) and the selenium bridge concept is being further pursued.

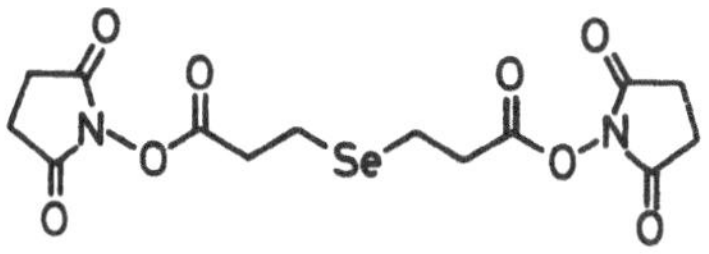

1

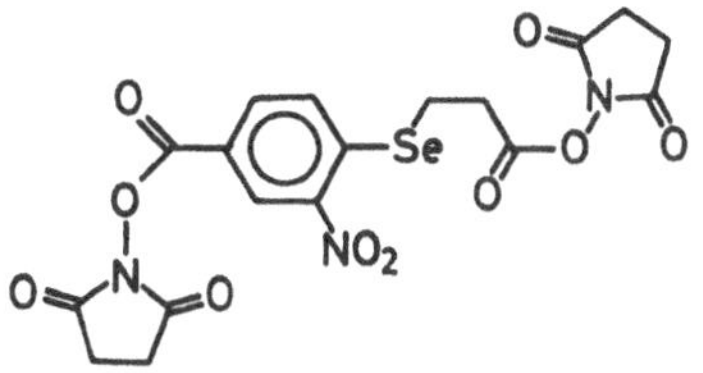

2

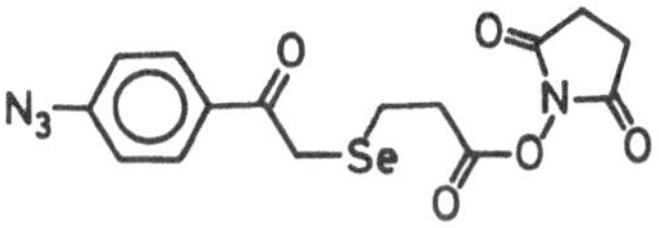

3

Scheme 4

6. Photobiotinylation reagents

Biotinylation is becoming an important alternative to radioactive labeling, especially with nucleic acid hybridization probes. Photobiotinylation reagents based on azido-nitro-anilino- or trimethylpsoralen ligands have been described [9-11] and are currently being used for biotinylation of nucleic acids.

Exploiting our experience with various DNA-photoprobes (cf. chapter 16, O. Buchardt et al.) we have developed several photobiotinylation reagents (Scheme 5) which photoreact efficiently with DNA (Table 2), and work is now in progress to optimize the conditions for their use in nucleic acid hybridization probes. Furthermore, reagent **6** is quite efficient for photobiotinylation of proteins.

Table 2

Reagent	Relative biotinylation of DNA	Relative biotinylation of Protein
4	0.65^{+}	<0.01
5	0.70	0.01
6	1.0	1.0
7	0.22	<0.01

$^{+}$Using a ^{3}H-derivative of **4** it was found that this biotinylation corresponds to an efficiency of ~ 25% of the added reagent.

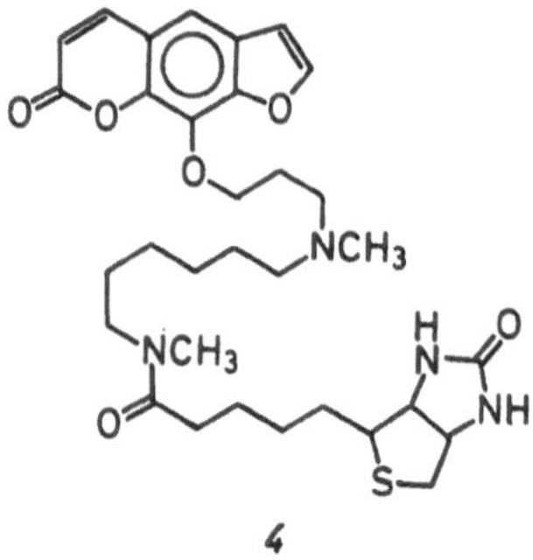

Scheme 5

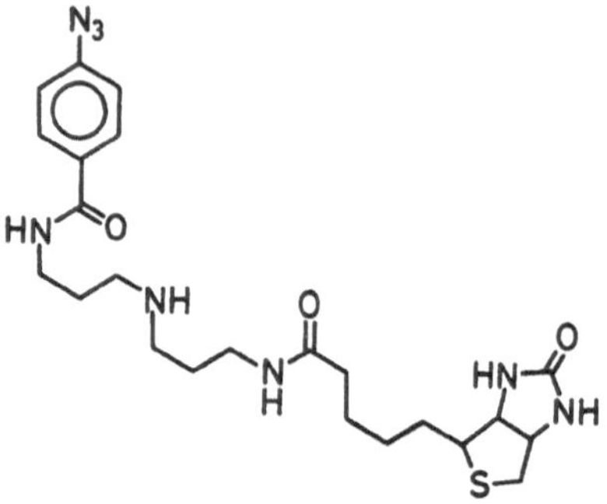

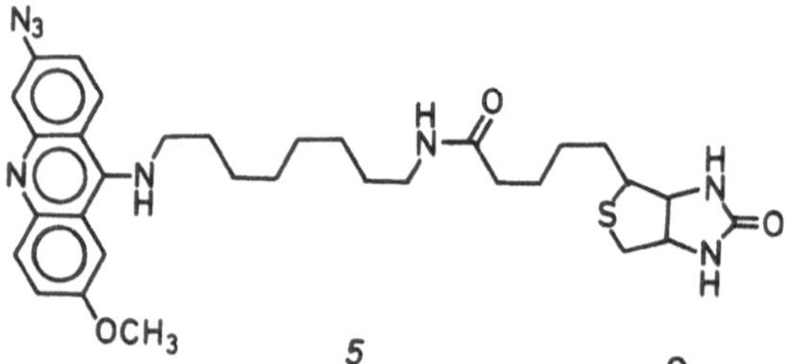

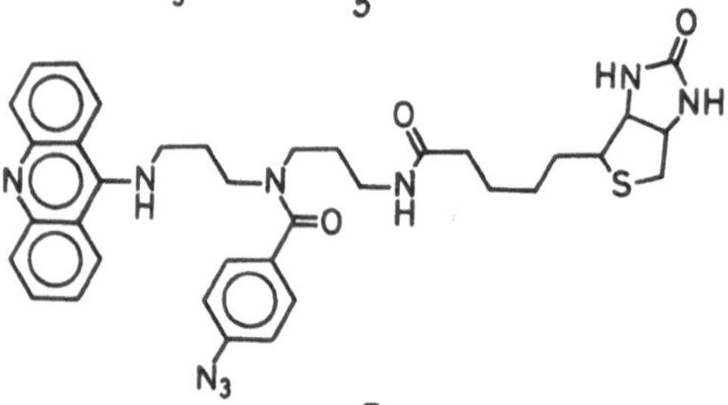

References

1. P. E. Nielsen and O. Buchardt, *Photochem. Photobiol.* **35**, 317-323 (1982).
2. Shields, C. J., Chrisope, D. R., Schuster, G. B., Dixon, A. J., Poliakoff, M. and Turner, J. J. (1987) *J. Am. Chem. Soc.* **109**, 4723-4726.
3. *"Diazo Compounds. Properties and Synthesis"*, ed. M. Regitz & G. Maas, Academic Press 1986.
4. P. E. Nielsen, J. B. Hansen, T. Thomsen, and O. Buchardt, *Experientia*, **39**, 1063-1072 (1983).
5. C. J. Shields, D. E. Falvey, G. B. Schuster, O. Buchardt and P. E. Nielsen, *J. Org. Chem.*, in press.
6. U. Henriksen, O. Buchardt, and P. E. Nielsen. In preparation.
7. E.g., D. Liotta, *"Organic Selenium Chemistry"*, Wiley, New York 1987.
8. O. Buchardt, H. I. Elsner, P. E. Nielsen, L. C. Petersen, and E. Suenson, *Anal. Biochem.* **158**, 87-92 (1986).

9.   Forster, A. C., McInnes, J. L., Skingle, D. C. and Symons, R. H. (1985) *Nucl. Acids Res.* **13**, 747-761.
10. Sheldon, E. L., Kellogg, D. E., Watson, R., Levenson, C. H. and Ehrlich, H. A. (1986) *Proc. Natl. Acad. Sci. USA* **83**, 9085-9089.
11. Saffran, W. A., Welsh, J. T., Knobler, R. M., Gasparo, F. P., Cantor, C. R. and Edelson, R. L. (1988) *Nucl. Acids Res.* **16**, 7221-7231.

NUCLEOPHILIC AROMATIC PHOTOSUBSTITUTIONS ON NITROPHENYL ETHERS. A NEW PHOTOAFFINITY LABELLING TECHNIQUE?

Jorge Marquet and Marcial Moreno-Mañas
Department of Chemistry.
Universidad Autónoma de Barcelona
Bellaterra. 08193-Barcelona.
Spain

ABSTRACT. The nucleophilic photosubstitution of alkoxy and other leaving groups in nitrophenyl ethers is a potentially useful reaction in photoaffinity labelling experiments. A mechanistic study of the photoreactions of several nitrophenyl ethers with different amines and hydroxide ion is presented. Three different mechanistic pathways have been identified for the photosubstitutions of 4-nitroveratrole. The photoreaction with hydroxide ion takes place in meta position with respect to the nitro group by a mechanism involving the excited triplet state ($S_N2^3Ar^*$). The photosubstitution with n-hexylamine occurs mainly in meta position through the singlet excited state ($S_N2^1Ar^*$). Finally, the photosubstitution with piperidine occurs in para position and the mechanism involves an electron transfer from the amine to the excited triplet state. The different regioselectivities observed with different amines depend upon the ionization potential of the amines.

1. INTRODUCTION

Photoaffinity labelling has been used for the identification of receptor sites in biological systems (1). The technique consists in the modification of the biologically active molecule (BAM) by means of an auxiliary group (G) that can be irreversibly linked to the receptor site (RS) and further localized. Two different strategies can be visualized (Scheme 1): 1) cross-linking in which a covalent bond is formed between the receptor and the auxiliary group in such a way that the BAM, the group G and the receptor site remain covalently linked and 2) label transfer in which the BAM is released in the act of photoaffinity labelling.

Some conditions should be fulfilled for the photoaffinity labelling technique to be successful: 1) the modified BAM must retain part of its original biological activity, 2) the reaction must be able to be triggered at will and 3) the label should be able to be detected in the modified receptor site or in one of its fragments. Photochemical reactions are very useful in fulfilling the second of the mentioned conditions

11

P. E. Nielsen (ed.), Photochemical Probes in Biochemistry, 11–29.
© 1989 by Kluwer Academic Publishers.

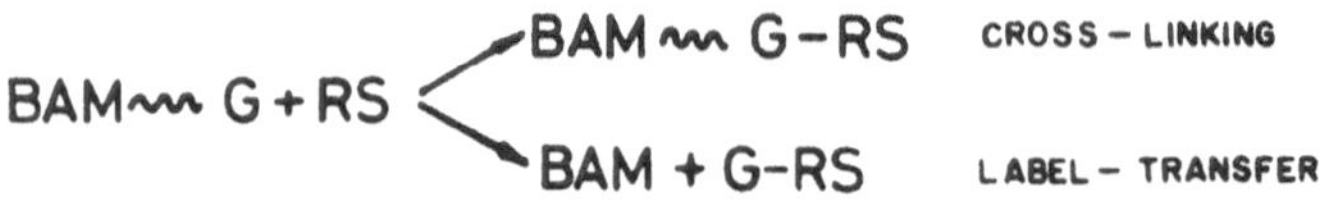

BAM∿ G + RS ⟨ BAM∿ G–RS   CROSS – LINKING
              BAM + G–RS   LABEL – TRANSFER

SCHEME 1.- General strategy for photoaffinity labelling.

The photochemically active auxiliary groups G so far used in the
cross-linking version are diazocompounds, azides and aromatic ketones.
However, new more selective and thermically stable groups are sought by
biochemists.

Havinga and his coworkers in Leiden described more than 20 years
ago several examples of the so called nucleophilic aromatic
photosubstitution on nitrophenyl ethers (2). Some of them are given in
scheme 2. Thus, 4-nitroanisole reacts under irradiation (> 300 nm) with
methylamine to afford chemically useful yields of N-methyl-4-
nitroaniline. 4-Nitroveratrole affords two different substitution
products: 2-methoxy-N-methyl-5-nitroaniline (major) and 2-methoxy-N-
methyl-4-nitroaniline (minor) in a 9:1 ratio. The major product comes
from substitution of the methoxy group in <u>meta</u> position with respect to
the nitro group whereas the minor product comes from substitution of the
methoxy group in <u>para</u> position.

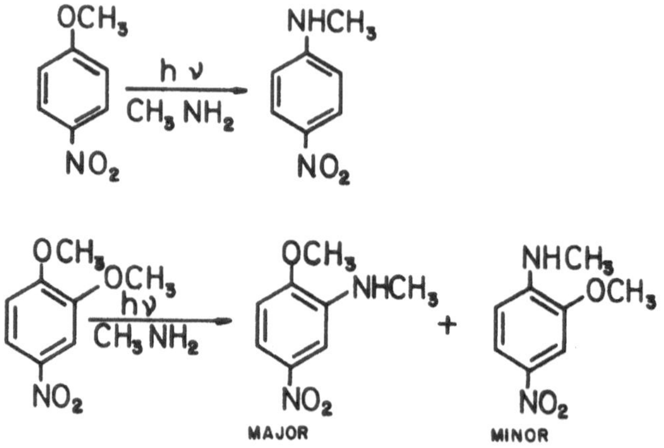

SCHEME 2.- Some examples of nucleophilic aromatic photosubstitution
described by the group of Havinga at Leiden.

These results originated the proposal by C. Cantor at Columbia University of using nitrophenyl ethers as possible photoaffinity labels (3). We have reformulated this proposal in scheme 3. Although the use of nitrophenyl ethers in photoaffinity labelling seems atractive it has not met with general acceptance (4).

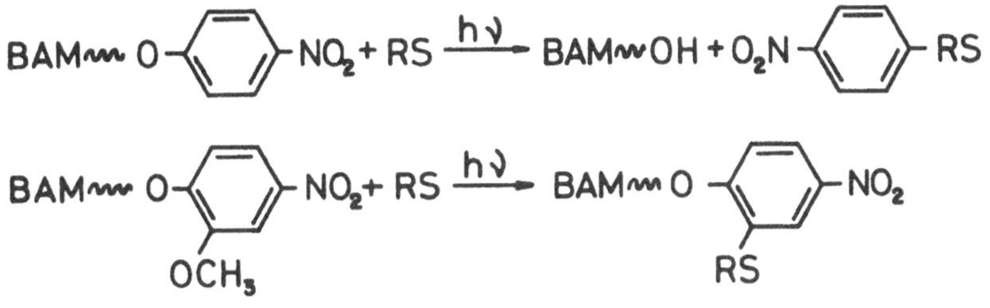

SCHEME 3.- Reformulation of the original Cantor proposal to use nitrophenyl ethers for photoaffinity labelling.

We became interested in this field through the approach of biochemists colleagues who suggested us to label the antibiotic cycloheximide for photoaffinity labelling purposes. We were successful in preparing some labelled derivatives of the antibiotic as well as of other active substrates such as the steroid estrone and the intercalating agent 9-aminoacridine (5).

However, at the beginning of our contribution we felt soon that a better general understanding of the nucleophilic aromatic photosubstitutions was required in order to use them for photoaffinity labelling purposes. Indeed, a revision of the literature revealed that the available information about the photochemical reactions of our model substrates: nitroanisoles, 4-nitroveratrole and the like in front of nucleophiles was clearly insufficient. Therefore, we undertook a research aimed at defining the scope and limitations of the nucleophilic aromatic photosubstitutions on nitrophenyl ethers with amines and the mechanistic aspects of the photoreactions involved.

2. RESULTS

2.1. The photoreactions of nitroanisoles and 4-nitroveratrole with amines. The regioselectivity problem

Our initial results (6) and those of the Havinga group at Leiden are collected in scheme 4. We first tried to extent the nucleophilic aromatic photosubstitution to a broader selection of amines. Havinga had reported the photosubstitutions in 3-nitroanisole with ammonia and with methylamine in aqueous medium as indicated. However, in our hands the photoreaction of 3—nitroanisole with n—butylamine afforded 3—nitrophenol as the only identified product. This compound arises from attack by the hydroxide ion present by hydrolysis of the amine.

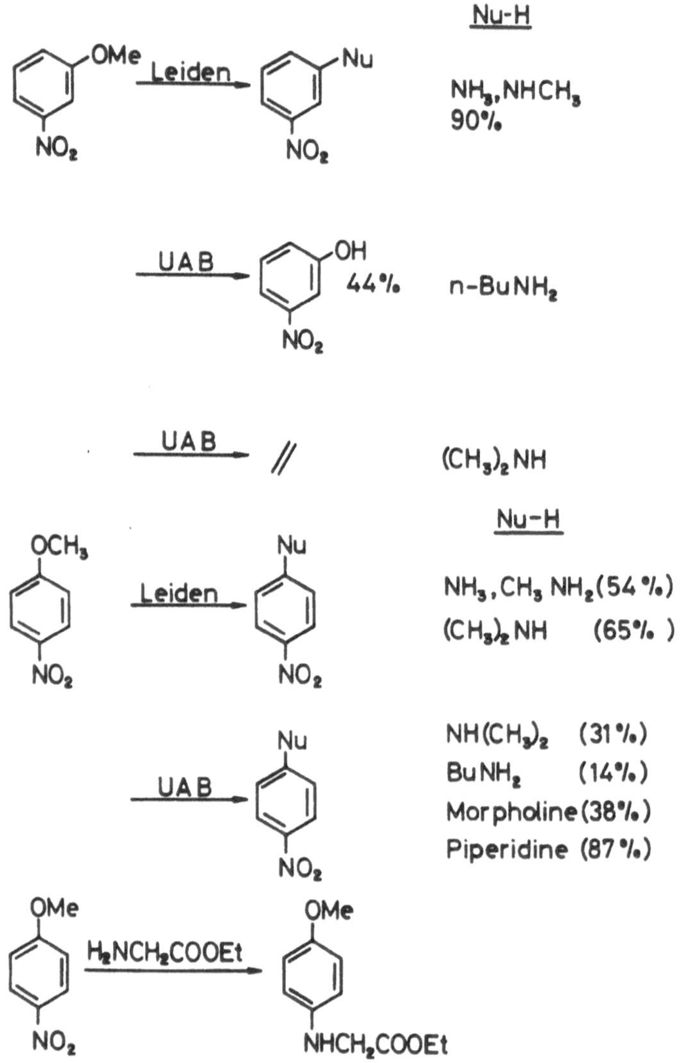

Nu-H

NH$_3$, NHCH$_3$
90%

UAB 44% n-BuNH$_2$

UAB // (CH$_3$)$_2$NH

Nu-H

NH$_3$, CH$_3$NH$_2$ (54%)
(CH$_3$)$_2$NH (65%)

NH(CH$_3$)$_2$ (31%)
BuNH$_2$ (14%)
Morpholine (38%)
Piperidine (87%)

SCHEME 4.- Nucleophilic aromatic photosubstitutions on 3- and 4-nitroanisoles with several amines. Results obtained at Leiden (Havinga) and at "Universidad Autónoma de Barcelona" (Marquet and Moreno-Mañas). General conditions: hν > 300nm; water or water-methanol or water-acetonitrile.

Moreover, no defined products were obtained from the reaction of dimethylamine with the same substrate.

When working with 4-nitroanisole we could reproduce the results of Havinga and coworkers with dimethylamine and also we observed clean photosubstitutions with n-butylamine, morpholine and piperidine (Scheme 4). No results were obtained with aniline and the use of ethyl glycinate led to an unexpected outcome: the substitution of the nitro group.

More interesting were the results obtained with 4-nitroveratrole (Scheme 5). Havinga and coworkers noticed that their photoreactions with ammonia and with methylamine produced regioselective meta-photosubstitution products and this was attributed to a charge control of the reactions, the charge densities in meta and para positions being inverted in the ground and in the excited state. Our own calculations using the semiempirical MINDO/3 method confirmed the inversion in charge densities (6). We could nicely reproduce the results of Havinga working with methylamine. However, when we performed experiments with other amines quite unexpected results were obtained. Thus, the photoreaction with morpholine produced only the para-photosubstitution product. Moreover, the reaction with dimethylamine produced only the aniline in which the dimethylamino group was in para position with respect to the nitro group. Some substitution in meta was also observed but the incoming nucleophile was the hydroxide ion always present as a result of the amine hydrolysis. Apparently we were dealing with two different types of amines: one type had propensity to attack the meta position with respect to the nitro group and a second type had propensity to attack the para position. The hydroxide ion can be categorized with the first type of amines. Finally, n-butylamine seemed to be in the borderline since its photoreaction with 4-nitroveratrole permited us to isolate both isomeric anilines in similar yields. It became soon clear that this effect was related to the ionization potential of the amines and of the nucleophiles in general. Thus, amines with a relatively high ionization potential produced meta attack as observed already by Havinga but amines with a relatively low ionization potential produced attack at the para position. This change in regioselectivity indicated a variation in the mechanism and we decided to study further this problem.

Alkoxy groups different from methoxy are going to be present in any photoaffinity probe. Therefore we studied also the photochemical reactions of several dialkoxynitrobenzenes with methylamine (Scheme 6). Clean meta photosubstitutions were obtained in all the cases studied (5).

As it was mentioned earlier we prepared some photoactive derivatives of the antibiotic cycloheximide and of the steroid estrone (Scheme 7). Their photoreactions with methylamine afforded again clean meta photosubstitutions (7). These photoreactions can be considered as models of the situation faced by the biologically active molecules in photoaffinity labelling experiments.

The use of 2-fluoro-4-nitroanisole produced similar results. Efficient photosubstitutions were observed with several amines although a borderline between two different orientations exists. n-Butylamine again gives rise to mixtures of photosubstitution products coming from meta and para attacks (Scheme 8).

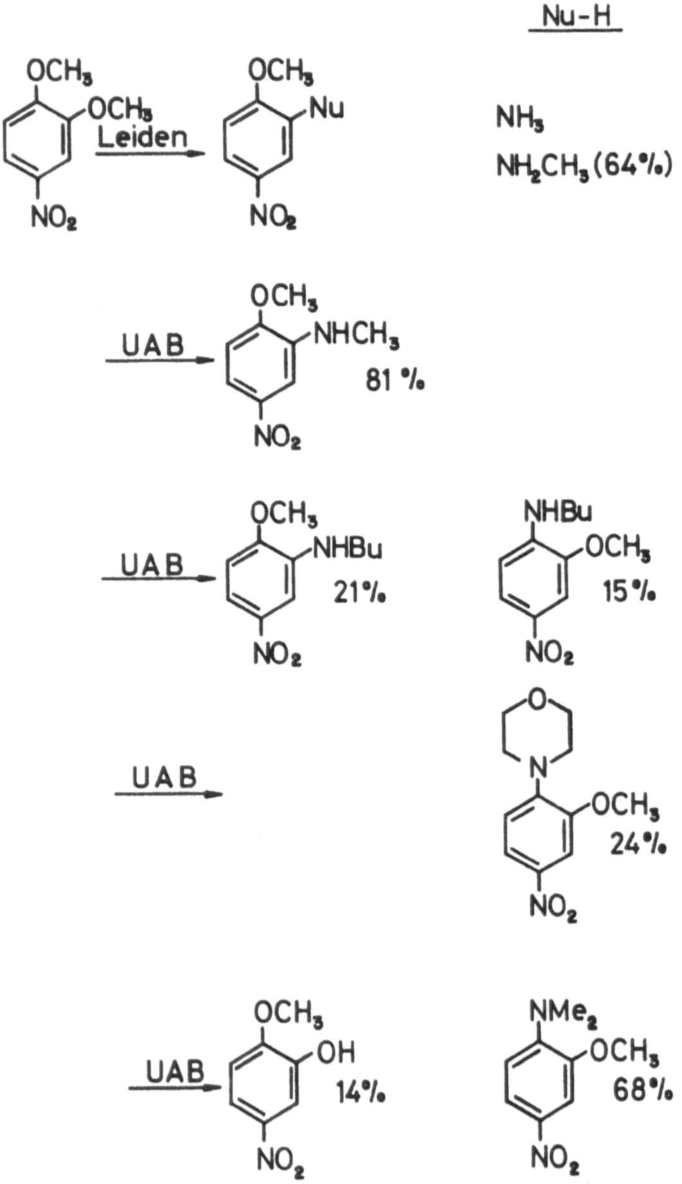

SCHEME 5.- Nucleophilic aromatic photosubstitutions on 4-nitroveratrole with several amines. Results obtained at Leiden (Havinga) and at "Universidad Autónoma de Barcelona" (Marquet and Moreno-Mañas). General conditions: hν > 300nm; water or water-methanol.

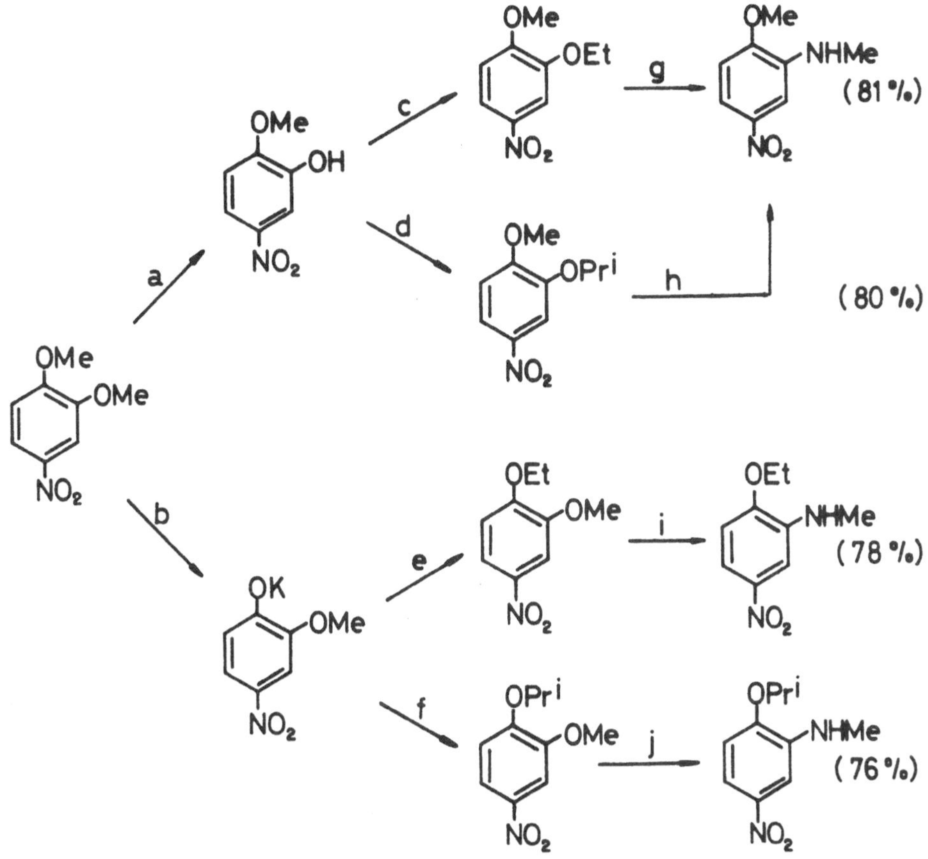

a: hν/NaOH/H₂O/15h.   b: KOH/H₂O/ref./19 h.   C: Br Et/Na₂CO₃/acetone/ref.

d: Br Pr^i/Na₂CO₃/acetone/ref.   e: Br Et/DMF/ref./4h.   f: Br Pr^i/DMF/ref./4 h.

g,h,i,j: H₂O/MeOH/hν/4h/MeNH₂

SCHEME 6.- Nucleophilic aromatic photosubstitutions on 3,4-dialkoxynitrobenzenes

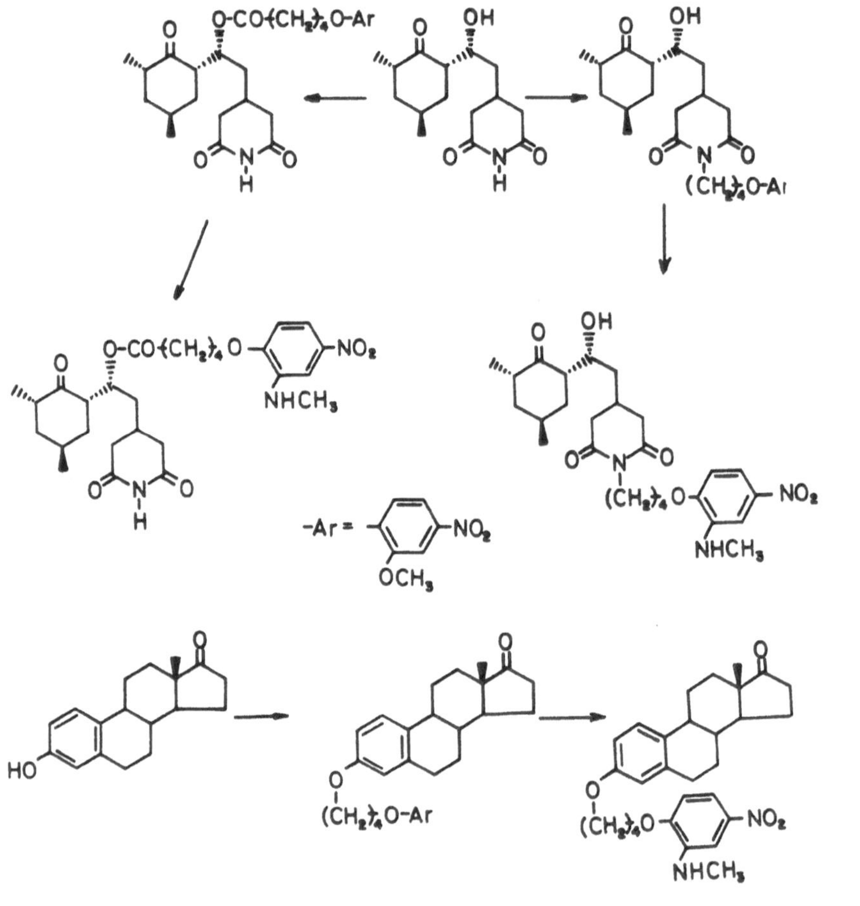

SCHEME 7.- Nucleophilic aromatic photosubstitutions on photoactive derivatives of cycloheximide (top) and estrone (bottom). General conditions: hν > 300nm; water-acetonitrile for cycloheximide derivatives and DMF-water for the estrone derivative.

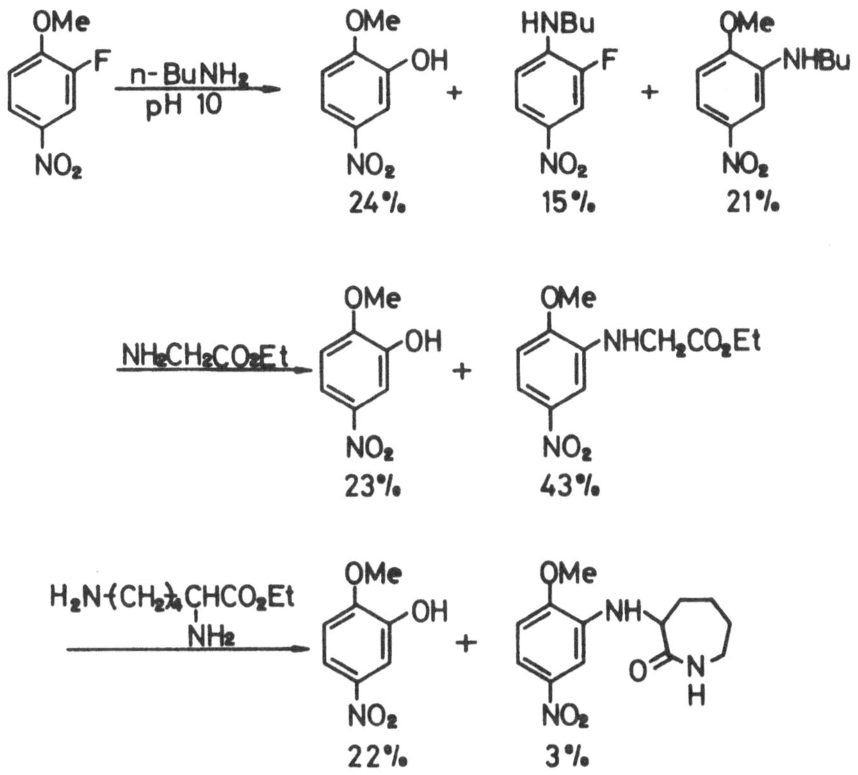

SCHEME 8.- Nucleophilic aromatic photosubstitutions on 2-fluoro-4-nitroanisole. General conditions: hν > 300nm; water-methanol or water-acetonitrile.

2.2 Photoreactions of 4,5-dinitroveratrole. The competence from photoreduction

In order to avoid the regioselectivity problems experienced with 4-nitroveratrole we next studied the photoreactions of 4,5-dinitroveratrole in which both methoxy groups have identical reactivity (8). The aromatic substrate was indeed active in photosubstitution reactions affording satisfactory yields of the corresponding anilines with metylamine and ethyl glycinate (Scheme 9 and Table I). However, the photoreaction with n-butylamine afforded not only the photosubstitution aniline but also a significant amount of reduction product 5,6-dimethoxybenzofurazane, **1**. When this experiment was repeated in the presence of benzophenone (triplet sensitizer) the ratio substitution/reduction products showed no variation.

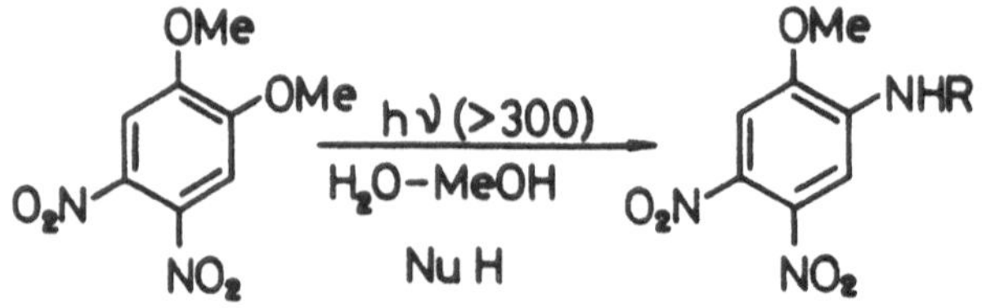

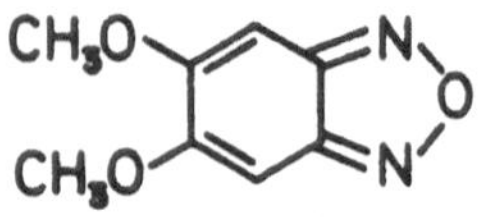

SCHEME 9- Nucleophilic aromatic photosubstitutions vs. photoreductions in 4,5-dinitroveratrole. See Table I.

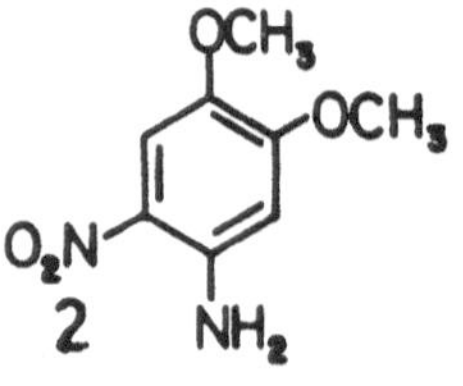

TABLE I. Reactions of 4,5-dinitroveratrole with amines[a]

Amine	Photosubstitution Products(%)[b]		Photoreduction Products(%)
Methylamine	R = CH$_3$	(36)	
Ethyl glycinate	R = CH$_2$COOEt	(80)	
n-Butylamine	R = n-Bu	(42)	1 (13)
Dimethylamine	———		1 (6), 2 (8)[c]
Trimethylamine	———		1 (6), 2 (6)

[a]General conditions: 400W medium pressure Hg lamp/ Pyrex filter/ water-methanol, pH 8, 4h. The experiment with ethyl glycinate was performed in acetonitrile-water. [b]Isolated yields calculated with respect to reacted starting material. [c] Complex mixture of reduction (major) and substitution products.

The photoreaction with dimethylamine afforded a complex mixture of reduction products from which 5,6-dimethoxybenzofurazane and 4,5-dimethoxy-2-nitroaniline, 2, could be isolated. Some others were tentatively characterized by means of GLC/MS analysis. The result with triethylamine with which no photosubstitution was expected was essentially the same.

The above experiments were the starting point for a better mechanistic understanding of the photoreactions. The amines that afforded clean photosubstitutions belong to the high ionization potential type and those affording photoreduction belong to the low ionization potential group (Table II). These facts pointed out to an electron transfer from the low ionization potential amines to the aromatic photoexcited substrate as the initial step for the observed photoreductions. Indeed, we performed photoreductions on several aromatic dinitrocompounds in the presence of benzophenone as triplet sensitizer (8). Also, the hypothesis emerged that photosubstitutions in para position of 4-nitroveratrole were initiated by an electron transfer from the low ionization potential amine to the triplet state of the aromatic substrate.

Table II. Ionization potential of amines (in eV) calculated by MINDO/3[a]

NH_3	H_2NCH_2COOMe	CH_3NH_2	$n-BuNH_2$	$(CH_3)_2NH$	Et_3N
10.071	9.503	9.206	9.190	8.578	8.347

[a]Since no experimental data for all the involved amines were available calculations were performed for all of them. Calculated and experimental data when the last were available follow the same trend

2.3. The mechanistic studies. A threefold mechanistic pathway.

A mechanistic study was performed on the photosubstitutions of 4-nitroveratrole with three selected nucleophiles in order to get useful information on the reaction pathways (9,10).

2.3.1. The photoreaction of 4-nitroveratrole with hydroxide ion. The preparative photoreaction of the title species in THF-water (1:9) afforded 2-methoxy-5-nitrophenol (96% yield based on consumed starting material) (Scheme 10). The isomeric 2-methoxy-4-nitrophenol could not be detected. The quantum yields for the production of the phenol were determined for different nucleophile concentrations, an increase of quantum yield being observed by increasing the hydroxide ion concentration. Also, a significant quantum yield decrease was observed at constant hydroxide ion concentration upon increasing potassium sorbate (triplet quencher) concentration. These initial experiments indicated that the studied reaction takes place through the excited triplet state of the nitroaromatic substrate (10). Further evidence for a single mechanistic pathway will be presented later on.

2.3.2. The photoreaction of 4-nitroveratrole with n-hexylamine. This photoreaction was performed in methanol-water (1:4) and led to a mixture of N-hexyl-2-methoxy-5-nitroaniline and N-hexyl-2-methoxy-4-nitroaniline in a ratio 4.4:1.0 (Scheme 10). N-Hexylamine was used as nucleophile since its low volatility makes its concentration control easy for quantitative measurements. Also the photosubstitution produces clean mixtures of both regioisomers. A set of experiments were performed in order to determine the effect of triplet quenchers as well as of radical scavengers (Table III). Some conclusions could be drawn from them: the addition of triplet quenchers such as potassium sorbate and 1,3-cyclohexadiene produced an increase in the meta/para ratio thus indicating that the meta isomer production was not affected or was much less affected than the para isomer production. Also the addition of radical scavengers which do not interact with the triplet excited state of 4-nitroveratrole (11) such as m-dinitrobenzene and methyl viologen interfere much more the production of the para isomer than the production of the meta if the last one is affected at all. The use of a much less polar solvent such as isopropanol also worked in the same direction. Finally, when the irradiation was performed in a solution saturated with argon (absence of oxygen, a triplet quencher) a small but significant effect in favour of the para isomer production was observed. Similar results were obtained in related experiments with methylamine (10). All the available experience from the data so far discussed is compatible with a hypothesis: the meta isomer is formed through the singlet excited state and the para isomer is formed by means of an electron transfer from the amine to the triplet excited state to form a radical cation-radical anion pair that evolves to the final product. Clearly, the available evidence is compatible with the indicated pathways as being predominant in the studied photosubstitutions. Additional evidences will be provided later on which point out that these pathways are indeed the only ones.

TABLE III. Effect of triplet quenchers and radical scavengers on the reactions of 4-nitroveratrole with n-hexylamine[a]

Atmosphere	Solvent	Additive	ϕ_{meta}/ϕ_{para}
Air	b	----	5.9
Air	b	Potassium sorbate[c]	16.3
Air	b	1,3-Cyclohexadiene[c]	>30
Argon	b	----	4.2
air	b	m-Dinitrobenzene[d]	>30
Argon	b	MV^{2+} $2Cl^-$ [e,f]	>30
Air	Isopropanol	----	22
Air	Hexane	----	g

[a]4-Nitroveratrole (2.5×10^{-3}M), n-hexylamine (2.5×10^{-1}M), 125W high pressure Hg lamp, pyrex filter, 1h irradiation. [b]Methanol-water (1:4). [c]5.1×10^{-2}M. [d]3.7×10^{-3}M. [e] A filter prepared with 0.005 g of 4-nitroanisole in 1 mL of methanol and 100 mL of water was used in addition to the pyrex filter to eliminate light under 320 nm and ensure MV^{+2} did not absorb. [f] 1.8×10^{-2}M. [g] Very complex mixture of products

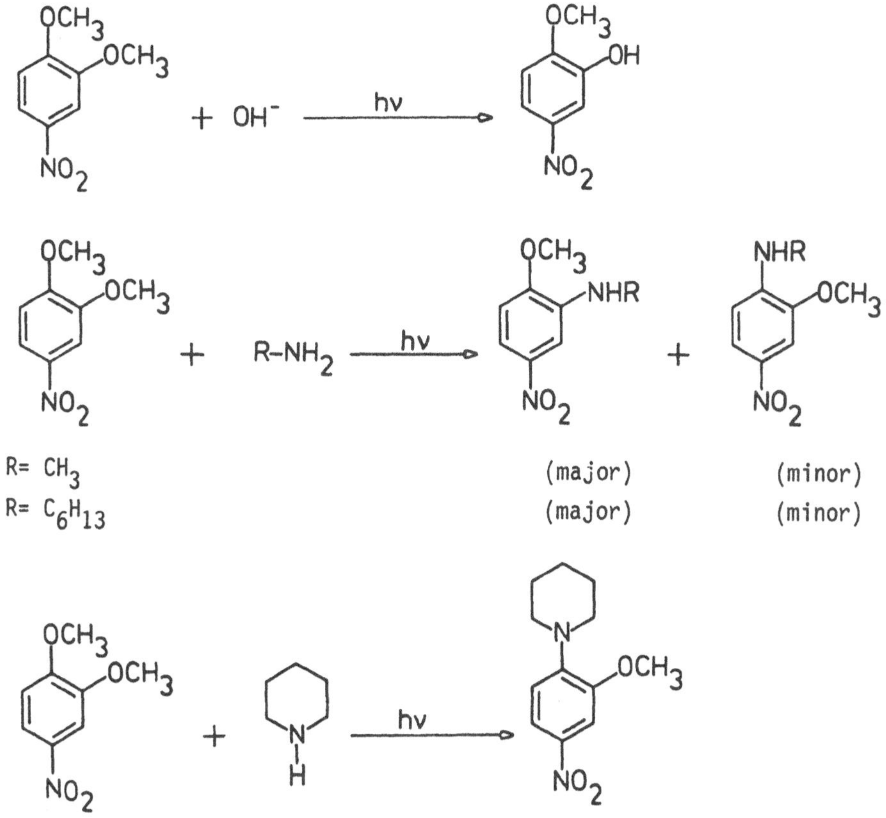

<u>Scheme 10.</u>- Nucleophilic aromatic photosubstitutions on 4-nitroveratrole with hydroxide ion, n-hexylamine and piperidine. General conditions: hν> 300 nm; THF-water (1:9) for hydroxide ion, methanol-water (1:4) for n-hexylamine and piperidine.

<u>2.3.3. The photoreaction of 4-nitroveratrole with piperidine.</u> The preparative photoreaction in the standard solvent (methanol-water 1:4) afforded N-(2-methoxy-4-nitrophenyl)piperidine as the only identified product in 25% yield based on consumed starting material (Scheme 10). A related reaction performed in isopropanol led to no photosubstitution at all being the starting 4-nitroveratrole completely consumed. We attribute this fact to the competence from photoreduction processes which are very favoured in hydrogen atom donor solvents as isopropanol.

The quantum yields of the photosubstitution exhibited an increase with increasing the nucleophile concentration and a decrease with increasing potassium sorbate (triplet quencher) concentration at constant piperidine concentration. Moreover, the addition of m-dinitrobenzene stopped completely the reaction. This additive has been shown to be a radical scavenger and to produce no interaction with the 4-nitroveratrole triplet excited state (11).

All the discussed data are consistent with a reaction pathway through a triplet excited state to which an electron is transferred from the nucleophile. In methanol-water (1:4) solvent the reaction evolves to photosubstitution. However, in isopropanol, a good hydrogen atom donor the radical anion of the aromatic substrate evolves to photoreduction products (8). Again the possibility arises of these mechanistic picture to be predominant or single. Arguments in favour of the uniqueness will be presented later on.

2.4. The general mechanistic scheme.

The experimental data so far discussed seem compatible with the general kinetic diagram represented in scheme 11. The photohydrolysis proceeds through the triplet state T^1 to a complex possibly in the ground state surface to afford finally the meta photohydrolysis product in a reaction which can be categorized as a "Nucleophilic Aromatic Photosubstitution through the Excited Triplet State" ($S_N2^3Ar^*$). Application of the steady-state approximation in the absence of quencher to this part of the diagram produces equations 1 and 2:

$$\phi = \phi_{isc} \times \frac{k_p}{k_p + k_d} \times \frac{k_3[Nu]}{k_4 + k_3[Nu]} \qquad (1)$$

Therefore:

$$\frac{1}{\phi} = \frac{1}{\phi_{isc}} \times \frac{k_p + k_d}{k_p} \times \left[1 + \frac{k_4}{k_3[Nu]} \right] \qquad (2)$$

A linear relationship between ϕ^{-1} and $[Nu]^{-1}$ has been observed (Scheme 12, line A) giving support to the hypothesis of the single mechanistic pathway through the triplet state for the photohydrolysis of 4-nitroveratrole in tetrahydrofuran-water (1:9). Further evidence comes from the Stern-Volmer plot of ϕ_0/ϕ vs. quencher concentration $[Q]$, where ϕ_0 is the quantum yield at $[Q] = 0$. A straight line was indeed obtained (Scheme 13) following equation 3.

$$\frac{\phi_0}{\phi} = 1 + \frac{k_q}{k_3[Nu] + k_4}[Q] \qquad (3)$$

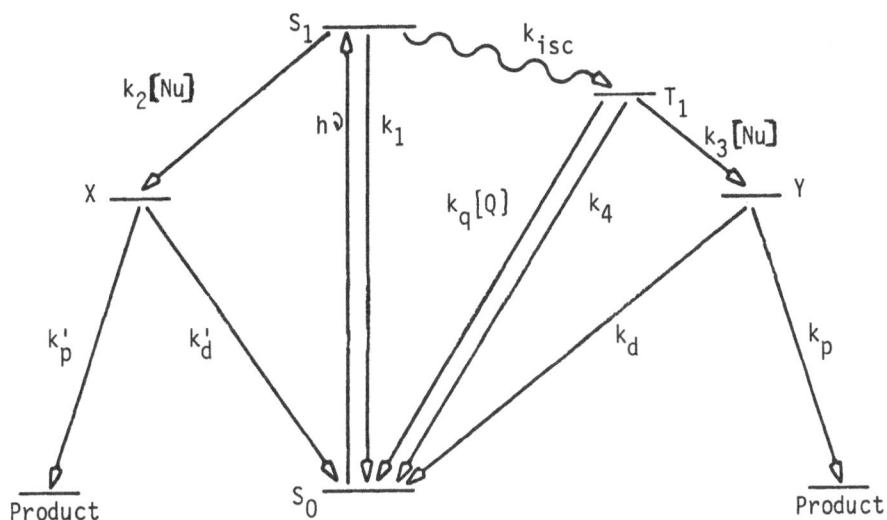

Scheme 11.- General mechanistic diagram for nucleophilic aromatic photosubstitutions.

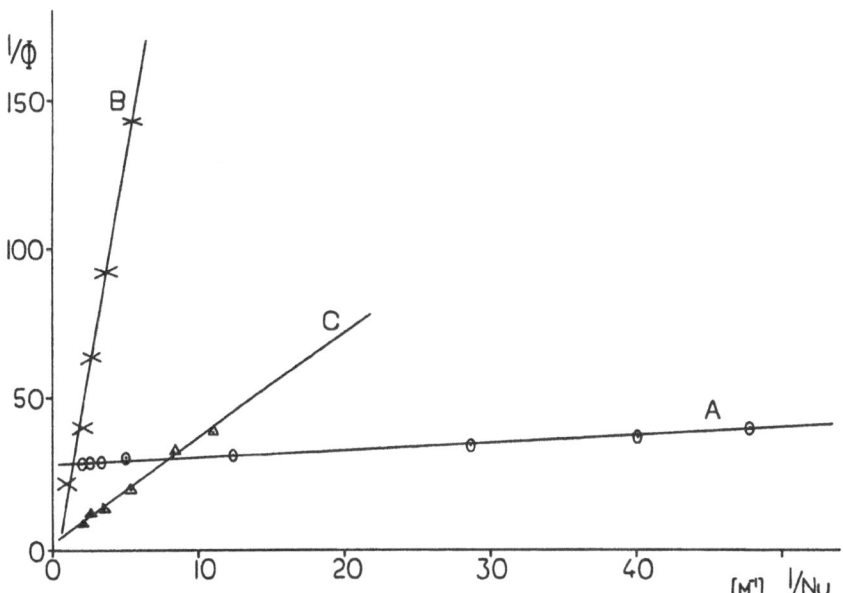

Scheme 12.- Plot of the inverse quantum yield vs. inverse nucleophile concentrations for A: hydroxide ion; B: piperidine; C: n-hexylamine.

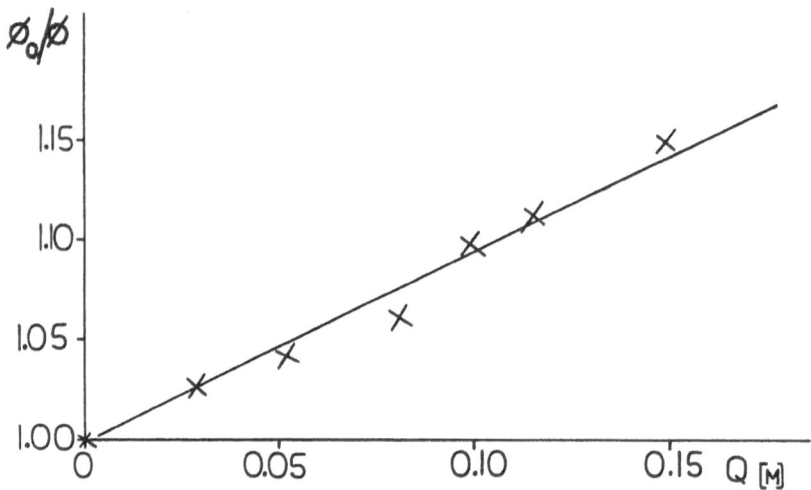

Scheme 13.- Plot of the inverse quantum yield vs. quencher concentration (potassium sorbate) for the nucleophilic aromatic photosubstitution of 4-nitroveratrole with hydroxide ion.

Similar results were obtained for piperidine by applying the same equations corresponding to the right hand side of the diagram (Scheme 12, line B and scheme 14). This results agree with the proposed reaction pathway through an excited triplet state involving electron transfer from the nucleophile.

Application of the steady-state approximation to the left part of the diagram produces equations 4 and 5:

$$\phi = \frac{k'_p}{k'_p + k'_d} \quad x \quad \frac{k_2[Nu]}{k_{isc} + k_1 + k_2[Nu]} \qquad (4)$$

$$\frac{1}{\phi} = \frac{k'_p + k'_d}{k'_p} \quad x \quad \left[1 + \frac{k_1 + k_{isc}}{k_2[Nu]}\right] \qquad (5)$$

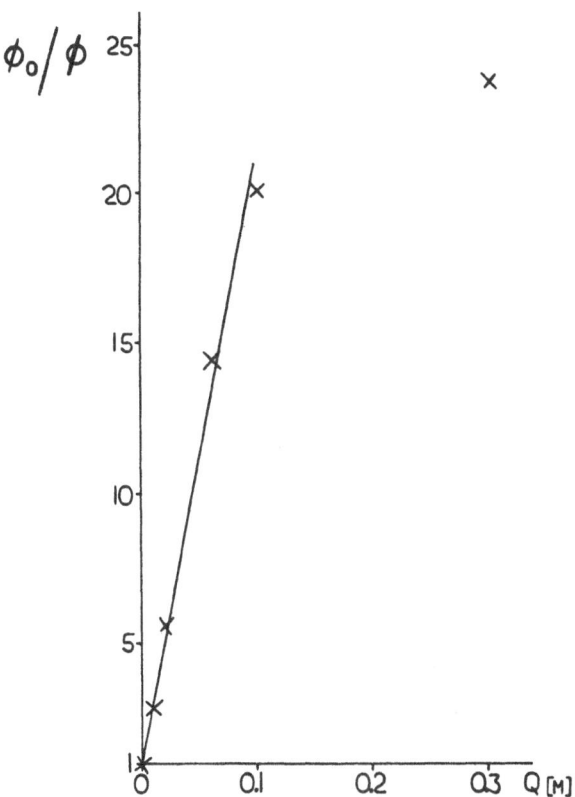

ϕ_0 / ϕ

Scheme 14.- Plot of the inverse quantum yield vs. quencher concentration (potassium sorbate) for the nucleophilic aromatic photosubstitution of 4-nitroveratrole with piperidine.

The plot ϕ^{-1} vs $\left[\text{Nu}\right]^{-1}$ gave a straight line for the production of the meta isomer in the reaction of nitroveratrole with n-hexylamine (See line C in Scheme 12). This is again a good indication of a single mechanistic pathway for that isomer. No linear relationship was found between ϕ_0/ϕ and quencher concentration (Stern-Volmer plot). All the available evidences suggest a mechanistic pathway for this photoreaction through the singlet excited state ($S_N2^1Ar^*$).

3. CONCLUDING REMARKS

Our work has emphasized the mechanistic complexity of the so called "Nucleophilic Aromatic Photosubstitution". This concept embraces a vast

array of reactions with very different mechanistic features. The
mechanistic knowledge obtained in our studies will be useful when
trying to apply Nucleophilic Aromatic Photosubstitution to fields of
science different from pure chemistry or photochemistry.

Can "Nucleophilic Aromatic Photosubstitutions" be useful for
photoaffinity labelling in biochemistry?. We believe that after the
contributions by us and other groups and although more work is needed
mainly with different nucleophiles the biochemists are now on much more
firm basis to use these old but now better understood reactions. The
decision is on their hands.

Although this is a review of our own contribution to the better
knowledge of Nucleophilic Aromatic Photosubstitutions the picture would
not be complete without mentioning other groups which in the last few
years have also contributed decisively to the better understanding of
the field. Apologizing in advance for any forget we would like to
mention the work by Mutai (12), Wubbels (13), Bunce (14) and Varma
(15). They results can be found in the indicated references and
essentially complement ours.

4. REFERENCES

(1) a.- J.R. Knowles; Acc. Chem. Res., 5, 155, (1972).
     b.- Photoaffinity Labelling, by H. Bailey and J.R. Knowles, in
     Methods in Enzymology, 46, 69, (1977). W.B. Jakoby and W. Wilchek,
     Eds. Academic Press.
(2) a.- M.E. Kronenberg, A. Van der Heyden, E. Havinga; Recl. Trav.
     Chim. Pays-Bas, 85, 56, (1966).
     b.- M.E. Kronenberg, A. Van der Heyden, E. Havinga; Recl. Trav.
     Chim. Pays-Bas, 86, 254, (1967).
     c.- J. Cornelisse, G.P. de Gunst, E. Havinga; Adv. Phys. Org.
     Chem., 11, 225, (1975).
     d.- J. Cornelisse, E. Havinga; Chem. Rev., 75, 353, (1975).
     e.- E. Havinga, M.E. Kronenberg; Pure Appl. Chem., 16, 137,
     (1968).
(3) P.C. Jelenc, C.R. Cantor, S.R. Simon; Proc. Natl. Acad. Sci. USA,
     75, 3564, (1978).
(4) H. Gozlan, V. Homburger, M. Lucas, J. Bockaert; Biochem.
     Pharmacol., 31, 2879, (1982).
(5) A. Castelló, J. Cervelló, J. Marquet, M. Moreno-Mañas, X. Sirera;
     Tetrahedron, 42, 4073, (1986).
(6) J. Cervelló, M. Figueredo, J. Marquet, M. Moreno-Mañas, J.
     Bertrán, J.M. Lluch; Tetrahedron Lett., 25, 4147, (1984).
(7) A. Castelló, J. Marquet, M. Moreno-Mañas, X. Sirera; Tetrahedron
     Lett., 26, 2489, (1985).
(8) J. Marquet, M. Moreno-Mañas, A. Vallribera, A. Virgili, J.
     Bertrán, A. González-Lafont, J.M. Lluch; Tetrahedron, 43, 351,
     (1987).
(9) A. Cantos, J. Marquet, M. Moreno-Mañas; Tetrahedron Lett., 28,
     4191, (1987).

(10) A. Cantos, J. Marquet, M. Moreno-Mañas, A. Castelló; Tetrahedron, 44, 2607, (1988).
(11) A.M.J. van Eijk, A.H. Huizer, C.A.G.O. Varma, J. Marquet; J. Am. Chem. Soc., in press.
(12) a.- K. Mutai, R. Nagaki, H. Takeda; Bull. Chem. Soc. Jpn., 58, 2066, (1985).
     b.- K. Mutai, K. Kobayashi, K. Yokoyama; Tetrahedron, 40, 1755, (1984).
(13) G.C. Wubbels, A.M. Halverson, J.D. Oxman; J. Am. Chem. Soc., 102, 4849, (1980).
(14) N.J. Bunce, S.R. Cater, J.C. Scaiano, L.J. Johnston; J. Org. Chem., 52, 4214, 1987.
(15) a.- C.A.G.O. Varma, J.J. Tamminga, J. Cornelisse; J.C.S. Faraday Trans. 2, 78, 265, (1982).
     b.- P.H.M. van Zeijl, L.M.J. van Eijk, C.A.G.O. Varma; J. Photochem., 29, 415, (1985).

5. ACKNOWLEDGEMENTS

Generous financial support from "CAICYT" ("Ministerio de Educación y Ciencia" of Spain through grants 845/81 and 343/84 is gratefully acknowledged.
    We want to express our gratitude to our past and present coworkers: J. Cervelló, Dr. M. Figueredo, A. Castelló, X. Sirera, A. Vallribera and in particular to Albert Cantos. Without their dedication, intelligence and skill this work would have never been possible.

PHOTOCHEMISTRY OF ARYL AND AROYL AZIDES: APPLICATION TO PHOTOLABELING OF BIOLOGICAL SYSTEMS.

Gary B. Schuster
Department of Chemistry
Box 58, Roger Adams Laboratory
1209 W. California Street
Urbana, IL 61801-3731

ABSTRACT. The major objective of this work is to provide the knowledge necessary for the sound application and interpretation of photolabeling experiments. A major goal in the study of living systems is to associate function with chemical structure. Several experimental approaches have been developed to facilitate this goal. Among the most useful of these are photolabeling procedures. In this approach, an activatable reagent is positioned adjacent to a target structure. The latent reactivity of the reagent is revealed by irradiation with light and the intermediate so formed binds irreversibly to some functional group at the site. Later, the biological structure is degraded and the labeled site identified. The aim of the work described herein is the definition and characterization of the chemical and physical properties of the high-reactivity intermediate. We detect these intermediates by time-resolved laser spectroscopy. The main focus of our effort has been aroyl and aryl nitrenes. Their properties are described herein and recommendations for their use in photolabeling experiments are made.

A major objective in the study of biological systems is to link chemical structure with function. Forming this link often requires the isolation and identification of a small, active portion of a larger molecule or system of molecules. For example, identification of the amino acid residues at the active site of an enzyme often precedes formulation of a hypothesis describing its mechanism of catalysis (1). Similarly, the mechanics of recognition between binding sites and hormones, or other agents, is often revealed after the chemical composition of the binding site is known (2). An important technique for identification of these critical segments of large structures is affinity labeling (3).

Affinity labeling takes advantage of the remarkable specificity of biological recognition to position a reactive chemical functionality uniquely at a target site. In a subsequent step, this position is marked for later detection by the formation of a covalent bond between the reactive functionality and appropriate residues at the target site. Photoaffinity labeling refers to a specific subset of affinity labeling techniques characterized by the use of light to activate latent reactivity in a specifically bound receptor molecule (4). Closely related to photoaffinity labeling procedures are photochemical cross-linking experiments (5) where the light-initiated reaction is employed to form a covalent bond between two macromolecules. This process can reveal the spatial relation between the target molecules in their natural state. A third photochemical technique used in biochemistry is topological labeling (6). These experiments do not probe a binding or recognition site, but instead report on the accessibility of a macromolecule from a specific region of a larger structure.

P. E. Nielsen (ed.), Photochemical Probes in Biochemistry, 31–41.
© 1989 by Kluwer Academic Publishers.

Topological labeling experiments have been particularly valuable in studies of membrane organization (7). Photoaffinity labeling, photocrosslinking, and phototopological experiments form a group that we refer to generally as photolabeling experiments.

Photoaffinity labeling was introduced about twenty years ago by Westheimer (8). In the intervening period, this technique has developed into a widely employed set of photolabeling procedures in the biological sciences. For example, photolabeling techniques have been used recently to study electrostatic and hydrophobic interactions in lipid vesicles (9), to identify components of human erythrocyte membranes (10), and to assign the glucagon receptor in rat liver membranes (11). Photolabeling procedures have gained wide acceptance in the biological community. The importance and utility of these techniques are clearly established and are well beyond question.

The widespread use of photolabeling techniques is easily traced to several key advantages that this process has over more conventional approaches. First, in the ideal case, the photochemical reagent retains a high affinity or specificity for a specific macromolecule or region so that the majority will be specifically bound or located even in dilute solution. Critically, prior to activation by light, the reagent typically is inert so that its association with the target site can be easily assessed. Second, the reactivity of the probe molecule can be revealed selectively after it has been placed in a particular environment. For example, photoaffinity labeling procedures provided identification of sites of nucleophilic reactivity in hydrophobic substructures in the presence of an abundance of nonspecific nucleophiles (12). Finally, the reactivity of the intermediates that are created photochemically are typically much greater than can be managed by conventional labeling procedures. In an idealized version, the key intermediate would possess globalreactivity. That is, its reaction with any chemical entity at the target site would be instantaneous and irreversible. Such behavior is especially important for identification of sites that do not have a catalytic function. In these cases, there is no reason to presuppose that presence of highly reactive groups at the binding site. None of the currently popular photoactivatable chemicals have demonstrated global reactivity. The bane of many photolabeling experiments is low efficiency of covalent bond formation after photolysis (13). This is particularly serious in photocrosslinking experiments where isolation and identification of coupled macromolecules is often a prerequisite to successful completion of the procedure.

Knowles' introduction of aryl azides as photoaffinity labeling agents in 1969 focused attention on the special properties of these chemicals (14). Aryl azides are the class of reagents most frequently used for biological photolabeling experiments. It is difficult to know certainly why this is so, but it is probably related to their ease of preparation particularly with radioactive substituents. Despite the great significance thus associated with photolysis of aryl azides, there was surprisingly little known about their photochemistry or about the chemical properties of the intermediates formed when they lose molecular nitrogen. In reports where the details of the chemical processes that form the heart of azide photolabeling are considered, the key intermediate is typically formulated as a nitrene. However, for the commonly used reagents, formation of nitrenes is probably a parasitic process that leads to low covalent bonding efficiancies and to non-specific labeling.

The prime aim of our work is the examination of the necessary chemical transformations associated with photolabeling experiments. The approach we follow in this examination relies heavily on the direct observation of the photogenerated reagent(s) by pulsed laser spectroscopy. This procedure provides the unique opportunity to generate the key intermediates of the photolabeling experiment in high enough concentration and short enough time to observe them directly by optical spectroscopy. In this way it is possible to explore such issues as the identity of the reactive intermediate(s), its lifetime, the pathways of its reactions, and its reactivity with functional groups typically found at biological receptor sites.

I. Photolysis of Unsubstituted Aromatic Azides

It is well-known that irradiation of aromatic azides (compounds that have the azido group bound directly to a carbon atom of an aromatic nucleus) with UV light leads to efficient loss of nitrogen and the generation of highly reactive intermediates (15). One of the intermediates typically considered to be formed in this process is a nitrene (univalent nitrogen compound). Subsequent reaction of the nitrene with functional groups located on the target macromolecule is routinely considered to generate the covalent bond required for labeling (16), Eq. 1.

$$ArN_3 \xrightarrow{h\nu, -N_2} Ar\text{-}N \xrightarrow{Target\text{-}H} Ar\text{---}\overset{\overset{\text{H}}{|}}{N}\text{---}Target \qquad (1)$$

Aromatic Azide   Nitrene   Labeled Macromolecule

Our work (17), and results from other groups (18), has revealed that this is not often the actual scenario. We have studied a series of aromatic azides that indicate quite a different reaction scheme. Generalization of our finding leads to some predictions concerning the effectiveness of some photolabeling reagents.

A. PHENYL AZIDE (PhN3).

Although PhN_3 is structurally the simplest aromatic azide, its photochemical behavior is more complicated than many of its substituted analogs and higher homologs. Despite many years of effort, some of the details of the reactions of PhN_3 are still mysterious. However, it is now possible to identify with some certainty the key processes of importance to photolabeling experiments. Moreover, these processes form a pattern that is useful for analyzing the reactions of other aromatic azides of more direct relevance to actual photolabeling reagents.

The chemicals outcome of the photolysis of PhN_3 depends dramatically on the conditions of the reaction. Irradiation of very dilute (ca. $10^{-4}\underline{M}$) solutions of PhN_3 in inert solvents with a continuous lamp gives azobenzene in reasonable yield. Azobenzene is a product that is characteristic of the dimerization of triplet phenyl nitrene (PhN^3), Eq. 2. However, the yield of azobenzene is sharply reduced when either the concentration of PhN_3 or the power of the light source is increased. This unusual behavior is a consequence of the reaction of PhN^3, or a precursor to PhN^3, with PhN_3, Eqs. 3 and 4. This discovery explains the disquietingly inconsistent results previously reported for this reaction in the literature (19, 20).

$$PhN_3 \xrightarrow{h\nu} (PhN_3)^{*1} \xrightarrow{-N_2} (PhN)^1 \xrightarrow{isc} PhN^3 \xrightarrow{PhN^3} Azobenzene \quad (2)$$

$$(PhN)^1 + (PhN_3)_n \xrightarrow{\hspace{2cm}} Oligomer \quad (3)$$

$$(PhN)^3 + (PhN_3)_n \xrightarrow{\hspace{2cm}} Oligomer \quad (4)$$

Irradiation of PhN_3 in a reactive, nucleophilic solvent such as diethylamine (DEA) gives only a small amount of azobenzene but produces a good yield of 2-diethylamino-3H-azepine (21). It is this class of reaction, covalent bonding to an external nucleophile at the target site, that is the key to the utility of aromatic azides as photolabeling reagents. This reaction occurs through a closed-shell intermediate that had been suggested to be either a dehydroazepine or a bicyclic azirine (22), Eq. 5. The dehydroazepine has been detected at 8 °K in an argon matrix (23). In contrast, triplet phenylnitrene, formed specifically by sensitization, reacts with DEA to give primarily aniline, Eq. 6. This process does not directly form a covalent bond to the nucleophile and thus appears to be less useful in photolabeling experiments. The radicals formed in this triplet process may in fact lead to pseudolabeling.

$$PhN_3 \xrightarrow[-N_2]{h\nu} \left[PhN^1 \text{ and/or } \text{Dehydroazepine} \text{ and/or } \text{Azirine} \right] \xrightarrow{(Et)_2NH} \tag{5}$$

$$PhN_3 \xrightarrow[-N_2]{(sens)^{*3}} PhN^3 \xrightarrow{(Et)_2NH} PhNH_2 \tag{6}$$

We examined the photochemistry of PhN_3 by pulsed laser techniques in inert and in nucleophilic solvents employing both direct irradiation and triplet-sensitized reaction techniques. These experiments helped to show that the formation of PhN^3 occurs slowly in comparison with the other reactions of its precursor. The immediate precursor to PhN^3 in our scheme is the singlet nitrene (PhN^1). However, our findings reveal that PhN^1 exists only in a microscopic concentration with a lifetime too short to participate in bimolecular processes. In the reaction sequence we adopt, the rate-limiting step in the formation of PhN^3 is the conversion of the relatively long-lived, closed-shell intermediate to PhN^1. This is an important concept to apply in evaluation of photolabeling agents.

In most cases the lifetime and chemical reactivity of the closed-shell intermediate determine the success of a photolabeling experiment employing an aryl azide. In the case of PhN_3, the intermediate is dehydroazepine identified in Eq. 5. Our findings show that this intermediate is highly reactive. In the presence of 1 M secondary amine, perhaps a model for the active site of an enzyme, this species has a lifetime of about 200 nsec and is efficiently trapped to form a lH-azepine. These chemical properties would appear to make PhN_3 a useful agent. Unfortunately, the optical properties of PhN_3 (it absorbs only in the deep UV) render it nearly useless in this regard. Modifications to the structure of PhN_3 that were intended primarily to shift its chromophore to more useful wavelengths also cause significant changes to the chemical and physical properties of the derived nitrenes and closed-shell intermediates.

B. 1-NAPHTHYL AZIDE (1-NaN₃) AND 2-NAPHTHYL AZIDE (2-NaN₃).

1-NaN$_3$ has been used several times as a photolabeling agent with some success (24). To the best of our knowledge, 1-NaN$_3$ has never been used in this application. It is ironic that our results indicate the 2-NaN$_3$ should be the superior agent in many circumstances. The central results of our study of the naphthyl azides are outlined below.

The ground state of the nitrenes formed from irradiation of 1-NaN$_3$ or 2-NaN$_3$ (*i.e.*, 1-NaN and 2-NaN) are shown by EPR spectroscopy to be triplets (25). However, irradiation of these azides in an argon matrix at 8 °K shows that an intermediate identified as an azirine is formed in both cases, Eqs. 7, 8 (26). It is therefore curious, and revealing, that only 2-NaN$_3$ gives products characteristic of nucleophilic trapping of the azirine, Eqs. 9, 10 (27). Our investigation points to the reason for this contrasting behavior and illuminates an important criterion for the selection of aryl azide photolabeling agents.

Pulsed laser irradiation of 1-NaN$_3$ generates a transient species identified by its spectral and chemical properties as 1-NaN^3. This transient is not formed immediately after photoexcitation of the azide, but appears with a risetime of 2.8 μsec. We identify the rate-limiting step in the formation of 1-NaN^3 as the conversion of 1-azirine to 1-NaN^1. The singlet nitrene is presumed to intersystem cross to 1-NaN^3 very rapidly and irreversibly.

In comparison, pulsed irradiation of 2-NaN$_3$ also generates 2-NaN^3, but for this compound the risetime of the triplet nitrene is much longer that it is in the case of 1-NaN^3. This means that the 2-azirine from 2-NaN$_3$ has as much longer lifetime than the corresponding species from 1-NaN$_3$. Since it is this intermediate that is trapped to give eventually the all-important nucleophile incorporation product, the longer lifetime of the 2-azirine allows sufficient time for this process to occur. However, from the point of view of the photolabeling experiments, the long lifetime of this intermediate may also permit its migration from the target site and concomitant pseudophotolabeling unless there is a reactive nucleophilic group at the binding site. Thus, on the basis of our findings, we would recommend use of 2-NaN$_3$ derivatives as photolabeling reagents only for those cases where the target site is believed to contain nucleophilic residues. The different behavior of 1-NaN$_3$ and 2-NaN$_3$ reveals a pattern that is repeated throughout aryl azide photochemistry.

C. 1-PYRENYL AZIDE (1-PyN$_3$) AND 2-PYRENYL AZIDE (2-PyN$_3$).

Photolabeling of carcinogen binding proteins by 1-PyN$_3$ has recently been attempted (28). In this work a new receptor site was identified that binds polycyclic aromatic hydrocarbons in a specific and saturable fashion. 1-PyN$_3$ has also been used in some phototopological labeling experiments (29).

We have studied the photochemistry of 1-PyN$_3$ and 2-PyN$_3$ in detail (17b). The general pattern of our findings points to quite similar behavior for the pyrenyl and naphthyl azides. Irradiation of either 1-PyN$_3$ or 2-PyN$_3$ leads to loss of nitrogen and formation of an intermediate. This intermediate can be trapped covalently with external nucleophiles only in the case of 2-PyN$_3$. The ground-state of the pyrenyl nitrenes in both cases is the triplet (30). Yet, 2-PyN^3 gives reaction products that are characteristic of the closed-shell singlet intermediate. We have proposed an explanation for this behavior that features reversible intersystem crossing from ground-state 2-PyN^3 to form the slightly higher energy singlet state. This proposal accounts for the ability of 2-PyN$_3$ to trap nucleophiles present at ca. 100-fold lower concentration than is required in the case of 2-NaN$_3$.

It is clear from our results that neither 1-PyN$_3$ or 2-PyN$_3$ is an ideal labeling agent. However, if there is a nucleophilic residue at the target site, 2-PyN$_3$ should be the far superior agent. On the other hand, if there is no proximate nucleopohilic residue, we predict that use of 2-PyN$_3$ will lead to serious pseudolabeling problems.

II.   Photolysis of Nitro-substituted Phenyl Azide.

The nitro-substituted phenyl azides are by a long margin the most popular photolabeling agents. Before our work there was only one serious attempt to examine the photochemical properties of these reagents as they relate to photolabeling experiments (31). We have examined in detail the properties of two of these compounds.

A. 4-NITROPHENYL AZIDE (4NPA).

One reason that 4NPA is attractive as a potential photolabeling agent is that it absorbs in a convenient spectral region. A second is the notion that the electron-withdrawing nitro group will increase the electrophilicity of the nitrene and thereby enhance its reactivity (32). We have found high reactivity from 4-nitrophenyl nitrene (4NPN), but not of the sort that is generally useful to photolabeling experiments.

The ground state of 4NPN is the triplet (33). We have detected this species by pulsed laser techniques and have observed its bimolecular reactions. In an inert, non-nucleophilic solvent 4NPN^3 dimerizes to form the azobenzene in high yield, Eq. 11. Irradiation of 4NPA in the presence of nucleophilic amines does not lead to covalent trapping products in significant yield. This reaction gives, in addition to residual azo-aromatic, 4-nitroaniline and oxidative demethylation of the amine, Eq. 12. This process is probably not useful in photolabeling applications.

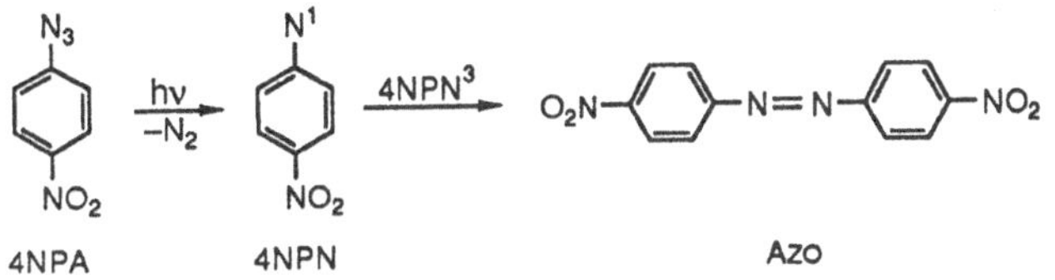

$$\underset{\text{4NPA}}{\underset{NO_2}{\underset{\bigcirc}{N_3}}} \xrightarrow[-N_2]{hv} \underset{\text{4NPN}}{\underset{NO_2}{\underset{\bigcirc}{N^1}}} \xrightarrow{\text{4NPN}^3} O_2N-\bigcirc-N=N-\bigcirc-NO_2 \quad \underset{\text{Azo}}{(11)}$$

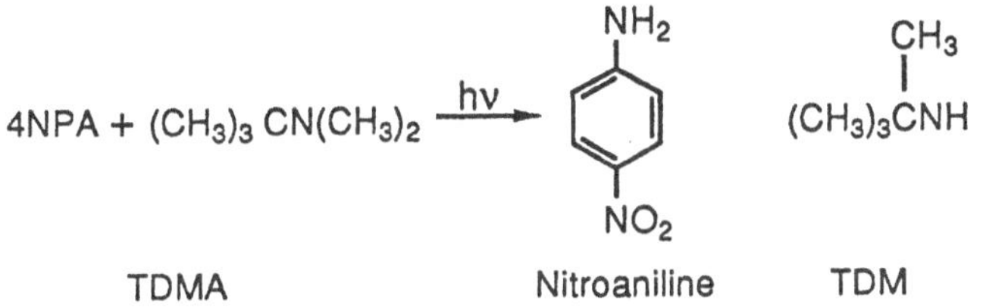

$$4NPA + (CH_3)_3\,CN(CH_3)_2 \xrightarrow{hv} \underset{\text{Nitroaniline}}{\underset{NO_2}{\underset{\bigcirc}{NH_2}}} \qquad \underset{\text{TDM}}{(CH_3)_3\overset{CH_3}{\overset{|}{C}}NH} \quad (12)$$

$$\underset{\text{TDMA}}{}$$

We studied the mechanism of the reaction of 4NPN[3] with amines by examining the product isotope effects for intramolecular and intermolecular competition reactions (34). Comparison of the deuterium content of t-butyl-N-methyl amine (TMA) formed from a 1:1 mixture of t-butyldimethyl amine (TDMA)-d_0 and -d_6 with that from TDMA-d_3 revealed that the initial step in the reaction of the nitrene with the amine is a single electron transfer. The reaction sequence supported by our finding is outlined in Scheme 1.

It is informative to compare 4NPA with the unsubstituted aromatic azides described earlier. We presume that the desired nucleophilic trapping reaction will proceed only from the closed-shell intermediate (azirine). The 4-nitro substituent evidently inhibits the formation of this species. This slows attack on the ring by the electrophilic singlet nitrene. So, the net effect of the substituent is to increase the efficiency of intersystem crossing to the non-electrophilic triplet state.

Scheme 1

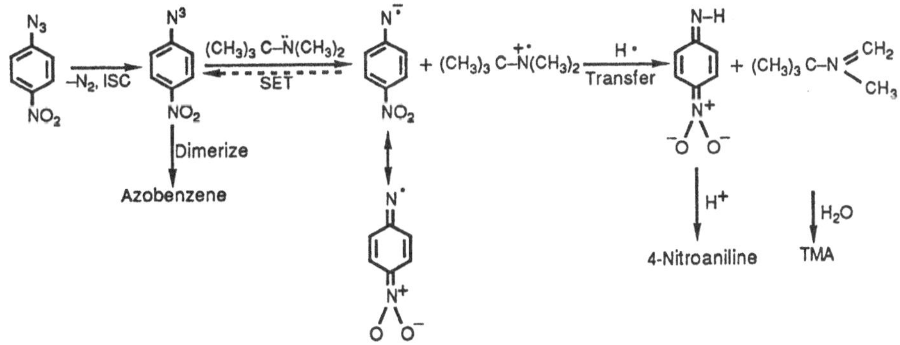

B. 3-NITROPHENYL AZIDE (3NPA).

This is a particularly important compound since a large fraction of the photolabeling experiments that employ azides use this agent (or a derivative) (3). Under appropriate conditions, irradiation of 3NPA gives products that incorporate external nucleophiles.

Irradiation of 3NPA in benzene solution with a continuous lamp or a pulsed laser gives the expected azo compound in high yield, Eq. 13. The laser spectroscopy shows that the azo compound is formed from an intermediate, presumed to be the triplet nitrene (3NPN^3), by a bimolecular process with a rate constant of ca. 1 x 10^9 M^{-1} s^{-1}. On a nanosecond time scale the growth of the azo compound obscures the nitrene.

(13)

3NPA   3NPN   AZO

Irradiation of 3NPA in solutions containing diethylamine gives three major products and a small amount of the azoaromatic. One of these products (ca. 40%) is 3-nitroaniline formed from reduction of the azide. The other two products are isomers with the molecular formula C$_{10}$H$_{15}$N$_3$O$_2$ (determined by mass spectroscopy). These two products are formed in a ratio of ca. 2:1 and account for 20% of the 3NPA consumed. The significance of finding these products is that they correspond to addition of DEA to the nitrene or its precursor, Eq. 14.

(14)

III. Design and Evaluation of New Photoactivatable Azides Useful as Photolabels.

The concept of global reactivity in photolabeling is an important one. Although specificity of reaction may sometimes be beneficial, the hope in most photolabeling experiments is that the reactive intermediate created by irradiation will combine immediately and irreversibly with whatever functionality happens to be present at the target site. In considering possible approaches to development of such an agent, we were led to a recent report that benzoyl nitrene inserts stereospecifically into unactivated carbon-hydrogen bonds whether it is generated by direct irradiation or by triplet sensitization, Eq 15 (35). This behavior suggests that this acyl nitrene might be a suitable global photolabeling agent. To accommodate the requirements of our laser experiments, we chose β-naphthoyl azide (BNA) for our initial investigation (36).

Direct irradiation of BNA in cyclohexane solution results in the formation of two products. One of these was identified as β-naphthylisocyanate (51%) by conversion to the carbamate with ethyl alcohol. The other, more significant, product is N-cyclohexyl-β-naphthamide (42%), Eq. 16. The amide is evidently formed by direct insertion of the β-naphthoylnitrene (BNN) into the unactivated carbon-hydrogen bond of the solvent.

(15)

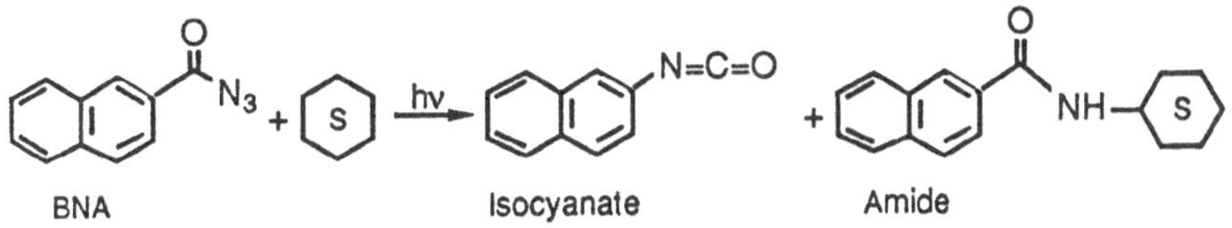

BNA Isocyanate Amide

Our results show that the isocyanate is not formed from the nitrene, but that it is created by rearrangement of the excited singlet azide. This finding is critically important since isocyanate formation is disastrous from the photolabeling point of view. That is, the isocyanate is a long-lived electrophile that will almost certainly cause unacceptable pseudolabeling to occur. We have discovered that we can completely suppress isocyanate formation while retaining acylnitrene reactivity by triplet sensitization.

By combining pulsed laser spectroscopy with conventional photochemical techniques, the photoinitiated reactions of BNA outlined in Scheme 2 were uncovered. Direct excitation of BNA creates an excited singlet state that either loses N_2 to make the nitrene, rear

Scheme 2

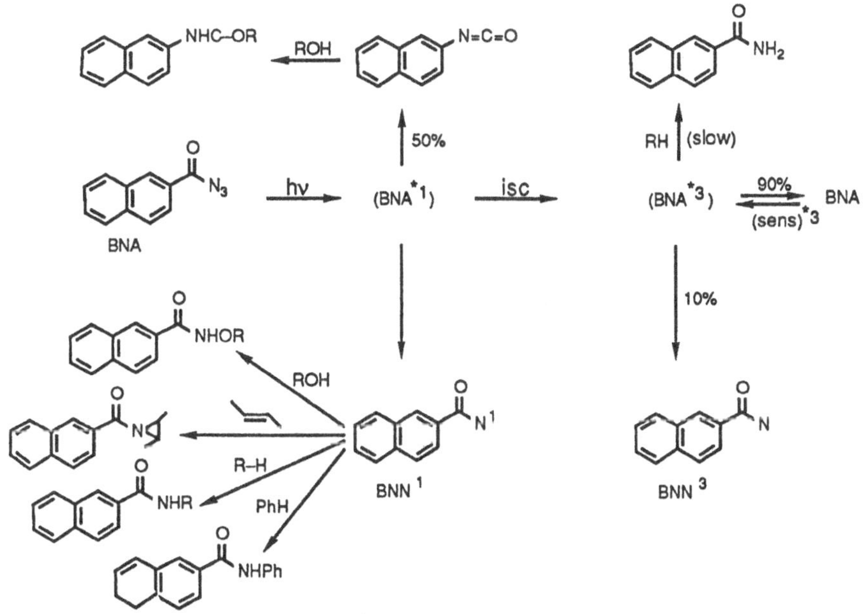

ranges to form the isocyanate, or intersystem crosses to the triplet azide. The triplet azide formed in this way, or by sensitization, either returns to BNA, very slowly abstracts hydrogen from the solvent, or loses nitrogen to form, presumably, the triplet nitrene (BNN[3]). We are unable to detect BNN[3] by low temperature (4 K) EPR or by optical spectroscopy or by pulsed laser techniques. Comparison of the products formed from direct and triplet-sensitized photolysis of BNA indicates that, contrary to computational results (37), the ground state of BNN is the singlet. This conclusion is very important for application of aroyl azides as photolabeling agents. It is the singlet nitrene that exhibits the global reactivity we hope to exploit.

As mentioned earlier, BNN inserts efficiently into the unactivated C-H bonds of cyclohexane. Since all conceivable target sites will contain carbon-hydrogen bonds, this feature alone insures global reactivity. Additionally, preliminary nanosecond time-resolved experiments indicate that the insertion reaction probably occurs in under 14 nsec.

As shown in Scheme 2, BNN[1] not only reacts with saturated hydrocarbons, but it also traps nucleophiles (we have examined only alcohols so far), olefins, and aromatic hydro carbons. Indeed, the key to our suggestion that the ground state of BNN is the singlet comes from an analysis of the stereochemistry of addition of BNN to cis- and trans-2-pentene under direct and triplet-sensitized conditions. In all cases the formation of aziridine occurs with total retention of configuration. This is precisely the result expected if the nitrene is a ground-state singlet. Moreover, this finding is opposite to those for azidoformates which appear to have triplet ground states (38).

IV. Conclusions.

Photolabeling is an important biochemical technique. Aryl azides will be effective photolabeling agents when a reactive nucleophile is present at the target site. Aroyl azides may provide reagents with sufficient reactivity to label unactivated sites.

V. Acknowledgement.

This work was supported by grants from the National Institute of Health and the National Science Foundation.

VI. References and Notes.

(1) Potter, R. L. and Haley, B. E. In *Methods in Enzymology*, Vol. 91, C. H. W. Hirs and S. N. Timasheff, Eds. Academic: New York, 1983.

(2) Barden, R. E.; Achenjang, F. M.; Adams, C. M. In *Methods in Enzymoloty* , Vol. 91, C. H. W. Hirs and S. N. Timasheff, Eds. Academic: New York, 1983.

(3) Bayley, H. in *Membranes and Transport* , Vol. 1, A. Martinosi, Ed., Plenum: New York, 1982.

(4) Tometsko, A. M.; Richards, F. M. 'Applications of Photochemistry in Probing Biological Targets,' *Ann. N. Y. Acad. Sci.* 1980,346.

(5) Ji, T. H., *Methods Enzymol* ., 1983, 46 , 580.

(6) Brunner, J., *Trends Biochem. Sci.* 1981, 6 , 44.

(7) Staros, J. U.; Richards, F. M.; Haley, B. E. *J. Biol. Chem.* 1975, 250 , 8174.

(8) Singh, A.; Thornton, E. R.; Westheimer, F. H. *J. Biol Chem.* 1962, 237 , PC3006.

(9) Based on a survey of Chemical Abstracts Online.

(10) Brunner, J.; Spiess, M.; Aggeler, R.; Huber, P.; Semenza, G. *Biochemistry* 1983, 22 , 3812.

(11) Aimoto, S.; Richards, F. M. *J. Biol. Chem.* **1981**, *256* , 5134.

(12) Johnson, G. L.; Vincent, I. M.; Pilch, P. F. *Proc. Natl. Acad Sci. USA* **1981**, *78* , 875.

(13) Fleet, G. W. J.; Porter, R. R.; Knowles, J. R. *Nature* **1969**, *224* , 511.

(14) Knowles, J. R. *Acc. Chem. Res.* **1972**, *3* , 155.

(15) *Azides and Nitrenes* , Scriven, E. F. V. Ed. Academic: New York, 1984.

(16) Bayley, H.; Knowles, J. R. *Methods Enzymol.* **1977**, *46* , 69.

(17) (a) Schrock, a. K.; Schuster, G. B. *J. Am. Che . Soc.* **1984**, 106, 5228. (b) Schrock, a. K.; Schuster, G. B. *J. Am. Chem. Soc.* **1984**, *106* , 5234.

(18) Nielsen, P. E.; Buchardt, O. *Photochem. Photobiol.* **1982**, *35* , 317.

(19) Horner, L.; Christman, A. *Chem. Ber.* **1963**, *96* , 399. DeGraff, B. A.; Gillespie, D. W.; Sundberg, R. J. *J. Am. Chem. Soc.* **1974**, *69* , 7491.

(20) Waddel, W. H.; Go, C. L. *J. Org. Chem.* **1983**, *48*, 2897.

(21) Doering, W. E.; Odum, R. A. *Tetrahedron*, **1966**, *22* , 81.

(22) Huisgen, R.; Vossius, D.; Appl, M. *Chem. Ber.* **1985**, *91* , 1.

(23) Chapman, O. L.; LeRoux, J.-P. *J. Am. Chem.Soc.* **1978**, *100*, 282.

(24) Bercovici, T.; Gitler, C. *Biochemistry* **1981**, *20* , 6872.

(25) Wasserman, E., *Prog. Phys. Org. Chem.* **1971**, *8* , 319.

(26) Dunkin, I. R.; Thompson, P. C. P. *J. Chem. Soc. Chem. Commun.* **1980**, 499.

(27) Carde, R. N.; Jones, G. *J. Chem. Soc. Perkin I* , **1975**, 519.

Collins, S.; Marletta, M. A. *Molecular Pharm.* **1984**, *26*, 353.

(29) Wolf, M. K.; Konisky, J. *J. Bacteriol* **1981**, *145* , 341.

(30) We were able to obtain EPR spectra that obeyed the Curie law at 1.2 K.

(31) Nakayama, H.; Nozawa, M.; Kanaoka, Y. *Chem. Pharm. Bull.* **1979**, *27* , 2775.

(32) Abramovitch, R. A.; Challand, S. R. *J. Chem. Soc. Chem. Comm.* **1972**, 964.

(33) Smirnov. V. A.; Brichkin, S. B. *Chem. Phys. Lett.* **1982**, *87* , 458.

(34) Liang, T.-Y.; Schuster, G. B. *J. Am. Chem. Soc.* **1987**, *109*, 7803.

(35) Inagaki, M.; Shingaki, T.; Nagai, T. *Chem. Lett.* **1982**, 9.

(36) Autrey, T.; Schuster, G. B. *J. Am. Chem. Soc.* **1987**, *109* , 5814.

(37) Poppinger, D.; Radom, L. *J. Am. Chem. Soc.* **1978**, *100* , 3674.

(38) McConaghy, J. C.; Lwowski, W. *J. Am. Chem. Soc.* **1967**, *89* , 2357, 4450.

MEMBRANE PROTEIN LABELLING WITH PHOTOREACTIVE PHOSPHOLIPID ANALOGUES

Cesare MONTECUCCO, Giampietro SCHIAVO, Emanuele PAPINI, Maurizio TOMASI[+]
and Roberto BISSON

C.N.R. Center for Biomembranes and Department of Biomedical Sciences,
University of Padua, Via Trieste, 35131 Padova, ITALY and
[+]Laboratory of Cell Biology, Istituto Superiore di Sanita', Viale Regina
Elena 29, 00161 Roma, ITALY

SUMMARY
Radioactive phospholipid analogues carrying a photoactivatable group
at different positions of the molecule have been synthesized to probe
lipid-protein interactions at different levels of the lipid bilayer. On
illumination, those protein regions exposed to lipids become labelled
and can be identified thus providing information on the membrane
topology of proteins and on transient lipid-protein interactions.

1. INTRODUCTION
Our present knowledge of the structure of membrane proteins is
poorer than that of soluble proteins, mainly because of the difficulties
in obtaining X-ray grade crystals of membrane proteins and in the
handling of hydrophobic peptides. It appears that membrane proteins have
a variety of sizes and shapes and that they show different organizations
with respect to the lipid bilayer (1). We can distinguish the following
broad groups of membrane proteins: A) integral proteins, B) periplasmic
proteins, C) proteins with transient lipid-protein interactions.
A) This group includes those proteins that translocate biological
objects in a vectorial reaction sometime coupled to an enzymatic
activity. These proteins generally have a large hydrophobic sector with
several peptide stretches transversing the lipid bilayer and they may be
formed by one or more subunits up to the twenty five of the
mitochondrial complex I (2). Most frequently the hydrophobic domain
involved in lipid-protein interactions is rich in α-helices running
across the lipid bilayer (3), though an integral membrane protein can
also be formed by different structures (4). These α-helices are 18-24

P. E. Nielsen (ed.), Photochemical Probes in Biochemistry, 43–58.
© 1989 by Kluwer Academic Publishers.

residues long, the lenght needed to span the membrane in such a conformation. The proportion between the water-exposed part(s) and the membrane-embedded sector shows a great variation from the ATP-synthase, with a largelly predominant mass projected out of the membrane (5), to bacteriorhodopsin, which is nearly completely comprised within the lipid bilayer (6).

This group also includes an heterogeneous set of proteins with the common feature of possessing a single transmembrane segment 20-24 residues long, most probably arranged in an α-helical conformation. In such a way the particular function proper of that protein is linked to a particular type of membrane. A common feature of these proteins is that most of their mass is external to the lipid bilayer. Many receptors and several membrane-bound enzymes fall in this category.

B) These proteins are localized at the membrane surface with no or very little parts protruding into the hydrophobic phase of the lipid bilayer. They vary for their mode of attachment to the membrane. A first group interacts with the polar head group of lipids and may show different selectivity for the various lipid classes (7) A second group is bound to the membrane via interactions with integral proteins as in the case of spectrin (8). Recently it has been shown that several proteins, including 5'-nucleotidase and the VSV antigen of Trypanosoma, are covalently bound to the inositol ring of phosphatidylinositol via a Cys residue and glycan and glucosamine moieties (9).

C) A particular case of lipid-protein interactions is shown by a group of proteins of great relevance in physiopathological processes such as perforins and complement (10), viral proteins (11) and protein toxins (12). These proteins are water-soluble and yet under certain conditions they become able to penetrate into the hydrophobic core of the lipid bilayer. This phenomenon involves a conformational rearragement of their structure that can be triggered by proteolysis, disulfide reduction, pH changes or other mechanisms (12-14).

2. MEMBRANE LABELLING

The need of detailed molecular information on the spatial arrangement of proteins with respect to the membrane and of methods suitable to follow transient lipid-protein interactions has driven efforts at the development of membrane labelling reagents.

An ideal probe for membrane studies should possess the following properties:

a) **labelling restricted to a selected region of the membrane.** This can be most conveniently achieved by anchoring the reactive group to a selected position of a phospholipid molecule.

b) **high reactivity and short lifetime,** in order to label also protein segments rich in unreactive aliphatic lateral chains and to follow rapid phenomena.

c) no chemical reactivity, to avoid labelling during the manipulation of the biological system required to reach the situation that is to be studied.

d) high sensitivity of detection, to permit the study of very minute amounts of materials without perturbing effects due to the need of using large amount of probe.

e) stability of reagents and of their products, to allow the performance of many experiments before the formation of by-products that may interfere with the phenomenon under study and to allow the identification of the sites of protein modification.

Requirements b), c) and e) are reasonably meet by nitrene and carbene precursors (15-20). These molecules may be induced to generate in situ and only when needed the highly reactive nitrene and carbene intermediates by illumination with long wave ultraviolet or visible light, that do not damage most proteins. The instability of the electron-deficient nitrene and carbene species leads to their reaction with neighbouring molecules or to their quenching by reaction with water or with the precursor or to their rearrangement. The relative rates of the different routes of reaction depend on the reactivities of the intermediate and of the partner molecule, on the concentration of the reagent in the membrane, on its tendency to rearrange. As a result the yield of formation of covalent derivatives with proteins in biological membranes is generally very low. Carbenes are generally more reactive than nitrenes and their patterns of labelling should be regarded as more faithful reports on the actual lipid-protein interaction experienced in the membrane under study. However this potential advantage is lessened in biological membranes because the major factor governing the result appears to be the reactivity of the membrane protein rather than that of the photoactive reagent (18,19). A further drawback of nitrenes and carbenes is that it is difficult to establish their lifetime in the biological membrane under study and this prevents their use in probing rapid phenomena.

Under red light, many nitrene and carbene precursors whithstand to nearly all biochemical manipulations except the treatment with thiols (21). They also are very stable to storage (years for nitroarylazides, our unpublished observations), even in the tritiated form, provided that they are stored at - 20° C in dilute solutions in absolute ethanol doped with toluene. Unfortunately much less is known about the chemical stability of their covalent derivatives with the various lateral residues of proteins. It appears that most derivatives are stable to the conditions required to perform SDS-PAGE, including boiling in the presence of reducing agents, and that they are sufficiently stable under the conditions required for amino acid sequence determination.

Requirement **d)** can be best meet by making the reagent radioactive. The higher is the specific radioactivity of the probe the smaller is the amount needed to provide a good signal to noise ratio and this is important in order to minimize the possibility of perturbations.

Requirement **a)** is a most important point to be fulfilled if detailed information on the membrane topology of a protein are required. A first generation of probes was that of small molecules whose membrane localization is due solely to their lipophilicity. Their usefulness is limited by serveral drawbacks. In fact their position within the lipid bilayer is undefined and hence they provide information with no spatial resolution. Moreover they are frequently adsorbed into hydrophobic pokets of the water-exposed portion of a protein thus leading to an erroneous attribution (18,19). The same result may be obtained if the photogenerated intermediate is less hydrophobic than its precursor and repartitions out of the membrane in the water phase. Moreover, when studying lipid-protein interactions, areas of strong lipid binding may be not accessible to a lipid-unrelated probe molecule.

These drawbacks can be overcome by anchoring the photoactive group to a selected position of the class of lipid molecule that one intends to study (22). If one wants to investigate unspecific lipid interactions, analogues of the most membrane aboundant lipid such as phosphatidylcoline and sphyngomielin should be prepared, while the interaction with a specific class of lipids is followed by preparing the correspondent photoactive analogue.

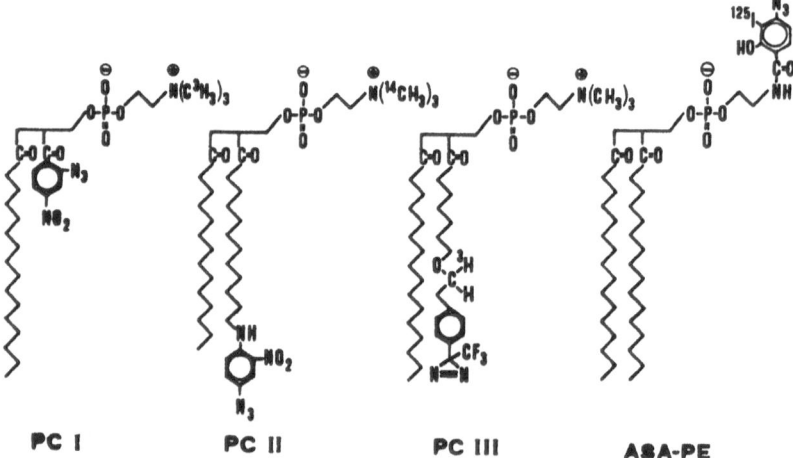

Fig. 1. Photoactive phospholipid analogues used in membrane labelling.

Fig. 1 shows the lipid analogues that we have used in different studies. PC I and PC II bear their photoactive group at two different levels of one fatty acid chain to probe different regions of the lipid bilayer.

Only the protein surface interacting with the polar head groups of phospholipids will be labeled by the tritiated PC I while protein regions intercalated into the deeper hydrophobic core of the membrane will be cross-linked to the $[^{14}C]$-labeled probe PC II. Hence by following the $^3H/^{14}C$ labeling ratio it is possible to obtain information on the varying depths of membrane penetration of different polypeptide chains or of different parts of the same protein molecule (17-20). A priori it cannot be excluded that the fatty acid chain of PC II may kink thus bringing the photoactive group near the membrane surface. However all the evidence available indicate that this does not occur significantly (14,23,24 and below). This possibility can be tested by including in the system under study glutathione, a membrane impermeant thiol reagent, that will quench immediately any water-exposed nitrene or carbene. Although at a superficial look PC I may resemble a lysophosphatidylcholine detergent molecule it does not behave as such because the OH in position 2 is not free, but it is esterified with an aromatic acid and hence it does not label proteins from the water phase as shown in several different systems (25-29).

PC III is similar to PC II, but contains the trifluoromethyl-diazirinylphenyl photoactive group that generates a carbene intermediate more reactive than the nitroarylnitrene formed by PC II (30).

ASA-PE has been recently introduced (31) to probe the very external surface of the membrane with its photoactive group coupled to the terminal amino group of dipalmitoylphosphatidylethanolamine and containing the amido and the OH groups both able to hydrogen bond water.

3. EXPERIMENTAL PROCEDURES

Several methods can be followed to perform a labeling experiment with photoactive phospholipids. They can be incorporated in the membrane by interdispersion among the other lipids during the usual procedures to form model membranes (17,32) or by incubation of a sonicated sample of the phospholipid reagent with the biological membrane to be probed (24,30,33-35). Only short illuminations (minutes) with a low wattage long wave ultraviolet lamp are sufficient to activate the probes depicted in fig. 1. The labelled proteins can be most conveniently identified by subjecting the sample to SDS-PAGE, which is very effective in removing all the unbound photoactivated phospholipids that run at the gel front. The amount of radioactivity bound to the resolved protein bands on the gel can be determined either by gel slicing and counting, as shown in figures 3 and 4, which permits the evaluation of double labelings or by fluorography and densitometry as in the case of table I.

4. HYDROPHOBIC PHOTOLABELLING OF CYTOCHROME C OXIDASE

Cytochrome c oxidase is the terminal component of most respiratory chains. In mitochondria it is a complex membrane-bound enzyme whose number of different subunits varies from thirteen in mammals to seven in lower eukaryotes. The three largest subunits are coded for by mitochondrial genes, while the others are synthesized on cytoplasmic ribosomes (36). A low resolution picture of the beef heart enzyme shows a Y-shaped molecule spanning the lipid bilayer with a major portion embedded in the membrane (37). As part of a study aimed to the definition of the topology of this enzyme complex, hydrophobic photolabeling was used to identify those subunits forming the lipid-protein boundary (38,39). Table I reports the relative amounts of labelling obtained with the two probes PC I and PC II with a detergent-solubilized, fully active, bovine heart cytochrome c oxidase.

table I

a comparison of the labeling of the subunits of yeast (Y) and bovine heart (B) cytochrome c oxidases with PC I and PC II in lauroyl maltoside. Comparisons are made for those subunits that show sequence homologies. Roman numbers identify the subunits of the bovine enzyme on the base of their order of migration in SDS-PAGE. Yeast subunits VIIa and VII comigrate in SDS-PAGE and hence labeling refers to their sum

polypeptide		residues		hydrophobic segments		PC I		PC II	
B	**Y**	**B**	**Y**	**B**	**Y**	**B**	**Y**	**B**	**Y**
I	I	514	510	12	12	++	+++	+++	+++
II	II	227	251	2	2	+	++++	++	++++
III	III	261	246	7	7	++	++++	++	+++++
IV	Va	147	153	1	1	++	++	++	++
V	VI	109	149	0	0	-	-	-	-
VI	IV	98	155	0	0	-	-	-	-
VII		85		0		-	-	-	-
VIII		84		1		++	-	++++	-
IX	VIIa	73	58	0	1?	-	}+++	-	}+
X	VII	56	57	1	1	+			
XI		50		?		++++		-	
XII	VIII	47	41	1	1	+	+	+++	-
XIII		46		1		-		-	

Several subunits of the bovine heart mitochondrial cytochrome c oxidase
are labelled and this is a indication that they are exposed to lipids
and form the lipid-protein boundary of this enzyme. Nothing can be said
about the location of the non labelled subunits because a lack of
labelling can be attributed to their being shielded or to an external
location with respect to the lipid bilayer as well as to a lack of
reactivity versus the nitrene intermediates. After this result was
obtained, the amino acid sequence became available and it is interesting
to note that all the labeled subunits have long uncharged stretches that
are believed to be arranged as transmembrane α-helices. Table I also
reports the result obtained under the same conditions with the yeast
enzyme, whose subunit composition is simpler. The three largest subunits
are labelled together with two smaller polypeptides. Subunit IV,
labelled in the bovine heart enzyme, is not labelled here, while subunit
Va is labelled in the yeast oxidase. This unexpected result can be
explained on the basis of their primary structures showing a close
similarity of bovine subunit IV with yeast subunit Va and of yeast
subunit IV with the unlabelled subunit VI of the bovine oxidase. As
noted above, both beef subunit IV and yeast subunit Va possess a single
long uncharged stretch that is likely to be arranged as a transmembrane
-helix. Another relevant point emerging from the above results is that
the patterns of labeling of PC I and PC II are different, further
indicating that they do report about regions of the membrane of
different depths. Although the amount of labeling cannot be strictly
correlated strictly to the amount of protein surface exposed to lipids
because of the different reactivity of the various amino acid residues,
the result suggests that subunit XI contributes to form the
lipid-protein boundary at the phospholipid head group level and shields
from the contact with lipids subunit XII that is exposed deeper in the
hydrophobic phase of the membrane.

The reliability of PC I in labelling membrane-embedded segments is
further emphasized by a detailed analysis carried out to identify the
amino acid residues cross-linked by PC I in cytochrome c oxidase subunit
II (28). Subunit II was fragmented and the pattern of fragmentation is
shown in the top panel of fig. 2. Fig. 2 also reports the elution
profile of the fragment mixture obtained by column cromatography (upper
trace). The determination of the radioactivity associated to the various
fractions (lower trace) shows that only those segments containing the
uncharged stretches of the lenght needed to span the membrane (black
boxes in the figure) are labelled. The bottom panel shows those labeled
residues that could be identified by Edman degradation.The labelled
residues lie within a short polar flanking region at the begin of the
first uncharged segment of subunit II.   These residues are those

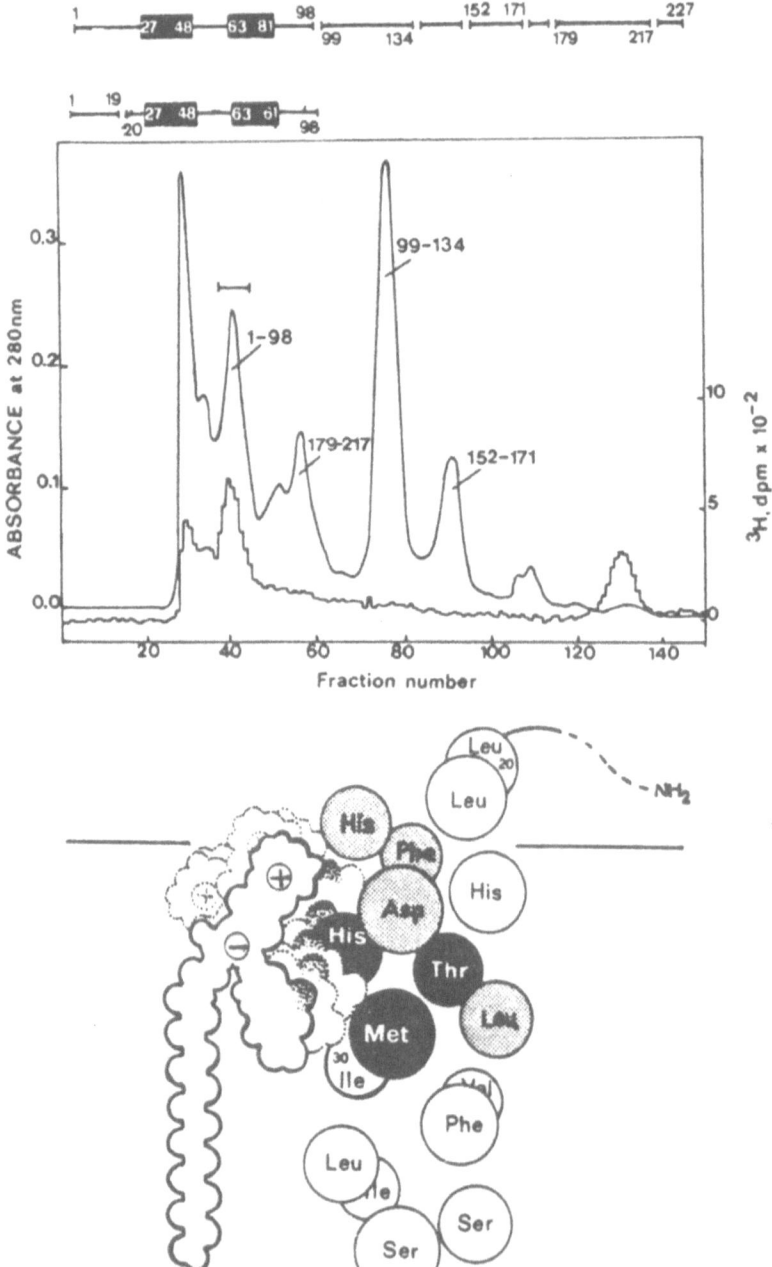

Fig. 2. Identification of sites of cross-linking of PC I with subunit II of bovine heart cytochrome c oxidase. The enzyme was labelled, and the

various subunits were separated by chromatography in the presence of SDS and subunit II was fragmented with trypsin. Fragments were loaded on a column of Bio-Gel P 30 and eluted with 70% acetic acid. The front peak is an aggregate of various fragments The second peak is fragment 1-98 and it was cut at residue 19 with S. aureus V protease; fragment 20-98 was isolated by chromatography on a Bio-Gel P 30 column eluted with 70 % acetic acid and sequenced by Edman degradation in a spinning cup Beckmann sequenator. The radioactivity released at each step was determined: black circles mark highly labelled residues, grey circles represent weaker labeling.

accessible to the photoactive group if the segment is arranged as a α -helix spanning the membrane. Unfortunately, for technical reasons, it was not possible to determine the other sites of cross-linking, which would have allowed to determine if the same arrangement is adopted also by the second uncharged segment.

5. HYDROPHILIC LABELLING OF MEMBRANE BOUND TETANUS NEUROTOXIN

Tetanus toxin is a dichain protein that binds specifically and very strongly to nerve cells. The membrane components involved in such binding have not been identified, but there are evidence that both a protein and lipids are involved (40). It has been proposed that polysialogangliosides, highly enriched in the nerve tissue, play a major role in tetanus toxin binding (41). We have investigated the interaction of tetanus toxin with lipid bilayers of different compositions containing or not containing gangliosides (42). The interaction was monitored with ASA-PE, interdispersed in minute amounts among the other lipids, by measuring the amount of reagent cross-linked to the two chains of the protein after illumination and removal of the unbound reagent with SDS-PAGE. Table I reports the amount of probe bound to the two chains of tetanus toxin after incubation with liposomes of different composition in a physiological solution and illumination. It appears that the toxin interacts strongly with vesicles with a net negative charge such as those made of the asolectin lipid mixture and very poorly with the zwitterionic eeg lecithin vesicles. The presence of purified mixed brain gangliosides does not lead to a relevant increase in toxin binding in both the asolectin and the egg lecithin vesicles. The highest labeling was obtained in the presence of liposomes made with a purified

table II

Labeling of tetanus toxin incubated with liposomes of the indicated compositions and containing the surface photoactive reagent $\begin{bmatrix} ^{125}I \end{bmatrix}$-ASA-PE. The amount of radioactivity bound to the two chains of the toxin is espressed as percentage of the radioactivity bound to the heavy chain of the toxin illuminated in the presence of asolectin vesicles containing gangliosides.

	Heavy chain	Light chain
Tetanus toxin with:		
Egg Lecithin	9–22	9–13
Egg Lecithin + Gangliosides	6–15	7–13
Asolectin	36–93	84–112
Asolectin + gangliosides	100	84–129
Bovine brain lipids	204	195

preparation of bovine brain lipids, whose major constituents were sulphatides, cerebrosides and gangliosides. These results indicate that tetanus toxin interacts with negatively charged membrane surfaces and that both its chains are involved. Ganglioside do not add much to this interaction and this is not in agreement with their proposed role as toxin acceptors. A possibility that emerges from this study is that a major contribution to the lipid fixation of tetanus toxin is provided by a unspecific interaction with negative lipids (that may include gangliosides) and that a minor role is played by a specific interaction with the oligosaccharide portion of gangliosides.

6. HYDROPHOBIC PHOTOLABELLING TO PROBE THE MEMBRANE INSERTION OF PROTEIN TOXINS

A large group of protein toxins relased by bacteria or plants bind to the cell surface and have intracellular targets (43); hence, somehow, they must be able to cross the hydrophobic membrane barrier. A first group of toxins, including cholera toxin and pertussis toxin, ADP-ribosylate proteins localized on the cytoplasmic face of the plasma membrane. A second group act on targets localized in the cytoplasm such as diphtheria toxin that ADP-ribosylates elongation factor 2 thereby blocking protein synthesis. We have investigated the mechanism of insertion of cholera and diphtheria toxins in lipid bilayers with PC I and PC II (13,14,44).

Cholera toxin is made of a binding part composed of five identical subunits termed B (11 kDa), of an enzymic subunit termed A1 (21 kDa) and of a joining fragment termed A2 (9 kDa). A1 is linked to A2 via a disulfide bond, whose reduction is required to freed the enzymatic activity of A1. Each B subunit possesses a single binding site for the oligosaccharide portion of ganglioside G_{M1}, the specific cellular receptor of cholera toxin.

The left panel of fig 3 shows the pattern of labeling of cholera toxin illuminated in the presence of G_{M1}-containing egg lecithin liposomes, doped with PC I and PC II.

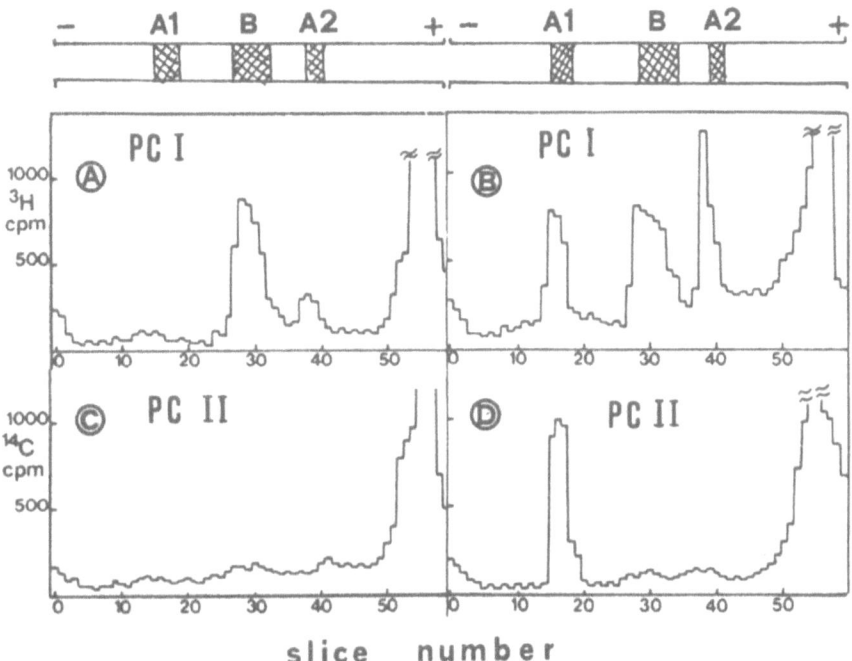

Fig. 3. Patterns of labeling with PC I (A and B) and with PC II (C and D) of cholera toxin illuminated in the presence of egg lecithin vesicles containing 1 % of the monosialoganglioside G_{M1} before (panels A and C) or after the reduction of the toxin with 20 mM reduced glutathion and dialysis (panels B and D). The amount of radioactivity bound to the various subunits was determined after SDS-PAGE, staining with Coomassie Blue (staining patterns are shown at the top of the figure) slicing and counting.

Only subunits B and A2 are labeled and only with the superficial probe PC I thus suggesting that cholera toxin binds to the liposome surface via the B subunits. The right panels show the patterns of labeling obtained after reduction of the liposome-bound toxin. While there is no change at the level of B, A2 is more labeled with PC I and subunit A2 is now labeled with both PC I and PC II.

The simplest interpretation of these results is that reduction of the single interchain disulfide bond is sufficient to bring about a gross conformational rearrangement of the toxin. Subunit A2 becomes hydrophobic and able to insert into the lipid bilayer to reach the other face of the membrane, where it exerts its activity. Our proposal that disulfide reduction is the trigger of the membrane insertion of cholera toxin (13) has been recently confirmed by 3-D image reconstruction of two-dimensional crystals of cholera toxin formed on G_{M1}-containing lipid monolayers before and after reduction (23).

Another important information obtained from this experiment is that the photoactive group of PC II does not appreciably fold back to the membrane surface because it does not label B. This important point is reinforced by the result obtained with cytochrome c oxidase and by the comparison of the patterns of labeling and of the sequences of the acetyl choline receptor from Torpedo marmorata and Torpedo californica (24), clearly showing that also in this system PC II does not label at the membrane surface.

Diphtheria toxin is made of two chains held together by a single interchain disulfide bond: chain A (21 kDa) is the enzymic part and chain B (37 kDa) is responsible for cell binding. Several evidence indicate that DT enters into cells via receptor-mediated endocytosis and escapes into the cytoplasma from an acidic intracellular compartment. The toxin undergoes a low pH-driven conformational change, that enables it to interact with the hydrocarbon chain of lipids (45) and alter the conductance of lipid bilayers (46). The pH range where the transition occurs is between 5 and 6 as determined by following the change of turbidity and the fusion of liposomes of DPPC-DPPA incubated with DT at different pHs (47). However, all these measurements could not discriminate between the role of the two toxin protomers. Figure 4 shows that this limitation does not hold for hydrophobic photolabelling, which is based on the analysis of the distribution of radioactivity bound to the toxin chains after their separation by SDS-PAGE. It clearly appears from fig. 4 that DT at low pH becomes able to interact with the hydrocarbon chains of phospholipids and that this interaction is mediated by both DT protomers. No labelling was found with PC II at neutral pH, while both chains were significantly labeled with PC I-tagged negatively charged phospholipid vesicles. The simplest interpretation of these results is that DT binds to the surface of negatively charged liposomes at neutral pH and that low pH both DT chains become hydrophobic and insert in the lipid bilayer. The present findings do not support the tunnel model put forward by Hoch et al (47)

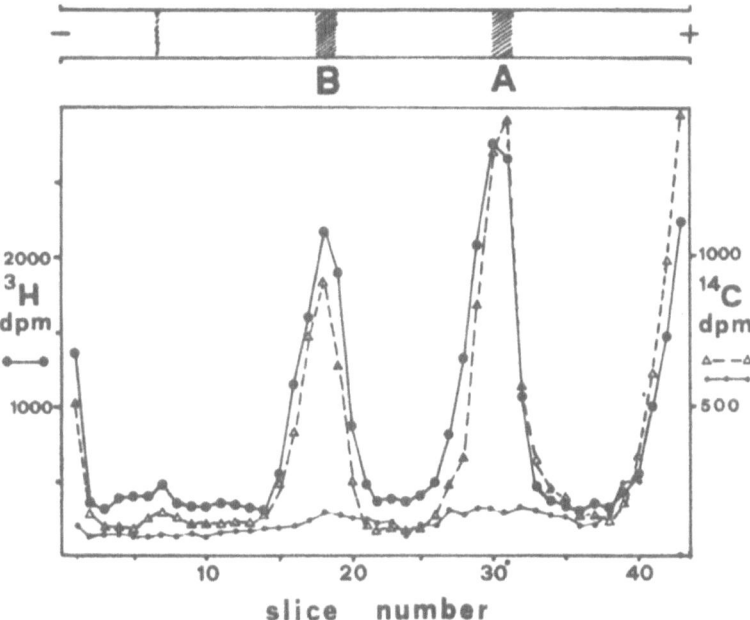

Fig. 4. Hydrophobic photolabelling of diphtheria toxin in the presence of asolectin vesicles. The top panel shows the Coomassie blue staining profile of the toxin electrophoresed on a SDS-gel after incubation with asolectin vesicles and illumination at neutral pH (PC II, •—•—•—•—•—) or at pH 4.5 with PC I (●   ●   ●   ) and PC II(—Δ—Δ—Δ—Δ—).

to account for the membrane translocation of diphtheria toxin. This model envisages that at acidic pHs fragment B inserts in the lipid bilayer and forms a hydrophilic tunnel large enough to allow the passage of the A chain in an extended form and without any contact with lipids. Here we have shown that chain A at low pH does interact with the fatty acid chains of phospholipids and we have discussed a different model (48).

CONCLUSIONS

The results described in the present paper indicate that photoactive phospholipid analogues can provide several important information on the structure and function of membranes. They are useful in determining which subunit and which segments of an integral membrane protein form the lipid-protein boundary. Moreover information on the depth of penetration in the lipid bilayer can be obtained. They can also be used advantageously to probe the insertion of proteins into membranes. An area where development is required is that of time-resolved photolysis to follow the very rapid phenomena that frequently occur in the two-dimensional solvent lipid bilayer.

AKNOWLEDGEMENTS:

We thank Prof. G.F. Azzone for encouragement and support. Part of the work described here has been supported by a grant from "Regione Veneto".

REFERENCES

1. Martonosi, A., Editor (1982) "Membranes and Transport", Plenum Press, New York, voll. 1-4.
2. DePierre, J.W. and Ernster, L. (1977) Annu. Rev. Biochem. **46**, 201-262
3. Wickner, W.T. and Lodish, H.F. (1985) Science **230**, 400-407
4. Rosenbush, J.P. (1974) J. Biol. Chem. **249**, 8019-8029
5. Senior, A.E. (1988) Physiol. Rev. **68**, 177-231
6. Henderson, R. and Unwin, P.N.T. (1975) Nature **257**, 28-32
7. Nielsen, N.C., Zahler, W.L. and Fleischer, S. (1973) J. Biol. Chem. **248**, 2556-2562
8. Bennet, V. (1985) Annu. Rev. Biochem. **54**, 273-304
9. Low, M.C., Ferguson, M.A.J., Futerman, A.H. and Silman, I. (1986) Trends Biochem. Sci. **11**, 212-215
10. E-Young, J.D., Cohn, Z.A. and Podack, E.R. (1986) Science **233**, 184-190
11. White, J., Kielan, M. and Helenius, A. (1983) Quart. Rev. Biophys. **16**, 151-185
12. Olsnes, S. and Sandvig, K. (1985) in "Endocytosis", Pastan, I and Willingham, M.C. Eds., Plenum Press, New York pp. 196-234
13. Tomasi, M. and Montecucco, C. (1981) J. Biol. Chem. **256**, 11177-11181
14. Montecucco, C., Schiavo, G. and Tomasi, M. (1985) Biochem. J. **231**, 123-128
15. Brunner, J. (1981) Trends Biochem. Sci. **6**, 44-46
16. Bayley, H. (1983) "Photogenerated Reagents in Biochemistry and Molecular Biology", Nort-Holland/Elsevier, Amsterdam
17. Bisson, R. and Montecucco, C. (1981) Biochem. J. **193**, 757-763
18. Bisson, R. and Montecucco, C. (1985) in "Progress in Protein-Lipid Interactions", Watts, A and DePont, J.J,H.H.M. Eds., Elsevier Science Publ., Amasterdam, pp. 259-287

19. Bisson, R. and Montecucco, C. (1986) in "Techniques for the Analysis of Membrane Proteins", Ragan, C.I. and Cherry, R.J. Eds., Chapman and Hall, London, pp. 153-184

20. Montecucco, C. (1988) Methods Enzymol. 165, 347-357

21. Staros, J.V., Bayley, H., Standring, D.N. and Knowles, J.R. (1978) Biochem. Biophys. Res. Commun. 80, 568-572

22. Chakrabarti, P. and Khorana, H.G. (1975) Biochemistry 14, 5021-5033

23. Ribi, H.D., Ludwig, D.S., Mercer, K.L., Schoolnik, G.K. and Kornberg, R.D. (1988) Science 239, 1272-1276

24. Giraudat, J., Montecucco, C., Bisson, R. and Changeux, J.P. (1985) Biochemistry 24, 3121-3127

25. Bisson, R., Montecucco, C. and Capaldi, R.A. (1979) F.E.B.S. Lett. 106, 317-320

26. Montecucco, C., Bisson, R., Dabbeni-Sala, F., Pitotti, A. and Gutweniger, H. (1980) J. Biol. Chem. 255, 10040-10043

27. Montecucco, C., Dabbeni-Sala, F., Friedl, P. and Galante, Y.M. (1983) Eur. J. Biochem. 132, 189-194

28. Bisson, R., Steffens, G.C.M. and Buse, G. (1982) J. Biol. Chem. 257, 6716-6720

29. Hoppe, J., Montecucco, C. and Friedl, P. (1983) J. Biol. Chem. 258, 2882-2885

30. Brunner, J., Spiess, M., Aggeler, R., Huber, P. and Semenza, G. (1983) Biochemistry 22, 3812-3820

31. Gao, Z. and Bauerlein, E. (1987) F.E.B.S. Lett. 223, 366-370

32. Montecucco, C., Schiavo, G., Brunner, J., Duflot, E., Boquet, P. and Roa, M. (1986) Biochemistry 25, 919-924

33. Gutweniger, H. and Montecucco, C. (184) Biochem. J. 220, 613-616

34. Volpe, P., Gutweniger, H. and Montecucco, C. (1987) Arch. Biochem. Biophys. 253, 18-145

35. Moscufo, N., Gallina, A., Schiavo, G., Montecucco, C. and Tomasi, M. (1987) J. Biol. Chem. 262, 11490-11496

36. Azzi, A. (1980) Biochim. Biophys. Acta 594, 231-252

37. Capaldi, R.A. (1982) Biochim. Biophys. Acta. 694, 291-306

38. Bisson, R., Montecucco, C., Gutweniger, H. and Azzi, A. (1979) J.Biol Chem. 254, 9962-65

39. Gutweniger, H., Bisson, R. and Montecucco, C. (1981) Biochim. Biophys. Acta 635, 187-193

40. Montecucco, C. (1986) Trends. Biochem. Sci. 11, 314-317

41. Mellanby, J. and Green, J. (1981) Neuroscience 6, 281-300

42. Montecucco, C., Schiavo, G., Gao, Z., Bauerlein, E., Boquet, P. and DasGupta, B.R. (1988) Biochem. J. 251, 379-383

43. Middlebrook, J.L. and Dorland, R.B. (1984) Microbiol. Rev. 48, 199-221

44. Papini, E., Schiavo, G., Tomasi, M., Colombatti, M., Rappuoli, R. and Montecucco, C. (1987) Eur. J. Biochem. **169**, 637-644
45. Blewitt, M.G., Chung, L.A. and London, E. (1985) Biochemistry **24**, 5458-5464
46. Hoch, D.H., Romero-Mira, M., Ehrlich, B., Finkelstein, A., DasGupta, B.R. and Simpson, L.L. (1985) Proc. Natl. Acad. Sci. U.S.A. **82**, 1692-1696
47. Papini, E., Colonna, R., Cusinato, F., Montecucco, C., Tomasi, M. and Rappuoli, R. (1987) Eur. J. Biochem. **169**, 629-635
48. Bisson, R. and Montecucco, C. (1987) Trends Biochem. Sci. **12**, 187-188

[125]J-ASA-PE, A PHOTOACTIVABLE, RADIOACTIVE PHOSPHOLIPID-ANALOGUE, DESIGNED FOR PHOTOLABELING PROTEINS IN CONTACT WITH PHOSPHOLIPID HEAD GROUPS

Selective photolabeling of subunits α in thermophilic ATPsynthase $TF_o \cdot F_1$

Zhan Gao and Edmund Baeuerlein
Max-Planck-Institut fuer Biochemie
D-8033 Martinsried Fed. Rep. Germany

ABSTRACT. To photolabel proteins, which are in close proximity to phospholipid head groups on membranes, N-(4-azidosalicyl)phosphatidylethanolamine, ASA-PE, was synthesized. Its special advantage is that it can be radioiodinated in the last step of the synthesis to the [[125]J]-iodo compound ([125]J-ASA-PE) with very high specific activity. This is necessary because the 4-azido-3[[125]J]iodo-2-hydroxybenzoyl group is located on the membrane surface and reacts during photolysis as well with proteins as with water. Additionally high specific activity allows to insert low amounts of [125]J-ASA-PE into proteoliposomes thus preserving biological activity of their proteins. To study structure and function of the thermophilic ATPsynthase $TF_o \cdot F_1$ of the thermophilic bacterium PS 3, it was first reconstituted with soybean phospholipids into liposomes and these were then incubated with [125]J-ASA-PE to provide preferred incorporation into the outside of the phospholipid bilayer. After photolysis subunits α of $TF_o \cdot F_1$ were labeled selectively. It is, therefore, possible that subunits α may be important in the interaction of TF_1 with TFo, to provide internal energy coupling between the proton channel TFo and ATP synthesis in TF_1.

INTRODUCTION

In the last ten years some photoreactive phospholipid analogues were used as "hydrophobic labeling reagents" to study the interaction of membrane proteins with the lipid bilayer. The photoactivable groups were anchored to selected positions of one fatty acid chain of a phospholipid. Photolabeling was, therefore, restricted to the hydrophobic membrane phase and allowed to determine regions of membrane proteins, which are exposed to the lipid layer (see for review (1)).

We introduce here the photoreactive radioactive phospholipid [125]J-ASA-PE (Fig. 2), which was developed for "hydrophilic photolabeling" on the membrane surface. The photoactivable, 4-azido-salicyloyl group, bound as amide to the amino function of the polar head group in phosphatidylethanolamine, may be radioiodinated to a specific activity up to 2000 Ci/mmol.

[125]J-ASA-PE was used to analyze the subunit structure and binding of the hydrophilic, catalytic moiety TF_1 to the membrane sector TFo of the thermophilic ATPsynthase $TF_o \cdot F_1$ (2).

MATERIALS and METHODS

ATPsynthase $TF_o \cdot F_1$ was prepared from thermophilic bacteria PS3 according to Kagawa and Sone (3). Activities were measured after addition of purified asolectin (ATPase activity: 13.7 - 15.2 U/mg protein at 56°C). Asolectin was purchased from Associated

P. E. Nielsen (ed.), Photochemical Probes in Biochemistry, 59–65.
© *1989 by Kluwer Academic Publishers.*

Concentrates, N.Y., Gramicidin D(Dubos) and 1,2-dipalmitoyl-sn-glycero-3-phospho-ethanolamine from Sigma, 1-aminonaphthalene-4-sulfonic acid, sodium salt from Aldrich and 2-amino-1,5-disulfonic acid from Fluka. The N-hydroxysuccinimideester of 4-azidosalicylic acid (ASA-NHS) was synthesized according to Ji and Ji (4). All operations with the photolabels ASA-PE and ^{125}J-ASA-PE were performed under red safety light.

Synthesis of the photolabel 1,2-dipalmitoyl-sn-glycero-3-phospho-N-(4-azido-2-hydroxybenzoyl)ethanolamine (ASA-PE).

100 mg (0.15 mmol) 1,2-dipalmitoyl-sn-glycero-3-phosphoethanolamine (PE) were dissolved in destilled chloroform and 90 mg (0.33 mmol) N-hydroxysuccinididyl-4-azidosalicylic acid (ASA-NHS) were added in the dark, followed by 300 μl triethylamine. The solution was stirred for 5 days at 20°C. The reaction was controlled by TLC analysis on silica gel (Merck 60), developed with chloroform/methanol/water/triethylamine (65:35:4:2). If most of PE had reacted to ASA-PE, the solvent was evaporated. The residue was resuspended in 1 ml 0.1 M sodium hydroxide. After 5 min 5 ml water and 3 ml of a saturated sodium chloride solution were added. The precipitate was collected by centrifugation and resuspended in destilled chloroform. This solution was dried with sodium sulfate, filtrated and evaporated. Yield: 40 mg (32%) of ASA-PE. Anal. Calc. for $C_{44}H_{76}N_4O_{10}PNa$: C, 60.39; H, 8.75; N, 6.40. Found: C, 60.71; H, 8.8; N, 5.92: ^{1}H-NMR (90 MHz, CDCl$_3$): δ/ppm: 7.58(d, 1H); 6.56(s, 1H); 6.49(d, 1H); 5.23(m, 2H); 4.18(m, 6H); 3.69(m, 2H); 2.31(m, 4H); 1.25(m, 52H); 0.74(m, 6H). IR-spectrum: 2100 cm^{-1} (azido group).

Synthesis of the radioactive photolabel 1,2-dipalmitoyl-sn-glycero-3-phospho-N-(4-azido-3-[^{125}] iodo-2-hydroxybenzoyl)ethanolamine (^{125}J-ASA-PE).

8 μg (10 nmol) ASA-PE and 2 mCi Na^{125}J (1 nmol) were dissolved in 200 μl of 0.1 M phosphate buffer, pH 7.4. Radioiodination of ASA-PE was initiated by the addition of 5 μl of 2 mM chloramin T and stopped after 10 min by the addition of 5 μl of 0.4% sodium hydrogensulfite. The product was precipitated by the addition of 200 μl of a saturated sodium chloride solution, extracted ten times with 200 μl chloroform. Yield of radioiodination: 1,78 mCi(spec.act: 200 Ci/mmol). ^{125}J-ASA-PE was analyzed by TLC on silica gel (Merck 60) and developed by chloroform/methanol/water/triethylamine (65:35:4:2); the products were made visible by iodine vapour. ASA-PE: Rf = 0.67; ^{125}J-ASA-PE: RF = 0.65. A single radioactive compound was detectable by a TLC scanner (Rf = 0.65).

Photolabeling of Gramicidin D by ^{125}J-ASA-PE.

2 mg Gramicidin D were sonified in 1 ml of a liposome solution, prepared with 50 mg asolectin by the cholate-dialysis method. 0.5 ml of these Gramicidin-liposomes were incubated with 0.1 μCi ^{125}J-ASA-PE for 2 h at 20°C. 0.5 ml of the unloaded liposomes were incubated with 0.1 μCi ^{125}J-ASA-PE in the same way. In 1 ml chloroform 1mg Gramicidin D and 0.1 μCi ^{125}J-ASA-PE were dissolved. Half of every solution was photolyzed with UV light and anlyzed by TLC as described below.

Photolabeling of 1-aminonaphthaline-4-sulfonic acid (ANS) and 2-aminonaphthaline-1,5-disulfonic acid (AND) by ^{125}J-ASA-PE.

Every 80 µl of liposomes, into which ^{125}J-ASA was incorporated, as described above, were purified by a Sepharose-4B column from free photolabel. The sodium salts of ANS or AND were added to these ^{125}J-ASA-PE liposomes to final concentrations of 0.1, 0.5, 1.0, 5.0 and 10 mM. For blanks ^{125}J-ASA-PE-liposomes alone and a solution of 50 µl 20 mM ANS and AND, resp. together with 0.5 µCi ^{125}J-ASA-PE were used. Half of every sample was photolyzed by UV light (> 300 nm) of a 125 W ultraviolett lamp (HPK 125W/L, Philips) for 15 min. 2 µl of every sample before and after photolysis were analysed by TLC on Silicagel 60 (Merck) using a solvent system of chloroform/methanol/water/triethylamine (5:35:4:2).

Photolabeling of the ATPsynthase $TF_o \cdot F_1$ reconstituted into liposomes.

$TF_o \cdot F_1$ proteoliposomes (protein/lipid ratio 1:100 w/w) were obtained by incubating preformed liposomes (5 mg asolectin) with $TF_o \cdot F_1$ (50 µg) as described in (2). 0.5 mCi of ^{125}J-ASA-PE (spec.act.: 200 Ci/mmol) which were dissolved in chloroform, were dried with a stream of nitrogen in a small cap. The $TF_o \cdot F_1$ proteoliposomes were added for an incubation time of 20 min at 37°C. Photolabeling was performed by irridation with UV light, as described above, for 10 min at 0°C. The samples were purified from free radioactive photolabel by centrifugation through 10% sucrose and analyzed by SPAGE as described in (2).

RESULTS AND DISCUSSION

A new photoreactive phospholipid, designed for hydrophilic photolabeling, was obtained by modifying 1,2-dipalmitoyl-phosphatidylethanolamine with an activated ester of 4-azido-salicylic acid to 1,2-dipalmitoyl-sn-glycero-3-phospho-N-(4-azido-2-hydroxy-benzoyl)ethanolamine (ASA-PE). This azido compound may be transformed to the highly reactive nitrene by ultraviolet light and during this reaction decreased the absorption maxima of ASA-PE at 282 nm and 312 nm (Fig. 1).
It is the advantage of ASA-PE that it may be easily radioiodinated to 1,2-dipalmitoyl-sn-glycero-3-phospho-N-(4-azido-3-[^{125}J]- iodo-2-hydroxybenzoyl)ethanolamine (^{125}J-ASA-PE, Fig. 2) up to a specific activity of about 2000 Ci/mmol. ^{125}J-ASA-PE may be incorporated into preformed liposomes almost completely by incubation for 90-120 min (Fig. 3).
During synthesis of ASA-PE disappeared the positive charge of the protonated amino group in phosphatidylethanolamine, because an amide bond was formed by acylation with azidosalicylic acid. Both this amide bond and the concomitantly introduced phenolic hydroxyl group were assumed to be together hydrophilic enough that the azidosalicylic group may be located on the surface of liposomes or natural membranes. This view was supported by studies with head group labeled fluorescent probes of phospha-tidylethanolamine, for example, N-(4-nitro- 2,1,3-benzoxadiazolyl) phosphatidylethanol-amine (NBD-PE) (5). Its fluorescence intensity was shown to be quenched by calcium, indicating its acessibility to hydrophilic molecules in the aqueous medium (6), though the fluorescent NBD-group was a heterocycle without hydrophilic functions. In addition to these considerations experiments were carried out to confirm the assumed location of the azidosalicylic acid function.
Gramicidin A, which is the main component (87,5 %) of Gramicidin D was introduced by Brunner and Richards (7) in the lipid-Gramicidin A model to test photoactivable

62

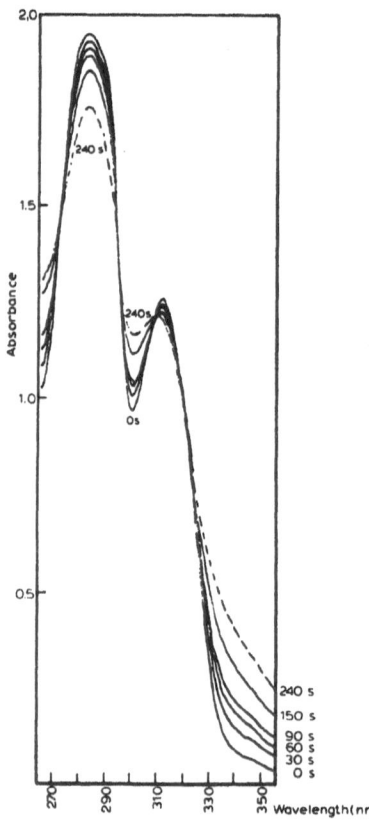

phospholipids for hydrophobic photolabeling. Gramicidin D was incorporated in our experiments into liposomes together with [125]J-ASA-PE. After irridation of these liposomes by ultraviolet light no cross-linking products of Gramicidin D with the photolabel could be detected (Fig. 4). Because dimers of the linear gramicidins are assumed to form transmembrane channels without hydrophilic, water-exposed domains, it was expected that Gramicidin D was not modified by the photolabel.

In the attempt to localize the photoreactive acidosalicylic group on the surface of membrane, an amphiphilic compound, 1-aminonaphthaline-4-sulfonate (ANS), as well as a hydrophilic one, 2-amino-naphthaline-1,5-disulfonate (AND), were used. [125]J-ASA-PE was incorporated into liposomes, which then were irridated with ultraviolet light in solution of various concentrations of ANS and AND, resp. The hydrophilic AND only was modified by [125]J-ASA-PE during photolysis (Fig.5), suggesting that the amphiphilic ANS was inserted mainly into the membrane and a smaller part of it located close to the phosphoryl group (8). From these results it appears to be very probable that the azidosalicylic group, bound to the amino group of phosphatidylethanolamine, extended to the bulk phase.

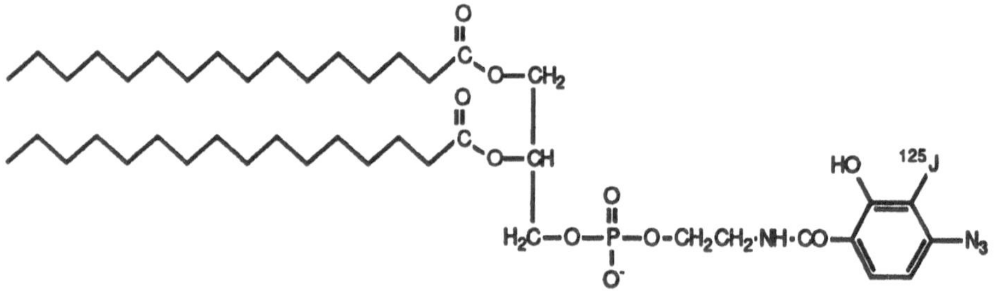

Fig. 1. UV spectra of ASA-PE after various times of irridation with ultraviolet light. Concentration: 280 mM ASA-PE in 50 mM Tricin-NaOH, pH 8.0.

Fig. 2. The radioactive and photoreactive phosphatidylethanolamine derivative [125]J-ASA-PE.

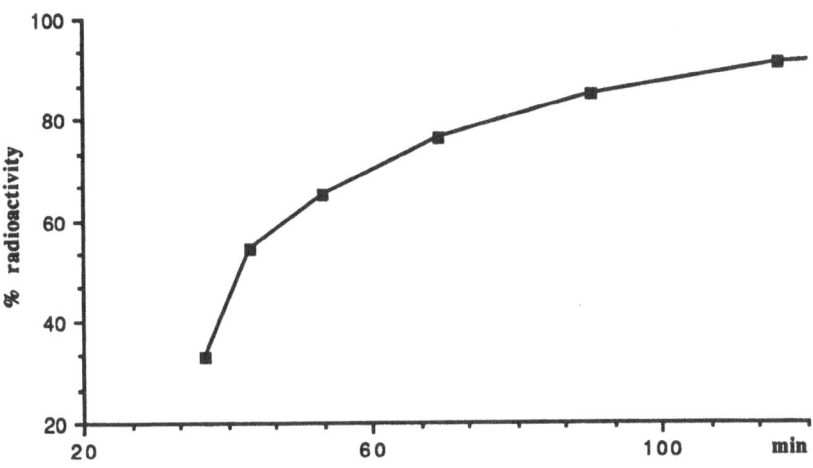

Fig. 3. Kinetics of the incorporation of ^{125}J-ASA-PE into liposomes. At various times 100 μl of a solution, in which 0.25μCi ^{125}J-ASA-PE in 2 μl of 0.1 M phosphate buffer pH 7.4 were incubated at 4°C with 2 ml of liposomes (50 mg asolectin), were separated by a Sepharose-4B column (0.5x10 cm) with 50 mM Tris-HCl pH 8.0. The radioactivity, incorporated into the liposomes, was determined.

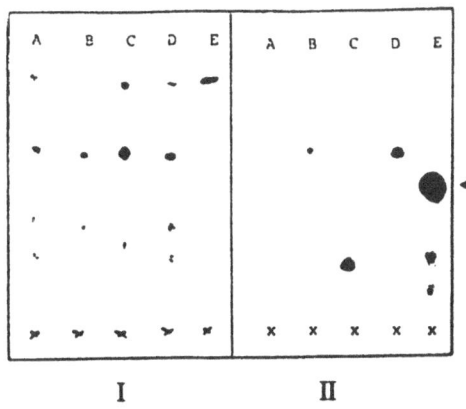

Fig. 4. Reaction of Gramicidin D with ^{125}J-ASA-PE under UV light. I. silica plate under UV light; II. autoradiopraphy of the same plate. A) Gramicidin in liposomes after irridation, B) ^{125}J-ASA-PE in liposomes after irridation, C) ^{125}J-ASA-PE and Gramicidin D without irridation, D) as C) but after irridation, E) Gramicidin and ^{125}J-ASA-PE in chloroform after irridation.Cross-linking product from Gramicidin and ^{125}J-ASA-PE as indicated by arrow. TLC analysis as described in Material and Methods.

^{125}J-ASA-PE was now used to study close to the membrane the subunit structure of the thermophilic ATPsynthase $TF_0 \cdot F_1$. The latter was incorporated into liposomes (lipid/protein ratio 100:1) and these preformed liposomes were added to ^{125}J-ASA-PE, which was distributed on the surface of a microvial, to provide preferred insertions of the photolabel into the outer surface of these proteoliposomes. After photolysis subunits α were labeled selectively (Fig. 6).

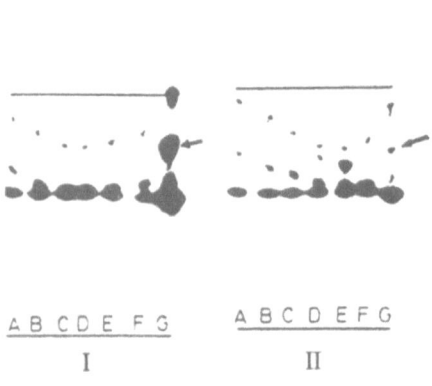

A B C D E F G   A B C D E F G

I   II

Fig. 5. Reaction of 1-aminonaphthalin-4-sulfonate (ANS) or 2-aminonaphthalin- 1,5-disulfonate (AND) with ^{125}J-ASA-PE under UV light. Autoradiography of TLC analysis as described in Material and Methods. Plate I: ANS. Plate II: AND. ^{125}J-ASA-PE was incorporated into liposomes, which were then incubated with solutions of 0.1 mM (A), 0.5 mM (B), 1 mM (C), 5 mM (D) or 10 mM (E) of ANS and AND, resp. and irridated by UV light. The reaction products of ANS and AND, resp., with ^{125}J-ASA-PE in water, without liposomes are indicated by an arrow (G).

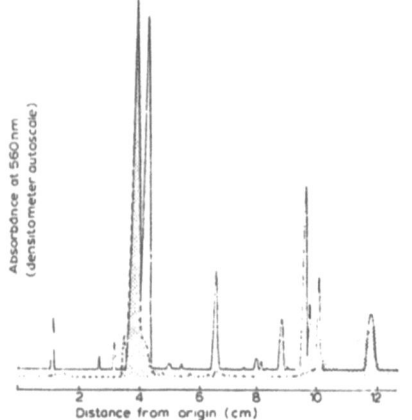

Distance from origin (cm)

Fig. 6. Photolabeling of thermophilic ATPsynthase $TF_0 \cdot F_1$ with ^{125}J-ASA-PE. (a) Coomassie Blue staining pattern of $TF_0 \cdot F_1$, (b) corresponding pattern after labeling with ^{125}J-ASA-PE, as determined by fluorographing (dotted line) SPAGE as described in (2).

Though hydrophilic domains of subunits a and b of TF_0 are exposed to the outside of the proteoliposomes and concomitantly to the TF_1 binding site, they were not modified in the photoreaction. It is, therefore, probable that subunits α are close to the surface of polar head groups of phospholipids on the proteoliposomes and constitute here the outer surface of TF_1, which is bound to TF_0. Subunits α appear to screen the core of the internal coupling of $TF_0 \cdot F_1$ between the proton channel TF_0 and the catalytic moiety TF_1, a coupling, which is suspected to be the interaction of subunit ε whith subunit β (Fig. 7)(9).

If a lipid to protein ratio of 2.4:1 (w/w) was used in the incubation procedure as described in (10), subunits α and ε of TF_1 and all three subunits of TF_0 were photolabeled. This possibly unspecific labeling may be obtained because liposomes were not formed for the reconstitution of $TF_0 \cdot F_1$ and phospholipids were bound to hydrophobic domains of the ATPsynthase, mainly to the TF_0 moiety. In addition to

125J-ASA-PE it could be confirmed recently by two independent methods (9), - a) by cross-linking reactions of a photoreactive, radioactive derivative of the ATPsynthase inhibitor dicyclohexylcarbodiimide (125J-A-DCCD-21), the ligand of which may be cleaved chemically after the photoreaction and b) by surface labeling with the fluorescent photolabel 1-azido-naphthaline-4-sulfonate - that subunits α are closer to the membrane than subunits β.

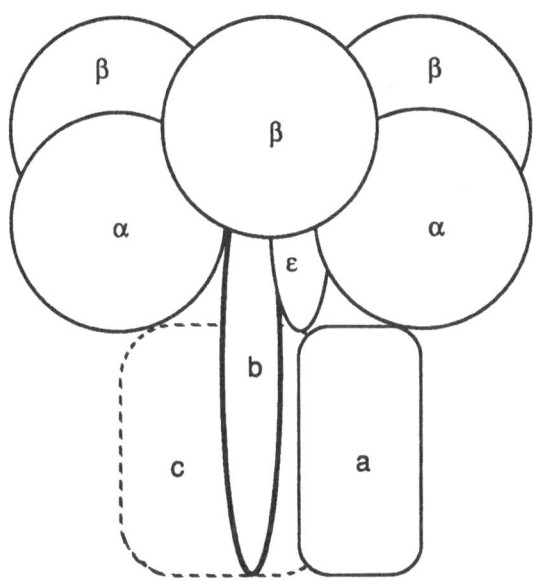

Fig. 7. Proposed model for $TF_o \cdot F_1$.

REFERENCES

(1)   Bisson. R. and Montecucco, C. in "Techiques for the Analysis of Membrane Proteins"; Ragan, C.I. and Cherry, R.J., eds.; Chapman and Hall, London, New York, 1986, pp.153-184

(2)   Gao, Z. and Baeuerlein, E. (1987) FEBS Lett. 223, 366-370

(3)   Kagawa, Y. and Sone, N. (1979) Methods Enzymol. 55, 364-372

(4)   Ji, T.H.and Ji, I. (1982) Anal. Biochem. 121, 286-289

(5)   Struck, D.K., Hockstra, D. and Pagano, R.E. (1981) Biochemistry 20, 4093-4099

(6)   Parente, R.A. and Lentz, B.R. (1986) Biochemistry 25, 1023-1026

(7)   Brunner, J. and Richards, F.M. (1980) J. Biol. Chem. 255, 3319-3329

(8)   Radda, G.K. and Vanderkooi, J. (1972) Biochem. Biophys. Acta 265, 509-549

(9)   Gao, Z. and Baeuerlein, E. (1988) 5. EBEC Short Report Vol. 5, 256-257 (B.Beechy ed.)

(10) Montecucco, C.,Dabbeni-Sala, F., Friedl, P. and Galante, Y.M. (1983) Eur. J. Biochem. 132, 189-194

PHOTOAFFINITY LABELLING OF RECEPTORS FOR MELANOCYTE-STIMULATING HORMONE

A.N. Eberle, F. Solca, W. Siegrist, T. Scimonelli, J. Girard
Department of Research (ZLF)
University Hospital and University Children's Hospital
Hebelstrasse 20, CH-4031 Basel, Switzerland

P.N.E. de Graan
Institute of Molecular Biology and Medical Biotechnology and
Rudolf Magnus Institute for Pharmacology, State University
Padualaan 8, NL-3508 TB Utrecht, The Netherlands

ABSTRACT. The photoaffinity labelling technique was applied to the study of receptors for α-melanocyte-stimulating hormone (α-MSH; α-melanotropin) on mouse and human melanoma cells as well as on Xenopus and Anolis melanophores. Twenty different α-MSH analogues containing one, two or three photolabels were synthesized and tested as photoprobes. A highly potent nitrene-generating azidophenyl compound, monoiodinated [4-Nle, 7-D-Phe, 9-Trp(Naps)]-α-MSH, was most suitable for the biochemical analysis of MSH receptors on melanoma cells whereas the physiological effects following photolabelling of intact cultured cells were best studied by using carbene-generating phenyldiazirine α-MSH analogues. Covalent linkage of α-MSH to its receptor on both melanoma cells and melanophores produced a long-lasting receptor stimulation. This irreversible agonism served as a new tool for the analysis of signal transduction mechanisms. The biochemical characterization of the covalent α-MSH-receptor complex of melanoma cells revealed a 45 kDa membrane glycoprotein, probably containing a relatively high content of sialic acid, thus explaining the acidic pI (4.7) of this receptor.

1. INTRODUCTION

α-MSH is a basic tridecapeptide with the following structure: Ac-Ser-Tyr-Ser-Met-Glu-His-Phe-Arg-Trp-Gly-Lys-Pro-Val-NH$_2$. The amino acid sequence of α-MSH corresponds to the N-terminal part of ACTH from which it is formed in melanotrophic cells of the pituitary pars intermedia as well as in hypothalamic neurons. α-MSH shares a common tetrapeptide, His-Phe-Arg-Trp, with β-MSH and γ-MSH which themselves are cleavage products of β-lipotropin (β-LPH) and, respectively, pro-γ-MSH (N-POMC). Like their immediate precursors (ACTH, β-LPH, pro-γ-MSH), α-, β- and γ-MSH are biosynthetically related and derived from a common prohormone, pro-opiomelanocortin (POMC) (Nakanishi et al., 1979).

P. E. Nielsen (ed.), Photochemical Probes in Biochemistry, 67–84.
© 1989 by Kluwer Academic Publishers.

    The 'classical' function of α-MSH is its action on pigment cells:
the peptide stimulates dispersion of the pigment granules (melanosomes)
in melanophores of fishes, frogs and lizards, and melanin formation in
melanocytes of lower and higher vertebrates, including man. It also
affects proliferation of melanocytes, melanin formation in melanoma
cells, sebum secretion in rats, lipolysis in rabbit fat cells and
aldosterone production under certain physiological conditions (reviewed
by Eberle, 1988). In mammals and man, however, the actions of α-MSH on
the central and peripheral nervous system are more important: the
peptide has been shown to affect performance in several behavioural
paradigms (de Wied and Jolles, 1982) and to stimulate regeneration of
damaged peripheral and central nerves (Gispen et al., 1987). In addition, α-MSH is the most potent antipyretic substance known and seems to
be involved in the control of fever (Lipton and Clark, 1986). It also
appears to have anti-inflammatory properties, modulating some of the
effects of interleukin-1 (Daynes et al., 1987). Another aspect of α-MSH
is its presence in the fetus and hence its possible influence on fetal
growth, in particular of the fetal adrenal gland (Swaab and Martin,
1981; Silman et al., 1981).
    MSH receptors in most mammalian target systems have not yet been
characterized, except for those on pigment cells which have been studied
extensively with a structure-activity approach (Eberle et al., 1985;
Hruby et al., 1984) and, more recently, with a receptor binding assay
(Lambert et al., 1985; Siegrist et al., 1988) and photoaffinity labelling experiments (Scimonelli and Eberle, 1987). The earlier studies on
melanophores of lower vertebrates showed that α-MSH contains two
hormonally 'active sites', i.e. a central (Glu-His-Phe-Arg-Trp) and a
C-terminal (Gly-Lys-Pro-Val-NH$_2$) message sequence (Eberle and Schwyzer,
1976) and an N-terminal potentiator sequence (Ac-Ser-Tyr-Ser-Met) (de
Graan et al., 1986). Quantitative information about the MSH receptor,
such as receptor numbers and affinities, has been obtained for melanoma
cells (Siegrist et al., 1988) but not for melanophores of lower vertebrates, mainly because this type of pigment cell cannot readily be
isolated. Therefore, photoaffinity labelling of MSH receptors on intact
melanophores was established as a method for studying the effects of
labelled receptors in a functional cellular system under various conditions and for a better quantitative characterization of the receptors
(de Graan and Eberle, 1980).
    This paper describes photoaffinity labelling experiments applied to
the characterization of MSH receptors on murine and human melanoma cells
and to the study of the physiological effects induced by MSH receptor
labelling of melanoma cells and melanophores.

2. MATERIALS AND METHODS

2.1. Photoreactive α-MSH Analogues

All peptides listed in table I were synthesized by a classical solution
approach (Eberle et al., 1981), except for [4-Nle,7-D-Phe,9-Trp(Naps)]-
α-MSH wich was prepared by modification of commerical [4-Nle, 7-D-Phe]-

Table I. Photoreactive α-MSH derivatives and their biological activities

Peptide[a]	Bioactivity[b]		Labelling efficiency[c]
	A	B	
Irreversible photolabels			
[1-(N$_2$Ac-Gly), 4-Nva]-α-MSH	15	–	10
[1-{N$_2$Ac(CF$_3$)-Ser}]-α-MSH	66	–	30
[1-(Tdbz-Ser)]-α-MSH	33	25	90
[1-(Apac-Ser)]-α-MSH	50	50	100
[1-(Npac-Ser)]-α-MSH	75	66	0
[1-(Apac-Ser), 9-Leu]-α-MSH	1	–	68
[13-Pap]-α-MSH	40	50	100
[9-Leu, 13-Pap]-α-MSH	1	–	20
[9-Trp(Naps)]-α-MSH	33	50	100
[4-Nle, 9-Trp(Naps)]-α-MSH	80	140	95
[4-Nle, 7-D-Phe, 9-Trp(Naps)]-α-MSH	1000	2500	100
Reversible photolabels			
[9-Trp(NapSS)]-α-MSH	30	–	90
[1-(ApSSpr-Ser)]-α-MSH	30	–	93
Multiple photolabels			
[1-(Tdbz-Ser), 13-Pap]-α-MSH	15	–	100
[9-Trp(Naps), 13-Pap]-α-MSH	20	–	100
[1-(ApSSpr-Ser), 9-Trp(Naps), 13-Pap]-α-MSH	10	–	20
Radioactive photolabels			
[1-D-Ala, 2-(3',5'-ditritio)Pap, 4-Nva]-α-MSH	8	–	65
[4-Nle, 2-{3'-(I-125)}Tyr, 9-Trp(Naps)]-α-MSH	60	–	–
[4-Nle, 2-{3'-(I-125)}Tyr, 7-D-Phe, 9-Trp(Naps)]-α-MSH	650	600	–

a Abbreviations for photolabels: Apac = 4-azidophenylacetyl; ApSSpr = (4-azidophenyl)-1,3'-dithiopropionyl; Naps = 2-nitro-4-azidophenylsulphenyl; N$_2$Ac= diazoacetyl; Npac = 4-nitrophenylacetyl; Pap = p-azidophenylalanine; Tdbz = 4-(3-trifluoromethyldiazirino)benzoyl.
b Percent biological activity (α-MSH = 100); A = Anolis melanophore assay; B = melanoma cell assay (S91 and B16).
c Percent stimulation of Anolis melanophores 100 min after the labelling experiment.

α-MSH (Bachem AG, Bubendorf, Switzerland) with freshly prepared 2-nitro-4-azidophenylsulphenylchloride (Naps-Cl) as described in detail by Eberle (1988). Iodination of [4-Nle, 7-D-Phe, 9-Trp(Naps)]-α-MSH was carried out by the chloramine T procedure followed by extensive purification on a reversed-phase minicolumn and by reversed-phase HPLC. This procedure yielded monoiodinated [2-{3'-(I-125)}Tyr, 4-Nle, 7-D-Phe, 9-Trp(Naps)]-α-MSH (referred to as photo-α-MSH tracer) which displayed a ~6-fold higher potency than α-MSH. The biological activity of photo-reactive α-MSH analogues and the corresponding radioactive peptides was assessed with the Anolis skin assay (Eberle and Girard, 1985) and the B16 melanoma cell in situ melanin assay (Siegrist et al., 1986) or the Cloudman S91 tyrosinase assay (Pomerantz, 1969; Fuller et al., 1987).

2.2. Photolabelling of Melanoma Cell Membranes

A suspension of 5 mg/ml B16 cell membranes was incubated for 90-120 min at 25°C with 10^6 cpm of photo-α-MSH tracer per mg protein, using a 50 mM Tris-HCl buffer, pH 7.4, containing 10 mM $MgCl_2$, 1 mM $CaCl_2$, 0.3 mM 1,10-phenanthroline (protease inhibitor) and 1 mM dithiothreitol. The sample was then UV-irradiated at 0-1°C for 5 min by applying the whole 310-550 nm spectrum of a mercury/xenon UV-lamp (180 mW/cm^2). Free tracer was first removed by ultracentrifugation and the pellet was solubilized and analyzed by SDS polyacrylamide gel electrophoresis followed by autoradiography.

2.3. Neuraminidase Treatment of Photolabelled MSH Receptor

Photolabelled melanoma cell membranes were resuspended in 100 mM Na-acetate buffer, pH 5, containing 30 mM KCl and 1 mM 1,10-phenanthroline, at a density of 5 mg/ml. Neuraminidase (0.1 mU/μg protein; type X, Sigma) was added and the suspension incubated for 0.5 to 7 h at room temperature with occasional shaking. The reaction was stopped by addition of electrophoresis buffer and the proteins were analyzed by SDS polyacrylamide gel electrophoresis followed by autoradiography.

2.4. Photolabelling of Cultured Melanoma Cells

Cloudman S91 melanoma cells attached to 24-well culture plates were incubated in the dark for 1-2 h in sterile medium containing the appropriate peptide concentrations. The dishes were placed on ice and irradiated under sterile conditions. The medium was exchanged and, after returning the cultures to 37°C, replaced 5 times in order to remove non-covalently bound hormone. The cultures were assayed for tyrosinase 24 h after the UV-irradiation.

2.5. Photolabelling of Melanophores

Four 2x2 mm pieces of dorsal skin of Anolis carolinensis were placed in a 5 ml pyrex beaker containing 2 ml of peptide solution, incubated in the dark for 5 min and then irradiated for 5 min. Controls were kept in a separate beaker in the dark. The skin pieces of both beakers were placed in fresh hormone solution, incubated for another 5 min in the dark, irradiated for 5 min (controls in the dark), washed by 4 buffer changes and the degree of pigment dispersion was recorded by visual assessment of the skin colour according to Eberle and Girard (1985).

3. RESULTS

3.1. Nitrene- and Carbene-Generating α-MSH Photolabels

Nitrene-generating azidophenyl-α-MSH photolabels were readily photolyzed at wavelengths of ≥ 310 nm with the energy of the UV-beam being in the order of 100-200 mW/cm^2. However, at longer wavelengths and lower intensities (e.g. 365 nm, 10-20 mW) these photolabels required a longer irradiation time for photolysis than carbene-generating diazirinophenyl-α-MSH photolabels. A comparison of two compounds containing the photolabel in the same position (N-terminal) is shown in table II. At 365 nm, the diazirinophenyl compound was photolyzed at an approximately twofold higher rate than the azidophenyl derivative. This is also reflected by a higher extent of receptor labelling with the diazirinophenyl compound, particularly when examined with melanoma cell monolayer cultures whose viability should not have decreased during and after photolabelling and hence could only be irradiated at 365 nm.

Table II. Comparison of photolysis times at λ = 365 nm and rate of incorporation of carbene-generating [1-(Tdbz-Ser)]-α-MSH (I) and of nitrene-generating [1-(Apac-Ser)]-α-MSH (II) into Anolis melanophore (A) and Cloudman S91 mouse melanoma cell (B) MSH receptors

Photo-label	Time of UV-irradiation	Extent of photolysis	Extent of labelling[a] A	B
I	1 min	50%	29%	
	2 min	73%	56%	40%
	5 min	93%	79%	
	10 min	100%		
II	1 min	11%	21%	
	2 min	21%	46%	20%
	5 min	42%	59%	
	10 min	63%		

a The extent of labelling was assessed by measuring the long-lasting response of Anolis melanophores (see 2.5) as well as the increase in tyrosinase activity of Cloudman S91 cells over controls treated with α-MSH for the same time period (see 2.4).

3.2. A Highly Potent Monoiodinated α-MSH Photolabel

The photoreactive analogue [4-Nle, 7-D-Phe, 9-Trp(Naps)]-α-MSH was about 10-fold more active than α-MSH in the Anolis melanophore assay and 25-fold in the B16 melanin assay (table I). Thus the compound proved to be the most potent MSH photolabel known to date.

Radioiodinated [4-Nle, 7-D-Phe, 9-Trp(Naps)] eluted as a single monoiodinated peptide peak from a C-18 HPLC column after prepurification on C-18 reversed-phase minicolumns (fig. 1). This photo-α-MSH tracer

72

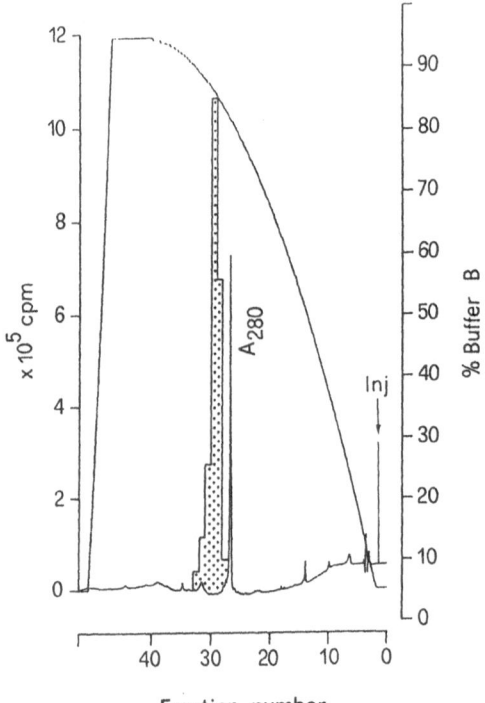

Figure 1. HPLC purification of monoiodinated [4-Nle, 7-D-Phe, 9-Trp(Naps)]-α-MSH tracer on a C-18 reversed-phase column. The peptide was eluted using an exponential gradient between 0.1% aqueous trifluoroacetic acid (buffer A) and 70% acetonitrile/30% water/0.1% trifluoroacetic acid (buffer B). The UV tracing shows the elution position of unlabelled peptide.

still displayed a ~6-fold higher bioactivity than α-MSH (table I). Binding analysis using photo-α-MSH tracer as radioligand produced virtually the same results (Solca et al., unpublished) as those obtained with monoiodinated α-MSH or [4-Nle]-α-MSH tracer (Siegrist et al., 1988). The tracer is therefore thought to interact with the same receptor site as the hormone with the natural sequence.

3.3. Analysis of MSH Receptors on Murine Melanoma Cells

The analysis by SDS polyacrylamide gel electrophoresis of the labelling pattern after UV-irradiation of B16-F1 melanoma cell plasma membranes in the presence of photo-α-MSH tracer revealed a single labelled band of approximately 45 kDa (fig. 2). The labelling was specific because the presence of excess α-MSH during incubation/photolysis abolished the insertion of tracer into this band (fig. 2). Gradual displacement of

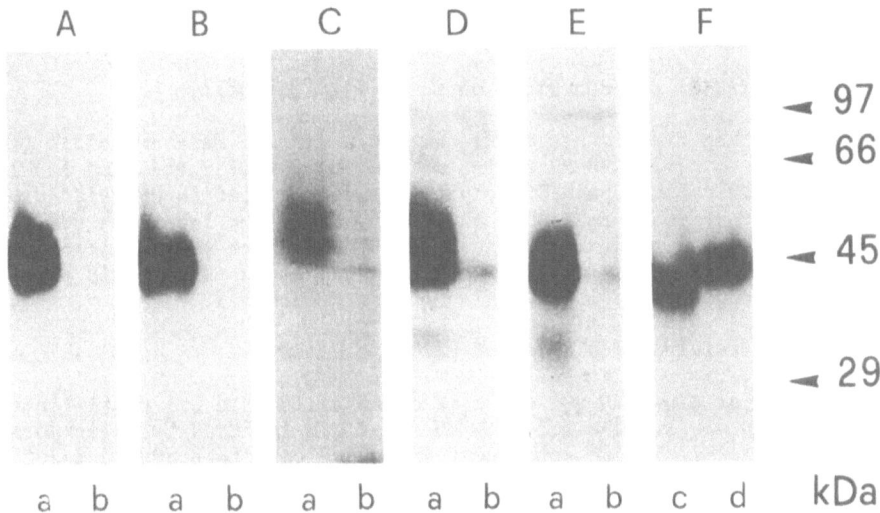

Figure 2. Labelling pattern of mouse and human melanoma cells after UV-irradiation in the presence of photo-α-MSH tracer (a) + a 1000-fold excess of α-MSH (b). Membrane proteins were separated by SDS polyacrylamide gel electrophoresis and analyzed by autoradiography. A: B16-F1 mouse cells; B: B16-F10 mouse cells; C: Cloudman S91 mouse cells; D: human D10 cells; E: human 205 cells; F: neuraminidase-treated B16-F1 cell membranes after (c) and before (d) treatment with the enzyme.

photo-α-MSH tracer with increasing concentrations of α-MSH decreases the extent of labelling in parallel with the displacement of the ligand in a normal binding assay (Scimonelli and Eberle, 1987). Photolabelling of Cloudman S91 cells produced a band with a slightly higher molecular weight (50 kDa) whereas no labelled band was observed with non-melanoma cells such as rat PC12 pheochromocytoma, dog MDCK II or human SK-N-MC neuroblastoma (Solca et al., unpublished).

Further characterization of the 45 kDa band of B16 melanoma cells by two-dimensional electroisofocussing/SDS gel electrophoresis revealed a relatively acidic pI for this protein (~4.7). The pattern indicated two spots of the same molecular weight but with slightly differing pI (Solca et al., unpublished). Chromatography over lectin columns showed that the protein is a glycoprotein because the radioactivity was retained by wheat germ agglutinin-agarose and specifically re-eluted by N-acetyl-glucosamine, suggesting the presence of, e.g., terminal sialic acids. Treatment of labelled membranes with neuraminidase, which cleaves N- and O-acylneuraminyl or N-acetylhexosamine residues, produced a new protein band with the molecular weight apparently reduced by 3 kDa (fig. 2). This supports the notion that the protein contains a relatively high content in sialic acid, thus explaining the acidic pI of the MSH receptor protein. Exposure of labelled membranes to glycopeptidase F produced a band at ~37 kDa which indicates that the MSH receptor may contain a relatively large glyco moiety (Solca et al., unpublished).

3.4. Analysis of MSH Receptors on Human Melanoma Cells

Photolabelling experiments using human D10 and 225 melanoma cells (fig. 2) as well as human LSD 22 cells (not shown) revealed the same labelled 45 kDa protein band upon SDS polyacrylamide gel electrophoresis/auto-radiography as found for mouse B16 cells. Since the labelling was also prevented by the presence of excess α-MSH and hence was specific for this band, it appears that the 45 kDa glycoprotein is not only part of the MSH receptor in mouse but also in human melanoma.

3.5. Irreversible Stimulation of Melanoma Cells

Incubation of Cloudman S91 mouse melanoma cells with 0.1 µM [1-(Apac-Ser)]-α-MSH or [1-(Tdbz-Ser)]-α-MSH for 1-2 h followed by UV-irradiation at 365 nm for 2 min and subsequent extensive washing produced a 20% or a 40% increase, respectively, in tyrosinase activity when determined 24 h later as compared with non-irradiated controls (table II). A further stimulation by a longer irradiation time was not possible because of cell damage produced by the UV light. Nevertheless, these results clear-ly demonstrate that labelled MSH receptors on Cloudman S91 melanoma cells produced a long-lasting signal which was sufficient to signifi-cantly increase the response of the cells.

3.6. Irreversible Stimulation of Melanophores

Irreversible stimulation of target cells by photoaffinity crosslinking of the hormone with its receptor was first shown for Xenopus melano-phores (de Graan and Eberle, 1980) and later for Anolis melanophores (Eberle, 1984). These systems soon became an important tool for mecha-nistic studies of receptor labelling on intact cells, mainly because the biological response (i.e. pigment dispersion and aggregation) is direct-ly visible and can be monitored for individual cells. Table III lists some of the effects studied and questions asked. The following para-graphs are not only a review of some of the most pertinent findings but also contain hitherto unpublished material.

3.6.1. Persistent stimulation of melanophores. Incubation of Xenopus tail-fin pieces or of Anolis dorsal skin pieces with [13-Pap]-α-MSH, [1-(Apac-Ser)]-α-MSH or [9-Trp(Naps)]-α-MSH in the dark followed by UV-irradiation produced long-lasting pigment dispersion (de Graan and Eberle, 1980; Eberle, 1984). MSH peptides without photolabel or photo-reactive non-MSH peptides did not induce this effect, and the UV-irradiation chosen did not affect the function of the melanophores. It seems that all sites of the α-MSH molecule carrying a photolabel, i.e. positions 1, 9 and 13, are in close contact with the MSH receptor. However, α-MSH derivatives whose Trp was replaced by Leu were effective only when the photolabel was attached at the N-terminal end but not when incorporated into the C-terminus (Eberle et al., 1983).

Table III. Application of photoaffinity labelling of MSH receptors on pigment cells for mechanistic studies

Question/finding	Test system	Ref.[a]
Long-lasting stimulation of target cells	Xenopus melanophores Anolis melanophores S91 melanoma cells	1 2 3
Reversibility of long-lasting effect by use of cleavable S-S containing MSH photolabel	Anolis melanophores	2
Temperature-dependence of receptor turnover	Xenopus melanophores	2
Extent of receptor labelling	Xenopus melanophores	1, 4
Ca^{2+} requirement for MSH-receptor binding	Xenopus melanophores Anolis melanophores B16 melanoma cells	5 6
Ca^{2+} requirement for signal transduction	Xenopus melanophores Anolis melanophores	5 6
Non-requirement of Mg^{2+} for receptor binding and signalling	Xenopus melanophores	7
Interaction of photolabelled MSH receptor with activated α_2-adrenergic receptor	Anolis melanophores	6
Interaction of C- and N-terminal α-MSH antibodies with covalent MSH-receptor complex	Anolis melanophores	see text
Interaction of photolabelled MSH receptor, depressed by activated α_2-adrenergic receptor, with forskolin	Anolis melanophores	6
Lack of effect of forskolin on MSH receptor labelling	Anolis melanophores	6

a References: (1) de Graan and Eberle, 1980; (2) Eberle, 1984; (3) Scimonelli and Eberle, 1987; (4) Eberle et al., 1981; (5) de Graan et al., 1981; (6) Eberle and Girard, 1985a; (7) de Graan et al., 1982.

3.6.2. Reversibility of long-lasting receptor stimulation.

Covalent attachment of the disulphide-containing [1-(ApSSpr-Ser)]-α-MSH to Anolis melanophores produced the same long-lasting stimulation as observed with [1-(Apac-Ser)]-α-MSH, but when the skin was washed in buffer containing 1% β-mercaptoethanol, the pigment slowly reaggregated (Eberle, 1984). This proved that the peptide was released from the receptor and that the MSH receptor probably remained on the cell surface during stimulation.

3.6.3. Receptor turnover. The molecular mechanism of the long-lasting stimulation is not yet clear but most probably the labelled MSH-receptor complex remains activated for several hours, generating a permanent intracellular signal. This hypothesis is supported by the finding that the long-lasting stimulation of Xenopus melanophores was strongly dependent on the temperature during the wash phase after UV-irradiation: it persisted with almost no loss in intensity when the temperature was kept at 12°C or 15°C whereas at 26°C the pigment reaggregated within 4 h. Thus in Xenopus melanophores, covalent MSH-receptor complexes seemed to be fairly stable at temperatures below 20°C but at 20°C receptor inactivation ranged between 20% and 30%, and at 26°C between 40% and 60% per h (de Graan and Eberle, 1980).

3.6.4. Dual role of calcium. It has long been known that MSH-induced pigment dispersion in melanophores depends on the presence of calcium in the medium. Incubation and UV-irradiation of Xenopus or Anolis melano- phores and photoreactive α-MSH in the absence of calcium did not produce long-lasting stimulation after readdition of calcium, indicating that the ion is essential for hormone-receptor binding (de Graan and Eberle, 1980; Eberle and Girard, 1985a).

When extracellular calcium was removed from the medium of Xenopus or Anolis melanophores after formation of a covalent MSH-receptor complex, the long-lasting stimulation was reversed (fig. 3). Readdition of a calcium-containing medium (without peptide) restored full pigment dispersion in both systems which showed that calcium is not only required for hormone-receptor binding but also for maintaining the long-lasting response. This view is supported by the finding that racemized α-MSH or [4-Nle, 7-D-Phe]-α-MSH, both of which produced a persistent activation of Xenopus melanophores in the presence of calcium, were displaced from the receptor after temporary removal of calcium (de Graan et al., 1982). In the Anolis system, [4-Nle, 7-D-Phe]-α-MSH was washed off only after repeated alternate incubation of the skin in calcium-free and calcium-containing buffer whereas pigment dispersion induced by photolabelling continued to persist in the presence of the ion (fig. 3).

Additional support for the existence of two calcium sites involved in hormonal stimulation of melanophores stems from experiments with calcium antagonists, such as D600 or verapamil. These compounds inhibted the formation of a long-lasting signal in Xenopus melanophores when present during UV-irradiation but were not effective after temporary removal of calcium once the covalent hormone-receptor complex had been formed (de Graan et al., 1982). The second calcium site seemed to precede the action of cAMP since this agent induced pigment dispersion in the absence of calcium. It seems therefore that the second calcium site is important for the signal transduction from the receptor to the catalytic unit of the adenylate cyclase and/or the activation of the latter.

3.6.5. Interaction with adrenergic receptors. Noradrenalin and dopamine (in a 100-fold higher concentration) reversibly inhibited the long-lasting stimulation of Anolis melanophores induced by covalently bound α-MSH (Eberle and Girard, 1985a). Removal of the catecholamines restored

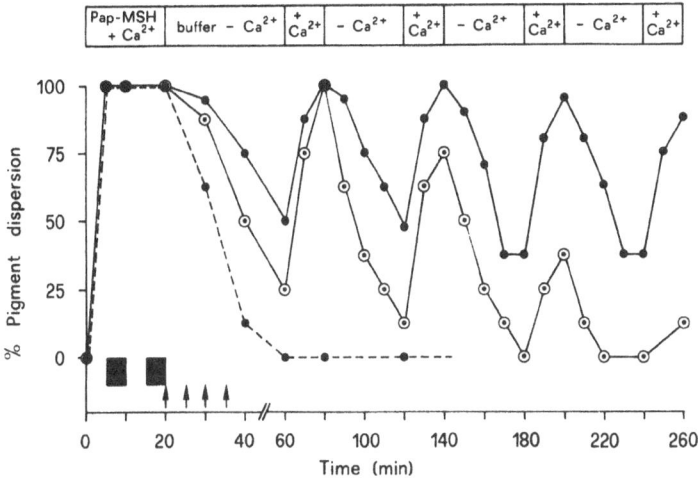

| Pap-MSH + Ca²⁺ | buffer – Ca²⁺ | + Ca²⁺ | – Ca²⁺ | + Ca²⁺ | – Ca²⁺ | + Ca²⁺ | – Ca²⁺ | + Ca²⁺ |

Figure 3. Reversible suppression of the long-lasting response in the absence of calcium and comparison of photolabelled and [4-Nle, 7-D-Phe]-α-MSH-'labelled' melanophores. Anolis skin pieces were photolabelled in the presence of 25 ng/ml of [13-Pap]-α-MSH (●) or incubated with 1 ng/ml of [4-Nle, 7-D-Phe]-α-MSH (☉). Repeated alternate transfer of the skin pieces between buffer with and without calcium slowly released non-covalently attached [4-Nle, 7-D-Phe]-α-MSH but not covalently bound photo-α-MSH. Non-irradiated controls (●--●). Black squares indicate time intervals of UV-irradiation and arrows are buffer changes.

the long-lasting signal. Repetitive reversible suppression of the long-lasting signal gave almost identical patterns.

The combination of an α_1-antagonist (prazosin) and an α_2-antagonist (yohimbine) with the catecholamines for switching off irreversibly stimulated Anolis melanophores revealed that the suppression of the long-lasting signal proceeded through activation of an α_2-adrenergic receptor (Eberle and Girard, 1985a). The binding of α-MSH to its receptor was however barely if at all affected by α_1- or α_2-agonists or -antagonists, and the turnover of MSH-receptor complexes did not appear to be elevated in the presence of catecholamines.

3.6.6. Effect of forskolin. The diterpene forskolin, which is known to increase intracellular cAMP by increasing adenylate cyclase activity (Seamon and Daly, 1981), induced pigment dispersion in Anolis and Xenopus melanophores with EC_{50}s of 40 nM and 700 nM, respectively. The effect of forskolin on Anolis melanophores was not influenced by the absence of extracellular calcium or magnesium but noradrenalin in a concentration of 100 nM significantly diminished the onset and extent of the forskolin-induced response. Forskolin did not alter receptor labelling when simultaneously present with photoreactive α-MSH. Interestingly, it rapidly restored the long-lasting signal after covalent receptor-labelling and subsequent inhibition by noradrenalin when added to the solution in the the presence of catecholamines. Thus it seems that forskolin and activated MSH receptors may act synergistically and may override the inhibitory signal of the α_2-receptors.

3.6.7. Effect of MSH antibodies on the long-lasting signal. The effect of N-and C-terminal antiserum on the photolabelling of MSH receptors was tested with both Xenopus and Anolis melanophores. When tail-fin pieces of Xenopus tadpoles were preincubated in normal medium containing C-terminal α-MSH antibodies (see Eberle, 1988) and subsequently incubated in the medium containing both [1-(Apac-Ser)]-α-MSH and the antibodies and then UV-irradiated, there was no long-lasting dispersion after washing of the tail-fin pieces. This indicates that the antibodies prevented covalent labelling of the MSH receptors (Eberle et al., 1982). Addition of the antibodies after the formation of the covalent hormone-receptor complex did not affect the irreversible dispersion, even if the incubation with the antiserum was performed in calcium-free medium.

Similar results were obtained with Anolis melanophores: application of N- or C-terminal antibodies to respectively C- or N-terminally cross-linked MSH-receptor complexes did not influence the long-lasting stimulation of these melanophores. If however the signal was 'turned off' temporarily by the application of noradrenalin (which possibly induced a 'weakening' of the MSH-receptor interaction) followed by the addition of the same antibodies, the removal of the noradrenalin-block led to a very marked decrease of the signal in the presence of C-terminal antibodies, to a less marked but significant decrease in the presence of N-terminal antibodies and to virtually no change in the presence of control serum (fig. 4.A). Detachment of non-covalently bound [4-Nle, 7-D-Phe]-α-MSH from the receptor by N- and C-terminal MSH antibodies showed the same pattern although part of the peptide dissociated already in the absence of specific antibodies (fig. 4.B).

These experiments indicate that (i) a free C-terminal end of α-MSH is more important for maintaining the long-lasting stimulation than a free N-terminal end of the hormone; (ii) α-MSH antibodies cannot easily loosen or detach crosslinked α-MSH from its receptor, and (iii) photo-labelled α-MSH is more tightly bound to its receptor than [4-Nle, 7-D-Phe]-α-MSH when exposed to various treatments, such as noradrenalin, MSH antibodies or calcium.

4. DISCUSSION

4.1. Biochemical Characterization of the MSH Receptor

The photoaffinity approach proved to be a useful tool for the study of MSH receptors on both melanoma cells and melanophores. There is now good evidence that the 45 kDa membrane glycoprotein labelled by the photo-MSH tracer is the MSH receptor or part of it: (i) labelling of intact melanoma cells produced a biological response, indicating that covalent

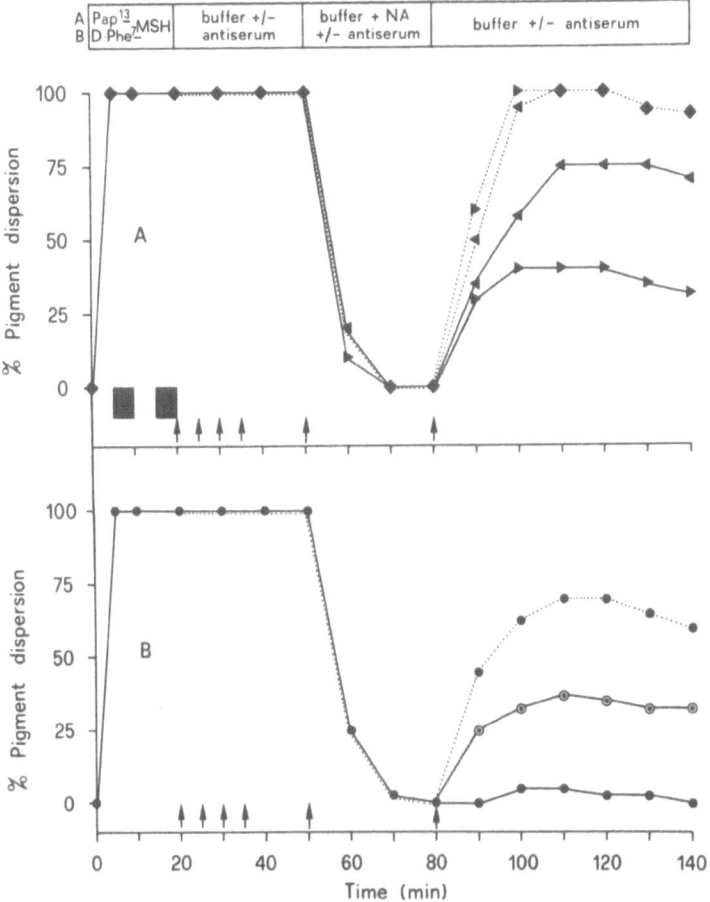

Figure 4. Effect of N- and C-terminal α-MSH antibodies on the long-lasting stimulation of <u>Anolis</u> melanophores after photo-labelling of MSH receptors. A: <u>Anolis</u> skin pieces were incubated in the presence of 25 ng/ml [1-(Apac-Ser)]-α-MSH (◆ ▶) or [13-Pap]-α-MSH (◆ ◀), UV-irradiated as explained in fig. 3, incubated with or without C- (▶●) or N-terminal (◀ ⊙) anti-bodies to which at t = 50 min 100 nM noradrenalin was added. At t = 80 min, the noradrenalin-block was released by trans-ferring the skins to buffer containing only the respective antibodies, and the response was recorded for another 60 min time interval. B: An identical experiment was performed with 1 ng/ml [4-Nle, 7-D-Phe]-α-MSH, except that the skins were not UV-irradiated (symbols used are indicated under A). The dotted lines are the controls without specific MSH antibodies.

MSH-receptor crosslinking did take place; (ii) non-irradiated MSH photolabel was specifically and reversibly bound by plasma membranes; (iii)
the MSH photolabel was only found in the 45 kDa band on SDS polyacrylamide gel electrophoresis, and its covalent insertion was competitively
inhibited by increasing concentrations of α-MSH, in parallel to the
curve obtained in the binding assay; (iv) the 45 kDa band was also found
after photolabelling of human melanoma cell lines.

The 'small' size of 45 kDa for the MSH receptor (subunit) is not
exceptional as the recent characterization of receptor systems for other
peptide hormones revealed molecular mass values of 30 kDa (renal V_2
vasopressin receptor; Fahrenholz et al., 1985), 30 and 38 kDa (hepatic
V_1 vasopressin receptor; Fahrenholz et al., 1986), 65 kDa (angiotensin
II receptor; Guillemette et al., 1986) and 43 kDa (substance K receptor,
without glyco moiety; Masu et al., 1987). The acidic nature of the MSH
receptor is not surprising either, but the low pI of 4.7 is probably
unusual, indicating the presence of a large number of sialic acid (and
perhaps sulphate) groups. Since neuraminidase treatment elicited a
remarkable shift on polyacrylamide gels, it is possible that sialic acid
residues play a dominant role in the negative overall charge of the
protein. The lack of formation of a small labelled fragment after glycopeptidase F treatment suggests that the photo-MSH tracer is inserted
into the protein moiety of the receptor.

4.2. Irreversible Stimulation of MSH Receptors

The labelling of MSH receptors on Xenopus and Anolis melanophores has
revealed that irreversible stimulation by formation of a covalent
hormone-receptor complex is a very useful approach for studying stimu-
lus-response coupling in intact cells. Its has been demonstrated for the
first time with melanophores that covalent binding of a peptide to its
receptor can lead to irreversible agonism, a concept which since then
has also found its application in many other hormonal systems. The
advantage of using permanently stimulated intact cells is not only that
receptor and adenylate cyclase are located in their different ionic
environment but also that an electrical potential is maintained across
the plasma membrane. Thus, the MSH-melanophore system offers the
possibility of studying, under almost physiological conditions, the
involvement of various factors, such as divalent ions and catecholamines, in receptor binding and membrane signal transduction, even when
the number of cells available is limited.

The photoaffinity labelling approach applied to intact melanophores
not only provided new information on temporal sequences in the stimulation of the cells, but was also useful for the study of the molecular
mechanism of peptide-receptor interaction. Recent preliminary experiments suggested that when α-MSH was 'tightly' linked to the receptor
simultaneously via two labels, located either on positions 9 and 13 or
on positions 1 and 13, the result was the same as with one label in
either position. If however a compound containing 3 photolabels was
used, producing an even 'tighter' interaction with the receptor, the
signal was less well preserved than with 1 or 2 labels. Release of the
reversible N-terminal label by incubation with β-mercaptoethanol

restored the signal to its full extent. This could indicate that the
stimulatory state of the MSH receptor requires some degree of flexibility between receptor and ligand.

The transmembrane signalling of the MSH receptor system is thought
to proceed in the same way as the well studied β-adrenergic receptor
system (reviewed by Levitzki, 1987). A striking difference, however, is
the involvement of calcium in the transduction of the α-MSH signal to
the adenylate cylcase. Although in most mammalian systems the coupling
proteins have been shown to contain magnesium- (or manganese-) binding
sites, there have been few reports on calcium-dependent adenylate
cyclase activation, such as found in brain tissue or adrenal cells
(reviewed by Bradham and Cheung, 1982). At present it is not yet clear
whether calcium binds directly to the N_s regulatory protein in <u>Anolis</u>
melanophores or whether a calcium-dependent regulatory protein is
involved in the activation of N_s. It is possible that a calmodulin-
related calcium-binding protein modulates hormone-receptor binding and
participates in the signal transfer to the adenylate cyclase (Gerst and
Salomon, 1987).

ACKNOWLEDGEMENTS

This study was supported by the Swiss National Science Foundation, the
Swiss and Basel Cancer Leagues, the Swiss Academy of Medical Sciences,
the Janggen-Pöhn-Stiftung (St.Gallen), the University Hospital and the
University Children's Hospital. We wish to thank Mrs. Sibylla Stutz and
Mrs. Roma Drozdz for expert technial assistance and Dr. Joyce B. Baumann
for helpful discussions.

REFERENCES

Bradham L.S. and Cheung W.Y. (1982) 'Nucleotide cyclases.' Prog. Nucleic
Acid Res. Mol. Biol. 27: 189-231.

Daynes R.A., Robertson B.A., Cho B.H., Burnham D.K. and Newton R. (1987)
'α-Melanocyte-stimulating hormone exhibits target cell selectivity in
its capacity to affect interleukin 1-inducible responses in vivo and in
vitro.' J. Immunol. 139: 103-109.

De Graan P.N.E and Eberle A.N. (1980) 'Irreversible stimulation of
Xenopus melanophores by photoaffinity labelling with p-azidophenyl-
alanine-13-α-melanotropin.' FEBS Lett. 116: 111-115.

De Graan P.N.E., Eberle A.N. and van de Veerdonk F.C.G. (1981) 'Photoaffinity labelling of MSH receptors reveals a dual role of calcium in
melanophore stimulation.' FEBS Lett. 129: 113-116.

De Graan P.N.E., Eberle A.N. and van de Veerdonk F.C.G. (1982) 'Calcium
sites in MSH stimulation of Xenopus melanophores: studies with photoreactive α-MSH.' Mol. Cell. Endocrinol. 26: 327-339.

De Graan P.N.E., Spruijt B.M., Eberle A.N., Girard J. and Gispen W.H.
(1986) 'ACTH(1-4) potentiates α-MSH-induced melanophore dispersion and
excessive grooming.' Peptides 7: 1-4.

De Wied D. and Jolles J. (1982) 'Neuropeptides derived from pro-opiomelanocortin: behavioral, physiological, and neurochemical effects.'
Physiol. Rev. 62: 976-1059.

Eberle A.N. (1984) 'Photoaffinity labelling of MSH receptors on Anolis
melanophores: irradiation technique and MSH photolabels for irreversible
stimulation.' J. Receptor Res. 4: 315-329.

Eberle A.N. (1988) The Melanotropins; Chemistry, Physiology and Mechanism of Action. Karger, Basel.

Eberle A.N. and Girard J. (1985) 'An Anolis skin melanophore assay
suitable for photoaffinity labeling studies with α-MSH.' Experientia 41:
654-656.

Eberle A.N. and Girard J. (1985a): 'Photoaffinity labelling of MSH
receptors on Anolis melanophores: effects of catecholamines, calcium and
forskolin.' J. Receptor Res. 5: 59-81 (1985).

Eberle A.N. and Schwyzer R. (1976) 'Hormone-receptor interactions. The
message sequence of α-melanotropin: demonstration of two active sites.'
Clin. Endocrinol. 5, suppl: 41s-48s.

Eberle A.N., de Graan P.N.E. and Hübscher W. (1981) 'Synthesis and biological properties of p-azidophenylalanine-13-α-melanotropin, a potent
photoaffinity label for MSH receptors.' Helv. Chim. Acta 64: 2645-2653.

Eberle A.N., de Graan P.N.E. and van de Veerdonk F.C.G. (1982) 'Photo-
affinity labelling of MSH receptors with [1-(Apac-Ser)]-α-MSH and [13-
Pap]-α-MSH.' In Neuroendocrinology of Vasopressin, Corticoliberin and
Opiomelanocortins (A.J. Baertschi and J.J. Dreifuss, eds.), pp. 231-238.
Academic Press, New York and London.

Eberle A.N., Girard J., de Graan P.N.E. and van de Veerdonk F.C.G.
(1983) 'Photoreactive [9-Leu]-α-MSH derivatives for receptor labelling.'
In Peptides 1982 (K. Bláha and P. Malon, eds.), pp. 589-592. De Gruyter,
Berlin.

Eberle A.N., de Graan P.N.E., Baumann J.B., Girard J., van Hees G. and
van de Veerdonk F.C.G. (1985) 'Structural requirements of α-MSH for the
stimulation of MSH receptors on different pigment cells.' In Pigment
Cell 1985 (J. Bagnara, S.N. Klaus, E. Paul and M. Schartl, eds.), pp.
191-196. University of Tokyo Press, Tokyo.

Fahrenholz F., Boer R., Crause P. and Tóth M.V. (1985) 'Photoaffinity
labelling of the renal V_2 vasopressin receptor. Identification and enrichment of vasopressin-binding subunit.' Eur. J. Biochem. 152: 589-595.

Fahrenholz F., Kojro E., Müller M., Boer R., Löhr R. and Grzonka Z. (1986) 'Iodinated photoreactive vasopressin antagonist: labelling of hepatic vasopressin receptor subunits.' Eur. J. Biochem. 161: 321-328.

Fuller B.B., Lunsford J.B. and Imam D.S. (1987) 'α-Melanocyte-stimulating hormone regulation of tyrosinase in Cloudman S91 mouse melanoma cell culture.' J. Biol. Chem. 262: 4024-4033.

Gerst J.E. and Salomon Y. (1987) 'Calcium and guanosine nucleotide modulation of melanotropin receptor function and adenylate cyclase in the M2R cell line.' In Membrane Receptors, Dynamics, and Energetics (K.W.A. Wirtz, ed.), Nato ASI Series, vol. 133, pp. 117-126. Plenum Press, New York.

Gispen W.H., de Koning P., Kuiters R.R.F., van der Zee C.E.E.M. and Verhaagen J. (1987) 'On the neurotrophic action of melanocortins.' Prog. Brain Res. 72: 319-331.

Guillemette G., Guillon G., Maire J., Balestre M.N., Escher E. and Jard S. (1986) 'High yield photoaffinity labeling of angiotensin II receptors.' Mol. Pharmacol. 30: 544-551.

Hruby V.J., Wilkes B.C., Cody W.L., Sawyer T.K. and Hadley M.E. (1984) 'Melanotropins: structural, conformational and biological considerations in the development of superpotent and superprolonged analogs.' In Peptide and Protein Reviews (M.T.W. Hearn, ed.), vol. 3, pp. 1-64. Marcel Dekker, New York.

Lambert D.T., Whitcombe P.E., Moellmann G.E. and Lerner A.B. (1985) 'Basic characterization of the receptor for MSH on Cloudman S91 melanoma cells.' In Pigment Cell 1985 (J. Bagnara, S.N. Klaus, E. Paul and M. Schartl, eds.) pp. 165-174. University of Tokyo Press, Tokyo.

Levitzki A. (1987) 'Regulation of hormone-sensitive adenylate cyclase.' Trends Pharmacol. Sci. 8: 299-303.

Lipton J.M. and Clark W.G. (1986) 'Neurotransmitters in temperature control.' Ann. Rev. Physiol. 48: 613-623.

Masu Y., Nakayama K., Tamaki H., Harada Y., Kuno M., Nakanishi S. (1987) 'cDNA cloning of bovine substance-K receptor through oocyte expression system.' Nature 329: 836-838.

Nakanishi S., Inoue A., Kita T., Nakamura M., Chang A.C.Y., Cohen S.N. and Numa S. (1979) 'Nucleotide sequence of cloned cDNA for bovine corticotropin-β-lipotropin precursor.' Nature 278: 423-427.

Pomerantz S.H. (1969) 'L-Tyrosine-3,5-^{3}H assay for tyrosinase development in skin of newborn hamsters.' Science 164: 838-839.

Scimonelli T. and Eberle A.N. (1987) 'Photoaffinity labelling of melanoma cell MSH receptors.' FEBS Lett. 226: 134-138.

Seamon K.B. and Daly J.W. (1981) Forskolin: a unique diterpene activator of cyclic AMP-generating systems. J. Cyclic Nucleotide Res. 7: 201-224.

Siegrist W. and Eberle A.N. (1986) 'In situ melanin assay for MSH using B16 melanoma cells in culture.' Anal. Biochem. 159: 191-197.

Siegrist W., Oestreicher M., Stutz S., Girard J. and Eberle A.N. (1988) 'Radioreceptor assay for α-MSH using mouse B16 melanoma cells.' J. Receptor Res. 8: 323-343.

Silman R.E., Street C., Holland D., Chard T., Falconer J. and Robinson J.S. (1981) 'The pars intermedia and the fetal pituitary-adrenal axis.' Ciba Found. Symp. 81: 180-190.

Swaab D.F. and Martin J.T. (1981) 'Functions of α-melanotropin and other opio-melanocortin peptides in labour, intrauterine growth and brain development.' Ciba Found. Symp. 81: 196-217.

TOPOGRAPHICAL ANALYSIS OF THE TORPEDO MARMORATA ACETYLCHO-
LINE RECEPTOR BY ENERGY TRANSFER PHOTOAFFINITY LABELING USING
ARYLDIAZONIUM DERIVATIVES.

F. Kotzyba-Hibert, A. Jaganathen, J. Langenbuch-Cachat, M. Goeldner, C. Hirth
Laboratoire de Chimie Bioorganique, Université Louis Pasteur, Faculté de Pharmacie, 67048 Strasbourg Cédex, France.

M. Dennis, J. Giraudat, J.L. Galzi, C. Mulle, J.P. Changeux
Neurobiologie Moléculaire et Laboratoire Associé au CNRS, Interactions Molécu-
laire et Cellulaire, Institut Pasteur, 25 rue du Dr Roux 75724 Paris Cédex 15,
France.

C. Bon
Unité des Venins, Institut Pasteur, 25 rue du Dr Roux, 75724 Paris Cédex 15,
France.

J.Y. Chang
Pharmaceutical Research Laboratories, Ciba-Geigy Ltd CH-4002 Basel, Switzerland.

C. Lazure, M. Chrétien
Laboratoire d'endocrinologie Moléculaire, Institut de Recherches Cliniques de
Montréal, Canada H2W 1R7.

ABSTRACT: The acetylcholine and the non competitive blockers binding sites of the native,
membrane-bound acetylcholine receptor from Torpedo Marmorata were covalently photolabeled by aryldiazonium derivatives. These derivatives are shown to be ligands of both
sites and to be selectively activated once bound by an energy transfer reaction involving
tryptophan residue(s) of the protein. The labeled subunits were analysed. The α subunit of
the acetylcholine receptor is essentially involved into the agonist binding site while each of
the five subunits is part of the non competitive blockers binding area. The α-chain residues Trp-149, Tyr-190, Cys-192, Cys-193 and an unidentified residue in the 80-105
segment are labeled in an agonist protectable manner, suggesting that they belong to the
agonist binding site. These amino-acids are located in the large amino-terminal hydrophilic
domain of the α-subunit primary structure and may lie in close proximity on the native
receptor.

P. E. Nielsen (ed.), Photochemical Probes in Biochemistry, 85–105.
© 1989 by Kluwer Academic Publishers.

86

INTRODUCTION

The peripheral nicotinic acetylcholine receptor (AcChoR) of vertebrates consists of four different subunits assembled into an heterologous $\alpha_2\beta\gamma\delta$ pentamer (Reviewed in Anholt et al., 1984, Changeux et al., 1984, Hucho, 1986). The cDNAs coding for the constitutive four chains of the AcChoR have been cloned and sequenced in Torpedo californica (Ballivet et al., 1982, Noda et al.,1982,1983 a,b,Claudio et al.,1983) , in Torpedo marmorata (for the α chain : Giraudat et al.,1982, Sumikawa et al.,1982, Devillers-Thiéry et al.,1983) and in several vertebrate species including humans (Boulter et al.,1986). Kinetic studies using radioactive (Neubig & al. 1982) or fluorescent ligands (Heidmann &Changeux, 1979,1980) indicate that the receptor might exist in at least four discrete conformations in equilibrium: the resting state (R), the active state (A), the intermediate state (I) and the desensitized state (D). In the synapse at rest, about 80% of the AcChoR population is in the (R) conformation while the desensitized state represents the remaining 20%. In the presence of acetylcholine, the equilibrium shifts from the resting to the transient active state (life time = 1 ms) in which the ionic channel is opened. These receptor conformations show an increasing affinity for the agonists and a decreasing one for the competitive antagonists going from the (R) to the (D) state. As an example, addition of agonists to the incubation mixture shifts the equilibrium towards the (A) state while addition of α-toxins will stabilise the (R) state.

The four subunits display striking sequence homologies and similar hydrophobicity profiles suggesting a transmembranous arrangement around a five fold pseudo-axis of symmetry. A central pit is proposed to contain the acetycholine-gated ion channel .This model has been confirmed by electron microscopy studies on two dimensional crystals of tubular AcChoR (fig. 1).

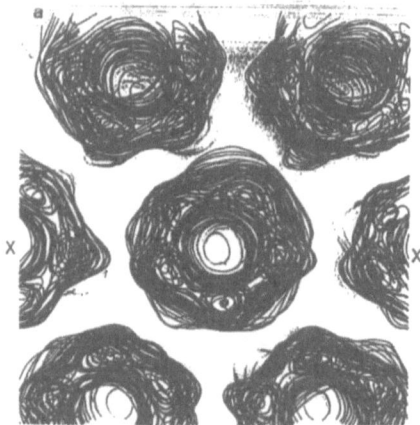 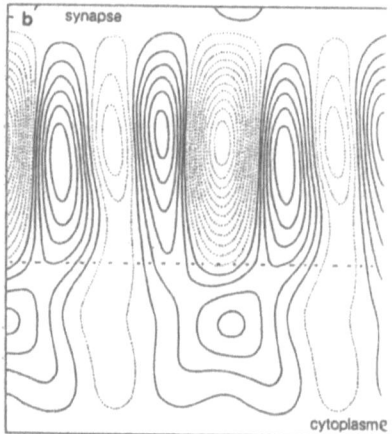

Figure 1 : AcChoR structure obtained by electron microscopy image analysis. Left : view of the receptor from the synapse showing the quasi-symetrical arrangement of the molecule around a 30 A diameter central pit. Right : cross-section of the receptor indicating that the molecule spans the membrane and that the central pit narrows sharply at the membrane level (Brisson & Unwin, 1985; with permission)

At least three significant pharmacological sites can be distinguished in the AcChoR complex: i) the acetylcholine binding site which is probably located on the α subunits and which binds other agonists and competitive antagonists such as α-bungarotoxin and d-tubocurarine; ii) the non-competitive blockers (NCB) binding area, able to bind histrionicotoxin, phencyclidine (PCP) and a large variety of molecules structurally unrelated; iii) the ionic channel which allows cation translocation through the membrane (Conti-Tronconi & Raftery,1982, Changeux et al.,1984).

Still much remain to be understood about the precise relationship between the function of the receptor and the molecular tridimentional organization of the different binding sites. In fact, one of the major goals in the present cholinergic receptor research, is to delineate on the receptor the different pharmacologically significant sites. In the absence of any structural information from radiocrystallographic experiments, a way to achieve this goal is to characterise which amino acids of the protein are involved in the ligand complexation. Photoaffinity labeling reagents are unmatched tools in that context since they are able to generate highly reactive species such as radicals, carbenes or aryl cations within the binding site, yielding irreversible binding with the amino acid residues located in the vicinity. The use of a radioactive probe allows further characterization of the labeled protein residues.

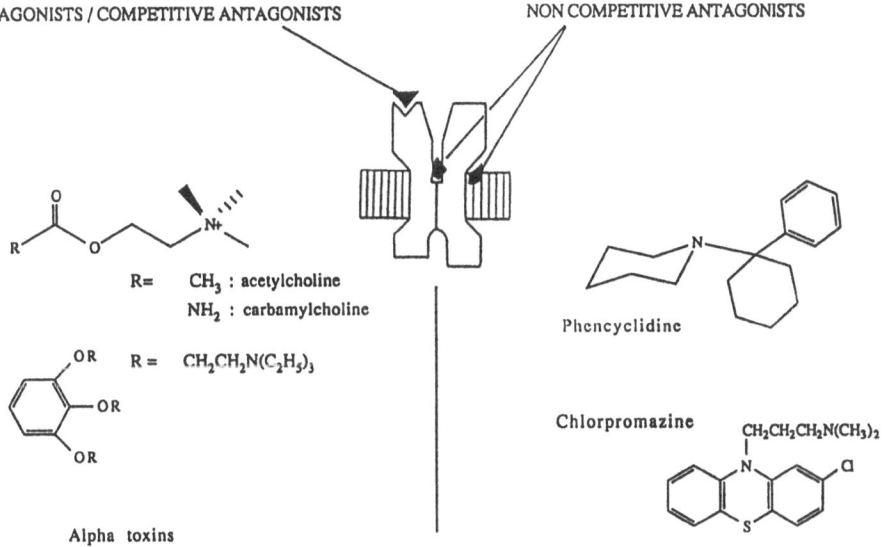

Figure 2 : Schematic localization of the different pharmacological sites on the receptor and structure of some prototypic ligands. A transection of the receptor is schematized, showing one of the two agonist binding sites: the high affinity NCB binding site located within the central pit and a secondary low affinity NCB binding site.

Labeling of specific sites on the AcChoR with radioactive affinity (Changeux et al., 1967, Silman & Karlin, 1969, Mautner & Bartels, 1970, Wieland et al., 1979, Kao et al., 1984) and photoaffinity (Witzemann et al., 1979, Nathanson et al., 1980, Oswald & Changeux, 1982) reagents have been used to identify labeled peptides of the active sites. Whilst the involvement of the α subunit in the acetylcholine binding site on the receptor complex is clearly demonstrated, localization of the NCB binding area is more complex to investigate. Photocoupling of various ligands to the receptor (Heidmann & Changeux, 1979; Giraudat et al.,1986; Hucho et al., 1986) indicate that the NCB binding site might be located within the central pit, involving the five subunits.

In the present communication we describe photoaffinity labeling of the different AcChoR binding sites with a new class of photoaffinity ligands: aryldiazonium salts (Kieffer et al. 1981). They are stable in the dark ($t_{1/2} > 2$ h) and generate, by irradiation, the corresponding aryl cation known to be a very reactive species ($t_{1/2} < 10^{-10}$ s in water). Furthermore, these salts can be preferentially photoactivated when bound to the target site, using a photosuicidal technique based on an energy transfer reaction with tryptophan residue(s) of the protein (Goeldner and Hirth 1980; Ehret et al., this volume).

MATERIAL AND METHODS

Live Torpedo marmorata were provided by the Institut Universitaire de Biologie Marine (Arcachon, France). Tritiated α–toxin obtained from Naja nigricollis was a gift of Drs Menez and Fromageot from the Commissariat à l'énergie atomique (Saclay, France). All other chemicals were of analytical grade. Fluorescence spectra were recorded with a Jobin-Yvon (JY 3C) spectrofluorimeter. Absorbance spectra were obtained with Kontron (uvikon 860) spectrophotometer. For irradiation experiment, monochromatic light was obtained from a 1000 W xenon-mercury lamp (Hanovia) connected to a grating monochromator (Kratos). The light intensity was measured with a thermopile (Kipp and Zonnen).

Electroplaque experiments. The pharmacological properties of DDF were examined by recording the membrane potential of an isolated electroplaque preparation from E. electricus (Schoffeniels & Nachmanson, 1957; Higman et al., 1963). The buffer solution used was 1.5 mM sodium phosphate buffer (pH 7), 150 mM NaCl, 5 mM KCl, 2 mM $CaCl_2$ and 2 mM $MgCl_2$.

Receptor-rich membrane preparation. AcChoR-rich membrane fragments (800-1800 nmol of α-toxin binding sites/g of protein) were prepared from fresh T. marmorata electric organs according to Saitoh et al., (1980).The concentration of α-toxin binding sites was determined under equilibrium conditions by DEAE filter assay (Fulpius et al.1972, Schmidt & Raftery, 1973). The association rate of the tritiated toxin to the receptor was measured by rapid filtration technique (Weber & Changeux 1974 a,b).

Irradiation experiments. Irradiation of electric eel electroplaque was carried out by a 40W microscope lamp normally used for organ dissection.In order to minimize the heating of the cell caused by the illumination, the lamp was turned on only 5 min every 15 min, and the preparation extensively washed by unexposed solution. Membranes irradiation were performed in a phosphate buffer (pH 7, 10 mM) at 10 °C in a 1 cm path length quartz cell, under gentle magnetic stirring. A monochromatic light beam (10 x 2 mm) was focused on the cell. We used either a visible light ($\lambda = 435$ nm, I = 40 μV) or UV light ($\lambda = 290$ nm, I = 60 μV), which correspond to the same number of incident photons in both conditions.We checked that DDF photodecomposes at the same rate at both wavelengths.

Purification of the α-chain. Solubilized membranes were subjected to preparative SDS-polyacrylamide gel electrophoresis and the bands corresponding to the different subunits eluted by diffusion (Giraudat et al., 1986). The eluates were dialysed against 0.1% SDS/0.01% thioglycol and then against water and finally lyophilised. The purified α-chain was carboxymethylated and precipitaded twice with acetone.

*Cleavage of the α–chain and purification of peptides.*The purified protein (2mg/ml) in 70% formic acid was incubated for 24 h at room temperature in the dark, in the presence of 0.06M CNBr and 4mM Trp. After lyophilisation, the residue was suspended in a water solution containing 0.1% trifluoracetic acid/6% n-propanol/3% acetonitrile/4 M guanidine hydrochloride.After centrifugation, the supernatant and the pellet (dissolved in pure formic acid) were subjected separately to HPLC separation, by reverse phase chromatography (Brownlee RP-300 column) followed by gel permeation (TSK 125 and 250).

Sequence analysis. Automated Edmann degradation was carried out on an Applied Biosystem gas-phase automated sequanator. Identification and quantification of amino acid derivatives were done by high performance liquid chromatography (HPLC; Knecht et al., 1983; Lazure et al., 1983)

RESULTS AND DISCUSSION

1. Acetylcholine binding site

Amongst the four different subunits of the AcChoR, the α subunit seems to be essentially involved in the agonist binding site. Several pieces of evidence strengthen this hypothesis i) affinity and photoaffinity labeling experiments of the acetylcholine binding site, using either agonists or α-toxins analogs, show a major incorporation of the probes into the α subunit; ii) the isolated α subunit is able to bind snake toxins with a good affinity (Merlie et al., 1983, Tzartos &Changeux, 1984, Smith et al., 1987); iii) mapping experiments of the toxin binding site by either monoclonal antibodies or synthetic peptides localize interacting peptides essentially on the α-subunit (Neumann et al., 1986, Lindstrom, 1986, Barkas et al. 1987).

Despite these results, few information are available regarding the precise part of the α-subunit primary sequence involved into the acetylcholine binding site. Clearly,agonists as well as toxins must bind to the synaptic part of the α-subunit. This binding area is probably located essentially between the N-terminal amino acid, proposed to be extracellular, and the first putative transmembrane fragment starting approximatively at Ile- 210. In fact, this sequence contains the glycosylation site (Asn 141), the "main immunogenic region" proposed to be located between α-46 and α-127 (see review Lindstrom, 1986) as well as the disulfide bridged adjacent cystein-192 cystein-193 pair previously labeled by a photoaffinity probe *after reduction of the receptor* (Kao et al., 1984, Dennis et al. 1986), a treatment known to alter its pharmacological properties.

To localize more precisely the amino acid residues involved in the agonists complexation,we developped a new class of photoaffinity labels of the acetylcholine binding site, derivated from aryldiazonium salts, DDF (I) being the prototypic drug. This compound is highly photosensitive and generates the corresponding arylcation able to react with a large variety of chemicals as shown in figure 1.

Besides its already "universal reactivity" the arylcation reacts with surrounding mole-cules at a rate higher than diffusion rate. Thus, if we generate it within the target binding site (by energy transfer reaction for instance, see this volume Ehret et al.), it will react with any surrounding residue, and analysis of labeled amino acids will give some insight into the receptor site topography.

1. 1. DDF AS A REVERSIBLE LIGAND

We studied first the pharmacological properties of DDF on the AcChoR. In the ab sence of light, DDF behaves as a competitive antagonist either _in vivo_ on Electrophorus Electricus electroplaque or _in vitro_ on receptor-rich Torpedo marmorata membranes (table I).

Furthermore, patch-clamp experiments on mouse myotubes, indicate that DDF behaves as a competitive antagonist of acetylcholine by shortening the opening frequency of the channel, and also possibly as a non-competitive blocker by reducing the channel mean open time. In vitro, DDF lowers the association rate of α-toxin to the receptor- rich membranes, indicating that these two molecules bind in a competitive manner to the acetylcholine site. From this experiment, mmolar range dissociation constant from the receptor was determined for DDF.

Table I: Pharmacological properties of DDF

isolated cells		
E. Electricus electroplaque[a]	Competitive antagonist	$K_d = 2\ 10^{-4}M$
C2 mouse myotube[b]	competitive antagonist non-competitive antagonist ?	
Receptor-rich-membranes[c]	competitive antagonist	$K_d = 10^{-3}M$

a) determined on isolated E. electricus electroplaque by measuring the influence of variable concentrations of DDF on the carbamoylcholine (30 μM) induced cell depolarisation amplitude. b) determined through the effect of DDF on AcCho-activated single-channel activity recording. c) DDF decreases the association rate of Naja nigricollis α-toxin to Torpedo marmorata receptor-rich membranes

---

Receptor intrinsic fluorescence quenching through DDF binding is an alternative way to quantify its affinity.

Figure 3 indicates that DDF binding leads to a quenching of the receptor fluorescence which can be partially restored by addition of acetylcholine to the medium. A more complete recovery of the initial fluorescence is achieved when phencyclidine, a ligand of the non-competitive blocker high affinity binding site, is added to the medium (not shown). This indicates that DDF is able to bind to the agonist as well as to the non-competitive blocker binding site.

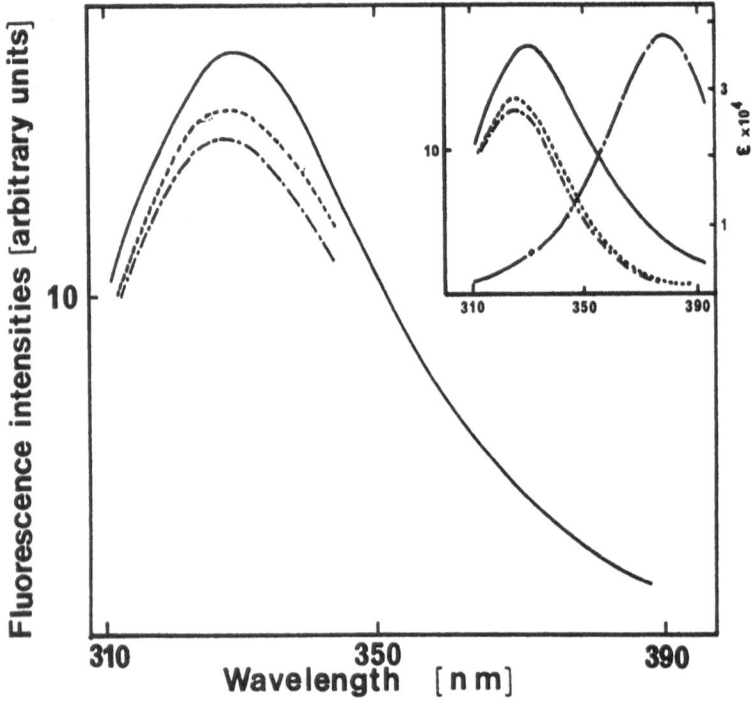

Figure 3: Corrected emission fluorescence spectra (λ_{ex} = 280 nm; T = 20 °C) of AcChoR rich membranes in the presence of DDF. Spectra were corrected for inner filter effect of DDF absorption at emission wavelength of the membranes; membranes alone (————), in the presence of 26 μM DDF and (—·—·—·—) in the presence of 26 μM DDF plus 10^{-4} M carbamoylcholine (- - - - - -). Insert : uncorrected fluorescence spectra (same symbols) juxtaposed to DDF ultraviolet absorp tion spectrum in phosphate buffer (——— · ———), (λ max 378 nm; ε = 38000 cm^{-1} M^{-1}).

1. 2. IRREVERSIBLE LABELING OF THE ACETYLCHOLINE BINDING SITE

In vivo, visible light irradiation of E. electricus electroplaque, in the presence of DDF, leads to an irreversible reduction of carbamoylcholine induced cell depolarisation amplitude. This effect is time- and DDF concentration-dependent and sensitive to d-tubocurarine (table II).

Table II : Irreversible blockade of E. Electricus electroplaque under
irradiation by 0.3 mM DDF

Number of successive irradiation	d-tubocurarine 2 μM	Membrane potential Amplitude response to 30 μM carbamylcholine
0	no	100%
1	no	85%
1	yes	100%
4	no	45%
4	yes	100%

Under irradiation, DDF reduces irreversibly the electroplaque depolarisation amplitude in response to application of 30μM carbamoylcholine. After every 5-min light exposure in the presence of 0.3mM DDF, the innervated face of the cell is extensively washed before the response to 30 mM carbamoylcholine is elicited.

In vitro, DDF reduces the α-toxin binding capacity of Torpedo marmorata receptor-rich membranes more efficiently, using energy transfer (irradiation wavelength 290 nm) than classical photoaffinity labeling conditions (irradiation wavelength 435 nm).

Table III : Irreversible loss of acetylcholine binding sites after classical (435 nm) or energy-transfer labeling (290 nm) by the same number of incident photons.

DDF(mM)	d-tubocurarine	hv.	acetylcholine sites available (%)
none	none		100
0.2	none		100
0.2	none	435 nm, 40 μV	79
0.2	2μM	435 nm, 40 μV	96
0.2	none	290 nm, 60 μV	40
0.2	10 μM	290 nm, 60 μV	97

AcChoR-rich membranes (0.9-1.4 μM α-toxin binding sites) in phosphate buffer 50mM, 150mM NaCl, are irradiated in the presence of 0.2 mM DDF during 30 min in the indicated conditions. After irradiation, the suspension is diluted 100x and the α-toxin binding capacity of the receptor measured under equilibrium conditions

For example, photoinactivation of the receptor, incubated in the presence of $2 \cdot 10^{-4}M$

DDF, leads to more than 50% site inactivation under energy transfer conditions, while about 15% of the sites are blocked when comparable direct photoaffinity conditions are used (table III).

These results and fluorescence quenching experiments indicate that at least one tryptophan residue must be located in the vicinity of the agonist binding site.

We checked that, under energy transfer as well as direct photoaffinity labeling conditions, DDF behaves as a good label of the acetylcholine binding site:

- in both conditions, the loss of α-toxin binding capacity (in vitro) as well as the diminution of the electroplaque depolarisation amplitude (in vivo) remain unchanged after extensive washings of the preparation.

- the extent of the inactivation is dependent upon both DDF concentration and the incident light intensity.

- it is possible to protect the receptor against DDF inactivation by adding either carbamoylcholine (agonist) or competitive antagonists such as d-tubocurarine or α-bungarotoxin in the incubation mixture.

- the tritiated probe incorporates irreversibly into the receptor, and the α–subunit is preferentially labeled (Fig 4). However about 50% of the radioactivity incorporated is sensitive to agonists or competitive antagonists indicating that DDF can irreversibly bind elsewere other than to the acetylcholine binding site. In fact, addition in the irradiation medium of the non-competitive blocker PCP, leads to a signal/noise ratio superior to 80%, indicating that DDF might also bind to the PCP binding site (Langenbuch-Cachat et al. 1988) This point will be discussed in the next chapter.

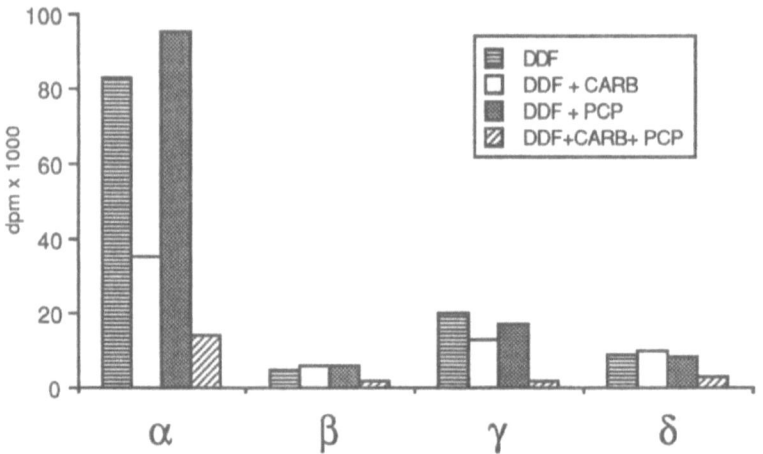

Figure 4: Effect of PCP and/or carbamoylcholine on tritiated DDF incorporation into the different AcChoR subunits under energy- transfer conditions. A solution of T. marmorata AcChoR-rich membranes (7.4 mM α-toxin binding sites) and 2×10^{-4} M ^{3}H DDF, in the absence or in the presence of 3×10^{-5} M PCP and/or 10^{-4} M carbamoylcholine, were irradiated under energy transfer conditions. Respectively 35% and 47% of the α-toxin-binding sites were inactivated in the absence and in the presence of PCP. Radioactivity incorporated into each subunit was measured after separation by SDS-gel electrophoresis. The four bars in each set represent respectively the following conditions: no further addition; 10^{-4} M carbamoylcholine; 3×10^{-5} M PCP; 10^{-4} M carbamoylcholine plus 3×10^{-5} M PCP.

There is a linear relationship between the ^{3}H DDF incorporation within the α subunit and the α-toxin binding sites inactivation, indicating an absence of site interaction. The stoechiometry of the inactivation reaction is close to unity, i.e. incorporation of one mole of DDF within the α-subunit, per mole of α-toxin binding site is sufficient to block it irreversibly (results not shown)

Altogether these results indicate that the DDF irreversible binding to the AcCho site is efficient (> 50% sites inactivated using DDF concentration lower than its K_d) and clean (stoechiometry of the inactivation close to the unity). On the other hand, we showed that the aryl cation generated within the binding site will react with any surrounding residues as well as water molecules. It is thus reasonable to assume that every labeled residue of the protein is in close vicinity to the bound diazonium during its photoexcitation. Analysis of these residues will give some information on the folding of the protein around the acetylcholine binding site.

We undertook this analysis on the α–subunit which is the major target of DDF incorporation. After labeling a large quantity of T. marmorata receptor rich membranes, in the presence of PCP, the subunit is purified by SDS gel electrophoresis, carboxymethylated and treated by cyanogen bromide. The HPLC reverse phase chromatography peptide map of the proteolysed subunit shows that the majority of the radioactivity is associated with three major peaks which account for approximately 70% of the total agonist protectable labeling (Fig 5).

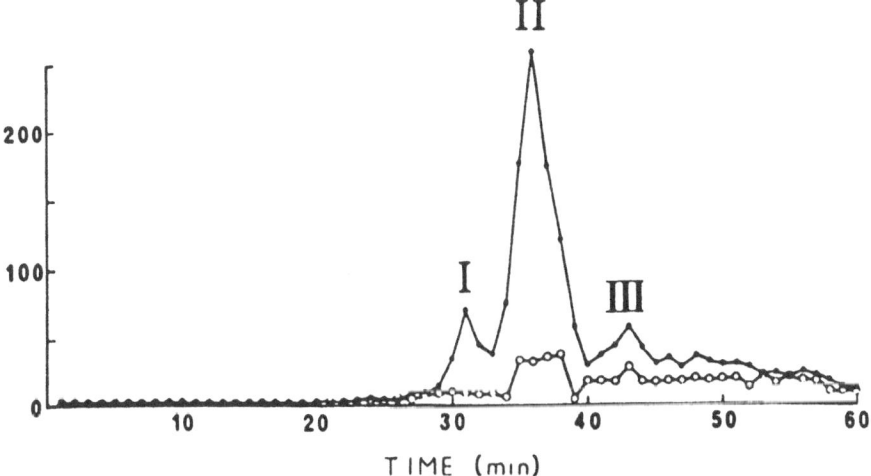

Figure 5: Radioactive profile of CNBr digested α-chain after photolabeling by ^{3}H DDF in the presence of 100 μM PCP and in the absence (•) or in the presence (o) of 100 μM carbamoylcholine.The reverse-phase HPLC column was equilibrated in 90% solvent A (0.1% trifluoracetic acid)/10% solvent B (0.1% trifluoracetic acid, 60% n-propanol, 30% acetonitrile).

Peak I contains approximately 5% of the total agonist sensitive labeling and after repurification reveals a single amino terminal sequence corresponding to the fragment starting at lysine-145. Measurements of the radioactivity at the sequanator output reveal that a clear release of tritium occurs at cycle 5 corresponding to α-Trp-149. This labeling disappears almost completely in the carbamoylcholine-protected sample.

Peaks II and peak III were purified and analysed in the same way. They contain respectively 60% and 5% of the carbamoylcholine-sensitive radioactivity. After purification, the N-terminal sequence analysis indicates that they correspond respectively to the cyanogen bromide fragments extending from α-Lys-179 and to the amino terminus of the mature α-subunit. Sequence analysis of peak II reveals that Tyr-190, Cys-192 and Cys-193 are labeled while peak III N-terminal sequencing does not allow labeled residue identification. Tryptic digestion of this 1-105 segment indicates that the labeled residue is probably located after residue α-Leu 80.

These experiments indicate that at least five residues are involved in the quaternary ammonium acetylcholine binding site. Except for the unknown amino acid residue in the sequence 80-105, which accounts for about 5% of the overall agonist sensitive DDF incorporation, no labeling of carboxylic residue is observed. This observation cannot be assigned to a loss of the labeling during the peptide purification, since we checked that the expected ester derivatives (p-alkoxydimethylaniline) is stable under CNBr cleavage and Edmann degradation. This result is surprising since it has been widely held that carboxylic residues must be involved in acetylcholine complexation (Luyten 1986). We cannot rule out that such carboxyl residues do exist within the acetylcholine binding site but ,if so, orientation of DDF within this site makes this reaction unfavorable. A long distance effect of the carboxyl residue in the ammonium complexation, a phenomenon already described in some antibiotic ionophores, may represent an alternative hypothesis. In agreement with this hypothesis, crystallographic studies show that hydroxyl groups can coordinate to quaternary ammonium (Rosenfield & Murray-Rust 1982).

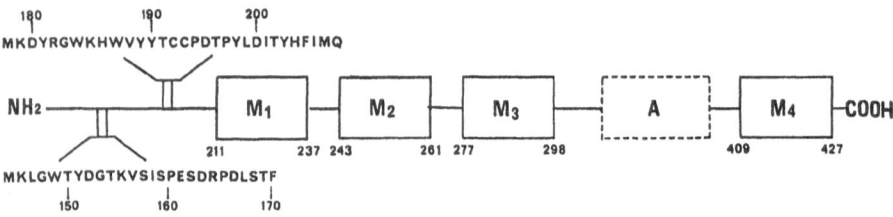

Figure 6: Schematic representation of the α-subunit primary sequence, showing the four putative transmembrane fragments (M1- M4), the glycosylation site Asn 141 and the disulfide bonds believed to link Cys-128-Cys-142 and Cys-192-193. The identified labeled residues in an agonist-sensitive manner are indicated by a filled star.

The five alkylated residues are located within three distinct regions of the large N-terminal hydrophilic sequence of the α-subunit (fig 6), and must probably be located in close vicinity to a single molecule of DDF, at the level of the acetylcholine binding site, in the native acetylcholine receptor. Since we do not know which orientation DDF can adopt when bound to the receptor, the simplest tridimentional representation of its binding site is a sphere whose diameter is the largest DDF dimension (approximately 12A).

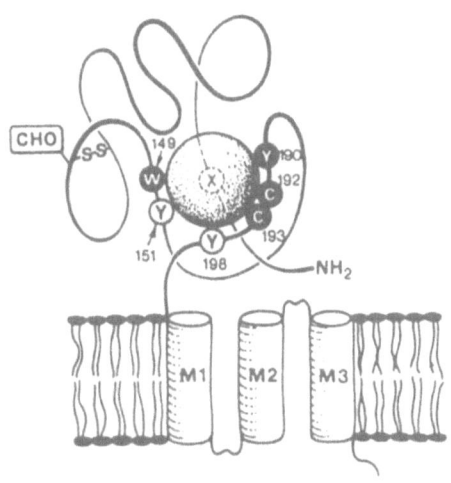

Figure 7: Schematic model for the DDF binding site. The sphere represents the space occupied by DDF in all its possible orientations within its site. Residues shown unanbiguously to react with DDF are indicated by a filled star, while open star denote residues for which suggestive evidence for labeling is obtained. The synaptic part of the α-chain was folded in such a way that all the labeled residues are in contact with the sphere surface. The three hydrophobic segments M1-M3 are putative transmembrane helices.

These results raise the question of the role of the labeled amino acids in the acetylcholine recognition by the receptor. It is reasonable to assume that they are probably involved in the acetylcholine ammonium complexation. In this respect, it is interesting to note that the labeled residues do not exist in the homologous positions on the β,γ, δ subunits, which are probably not involved in acetylcholine complexation. On the other hand, a good indication that they play an essential role in the acetylcholine receptor function (fig 8), is that they are conserved in α subunits of acetylcholine receptor isolated from the nervous system of different species, as indicated on the following scheme:

MOTOR ENPLATE:
Torpedo
K L G I W....K D Y R G W K H W V Y Y T C C P D T P Y
Mouse   α-subunit
K L G T W....K E A R G W K H T V F Y S C C P T T P Y
                                        γ-subunit
N C S L I......R H R P A K M L L D V A P A E E A T N V L
NEURONAL:
Chick
K F G S W........I K A P G Y K H E I K Y N C C E E I Y
Drosophila
K F G S W......M R V P A V R N E K F Y S C C E E P Y

figure 8: homologous sequences of the α-subunit AcChoR from different species. bold letters indicate the DDF labeled amino acid residues.

2. non-competitive blockers binding area

This site binds several molecules, structurally unrelated, leading to a blockade of the ion flux triggered by acetylcholine, without preventing the agonists/competitive antagonists binding. It exists in one copy per molecule of receptor and is probably distant from the agonist binding site by about 45 Å (Hertz et al., 1987), both sites being allosterically coupled.

Localization of this non-competitive blocker binding area using photolabeling techniques, has been investigated using chlorpromazine (Giraudat et al. 1986, 1987) and triphenylmethylphosphonium ion (Hucho et al., 1986) as photosensitized probes.These experiments lead to the proposal that the NCB site is a central one, delimited by homologous regions of the five subunits, since serines homologous to serine δ 262 are labeled on α and β subunits.However,the chemical reactivity of the generated species is not well understood and probably all the residues involved could not be detected. This is the reason why we use aryldiazonium derivatives to investigate in greater details the NCB binding area. As for the agonist binding site, these salts behave as reversible ligands in the dark and as efficient labeling reagents under irradiation.

2.1 Aryldiazonium salts as reversible ligands

Amongst several aryldiazonium derivatives,we selected three compounds, two aryldiazonium derivatives (I) and (II), differing by the N-alkyl chains length, and a derivative of imidazole (III), a smaller molecule existing essentially in a neutral diazo form at physiological pH:

$$ I : R = CH_3 $$
$$ II : R = C_4H_9 $$

$$ pK_a = 2,6 $$

Their affinity constants for the NCB binding area are determined either in the presence of α–toxin which stabilizes the native (R) state of the receptor or in the presence of carbamoylcholine which shifts the receptor to the desensitized (D) state.These compunds show affinity for the NCB binding site and are ten times higher than for the acetycholine binding site (not shown). Affinity of the aryldiazonium derivatives increases with the chain length, a phenomenon partially related to their hydrophobicity (table IV, Kotzyba-Hibert et al. 1985).Very few changes in the affinity for the (R) and (D) states are observed with derivatives (I) and (III),while changing the state of the receptor from (R) to (D) leads to a slight increase in the affinity of ligand (II). Such a phenomenon is frequently observed for NCB (Heidmann et al. 1984)

LIGAND	Table IV : Reversible binding characteristics of diazonium salts to membrane-bound AcChoR	
	Ki (M) against PCP	
	+ carbamoylcholine (D state)	+Toxin α (R state)
(I)	$2,5 \ 10^{-4}$	$1,7 \ 10^{-4}$
(II)	$1,6 \ 10^{-6}$	$4 \ 10^{-6}$
(III)	$1.5 \ 10^{-3}$	$1,8 \ 10^{-3}$

Table 4 : Ki against PCP were obtained by competition with ^{3}H -PCP according to Eldefrawi et al (1982). The native membranes were preincubated either with 2 molar excess of toxin α for 2 hours or with carbamoylcholine (10^{-4} M) for 10 min. prior to affinity measurement.

2. 2 Irreversible blockade of the PCP binding site

Table V shows the results of irreversible inactivation of the PCP binding site obtained after irradiation of the receptor-rich membranes in the presence of the different diazonium salts, when the receptor is maintained either in the (R) or in the (D) state.

Table V: aryldiazonium irreversible photoinactivation of the NCB binding site

Ligand	concentration	% irreversible binding	
		(D)state	(R) state
(I)[a]	$2 \ 10^{-4}$	40	40
(II)[a]	$2 \ 10^{-6}$	44	34
(III)[b]	$2 \ 10^{-3}$	40	0

AcChoR-rich membranes (1-1.6 μM α-toxin binding site) in phosphate buffer pH 7.2, 10 mM, 150 mM NaCl, are preincubated with α-toxin (2 molar excess) for two hours, or with carbamoylcholine (10^{-4} M) for 10 min. The diazonium derivative is then added and the solution is irradiated for 30 min. a) Irradiation conditions used: λ exc = 290 nm, C = 2 Ki; b) λ exc =313 nm; C = 2 Ki.

Comparison of the inactivation amplitude between the desensitized and the native states of the receptor by DDF shows no drastic differences. On the other hand, compound (II) blocks the NCB binding site preferentially in the desensitized state with a rather high yield (65% inactivation at 2 Ki). This observation can be accounted by a better site occupancy by the diazonium derivative in the (D) state compared to the (R) state, due to affinity constant variation between the two states. More surprising, is the result obtained with the imidazole diazonium derivative (III). Activated in direct photoaffinity labeling conditions (λ exc =313 nm) it gives a very high yield of inactivation (40%) when the receptor is in the (D) state, but does not block the receptor in the (R) state.This result, added to the absence of any detectable variation of its affinity to the NCB binding site, when the receptor shifts from the native state to the desensitized state, indicate that (III) may be considered as a conformation sensitive probe.The simplest interpretation of this result is that the diazonium group detects a much more hydrated environment in the (R) state than in the (D) state.

2.3 Polypeptides involved in the NCB binding area

To determine which polypeptide chains of the AchoR are involved in the NCB binding area, and ultimately which aminoacids are labeled with these photoactive ligands, we synthesized tritiated compounds (I) and (II). The membranous AcChoR, in the desensitized form, was irradiated in the presence of each radioactive diazonium derivative and tritium incorporation into every subunit was measured. With both probes, radioactivity incorporation occurs in the four chains α, β, γ, δ. However, prior incubation with unlabeled PCP lowers the radioactivity only by about 50% for DDF but by about 80% for (II).In both cases, the PCP sensitive radioactivity incorporates equally well into the five subunits. This is in agreement with the hypothesis that the NCB binding area is located within the central pit, in which the ligand bind with several quasi-equivalent orientations (Heidmann & Changeux, 1979).

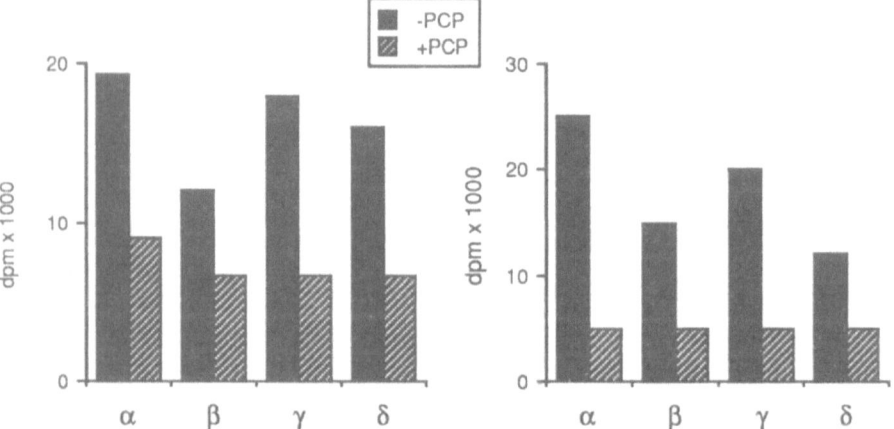

Figure 9: ^{3}H incorporation into the different receptor subunits, after irradiation of receptor-rich membranes in the presence of 10^{-4} M carbamoylcholine and (left) 2 10^{-4} M ^{3}H DDF or (right) 3 10^{-6} M ^{3}H (II).

H^3 ligand (II) seems to be a good candidate to detect the aminoacids involved in the NCB binding area since it labels very efficiently this site in energy transfer irradiation conditions (65% irreversible site occupancy, using a 2 Kd concentration) with a very good specificity (more than 80% of the ligand incorporation is prevented by PCP). The universal reactivity of the generated arylcation may be extented to other amino acid residues the localization of the NCB site shown to be probably located on the residues β-Ser-257 and β-Leu-257 of the M2 helix either by photolabeling experiments (Giraudat et al., 1986, 1987, Hucho et al., 1986) or by protein engineering experiments (Imoto et al., 1986).

CONCLUSION

In this paper, we demonstrate that simple aryldiazonium ions behave as good photoactivatable analogs either of acetylcholine or of non-competitive blockers of the acetylcholine receptor. In particular, we are able to selectively block each site without labeling the other, despite their low affinities. This is possible since we have efficient protecting drugs for both sites. The analogy of DDF with ammonium groups is found to apply to other receptor sites such as the active site of the acetylcholinesterase (Kieffer et al 1981) and the binding site of the muscarinic receptors (B. Ilien unpublished results).The great efficiency of the aryldiazonium ions as photoactivable analogs of ammonium groups is due to the conjunction of several favorable factors:
- Both structures have a positively charged group, with the charge distributed over several atoms
- the charged moiety (N_2^+) is also the reactive part of the photoactivated molecule.
- the generated aryl cation is a very reactive species with reaction rates probably higher than diffusion rate in the medium. It shows also little reaction specificity . Thus, it is reasonable to assume that it will react more rapidly with any surrounding residue of the binding site before leaving it (Ehret et al., this volume).
- some of them show spectral characteristics allowing selective photoactivation of the bound molecule through an energy transfer reaction with tryptophan residues located in the vicinity of the binding site.

These characteristics allow topographical analysis of active sites by characterizing the labeled subunits and the labeled amino acids on each target receptor subunit. In this paper we partially achieved this goal showing that the five acetylcholine receptor subunits are equally labeled when the probe is photodecomposed in the non-competitive binding area. This indicates that this site may be located within the central pit. A more precise analysis was carried out on the acetylcholine binding site, which suggests that at least five different amino acids, located in different regions of the synaptic part of the α-subunit, may lie in a close vicinity on the native receptor. It must however be kept in mind that, under our experimental conditions ,the two acetylcholine binding sites could be labeled and that no discrimination between the different states of the receptor occurs. Thus the cartoon showing the α-subunit folding integrates these spatial and temporal factors.

It will be interesting to generalize this approach to other biological systems in order to check the efficiency of aryldiazonium ions as photoprobes. It is reasonable to assume that substitution, on a defined ligand structure, of an amino group by aryldiazonium is the most promising way to achieve this goal.

ACKNOWLEDGMENTS

We thank Drs Menez and Fromageot for the gift of Naja nigricollis ^{3}H-α-toxin, Dr C. Pinset for C2 cell line, R. Knecht, O. Seegmüller and N. Franco for performing sequence analysis and Dr B. Ilien for its critical reading of the manuscript. This work was supported by the Muscular Dystrophy Association of America, the Collège de France, the Ministère de la Recherche et de l'Enseignement Supérieur, the Centre National de la Recherche Scientifique and the Commissariat à l'Energie Atomique.

REFERENCES

Anholt, R., Lindstrom, J. and Montal, M. (1984) Enzymes Biol. Membr. 3, 335-401.

Ballivet, M., Patrick, J., Lee, J. and Heinemann, S. (1982) Proc. Natl. Acad. Sci. USA 79, 4466-4470.

Barkas, T., Mauron, A., Roth, B., Alliod, C., Tzartos, S.J. and Ballivet, M. (1987) Science 235, 77-80.

Boulter, J., Evans, K., Goldman, D., Martin, G., Treco, D., Heinemann, S. and Patrick, J. (1986) Nature 319, 368-374.

Brisson, A. and Unwin, P.N.T. (1985) Nature 315, 474-477.

Changeux, J.P., Podleski, T.R. and Wofsy, L. (1967) Proc. Natl. Acad. Sci. USA 58, 2063-2070.

Changeux, J.P., Devillers-Thiéry, A. and Chemouilli, P. (1984) Science 225, 1335-1345.

Claudio, T., Ballivet, M., Patrick, J. and Heinemann, S. (1983) Proc. Natl. Acad. Sci. USA, 80 , 1111-1115.

Conti-Tronconi, B.M. and Raftery, M.A. (1982) Ann. Rev. Biochem. 51, 491-530.

Dennis, M., Giraudat, J., Kotzyba-Hibert, F., Goeldner, M., Hirth, C., Chang, J.Y. and Changeux, J.P. (1986) FEBS Lett. 207, 243-249.

Dennis, M., Giraudat, J., Kotzyba-Hibert, F., Goeldner, M.P., Hirth, C.G., Chang, J.Y., Lazure, C., Chrétien, M. and Changeux, J.P. (1988) Biochemistry 27 , 2346-2356.

Devillers-Thiéry, A., Giraudat, J., Bentaboulet, M. and Changeux, J.P. (1983) Proc. Natl. Acad. Sci. USA 80 , 2067-2071.

Eldefrawi, A.T., Miller, E.R., Murphy, D.L. and Eldefrawi, M.E. (1982) Mol. Pharmacol. **22**, 72-81.

Fulpius, B., Cha, S., Klett, R. and Reich, E. (1972) FEBS Lett. **24**, 323-326.

Giraudat, J., Devillers-Thiéry, A., Auffray, C., Rougeon, F., and Changeux, J.P. (1982) EMBO J. **1**, 713-717.

Giraudat, J., Dennis, M., Heidmann, T., Chang, J.Y. and Changeux, J.P. (1986) Proc. Natl. Acad. Sci. USA **83**, 2719-2723.

Giraudat, J., Dennis, M., Heidmann, T., Haumont, P.Y., Lederer, F. and Changeux, J.P. (1987) Biochemistry **26**, 2410-2418.

Goeldner, M.P. and Hirth, C.G. (1980) Proc. Natl. Acad. Sci. USA **77**, 6439-6442.

Heidmann, T. and Changeux, J.P. (1979) Eur. J. Biochem. **94**, 281-296.

Heidmann, T. and Changeux, J.P. (1980) Biochem. Biophys. Res. Comm. **97**, 889-896.

Heidmann, T. and Changeux, J.P. (1984) Proc. Natl. Acad. Sci. USA **81**, 1897-1901.

Hertz, J.M., Johnson, D.A. and Taylor P. (1987) J. Biol. Chem. **262**, 7238-7247.

Higman, H., Podleski, T.R. and Bartels, E. (1963) Biochim. Biophys. Acta **75**, 187-193.

Hucho, F., Oberthür, W. and Lottspeich, F. (1986) FEBS lett. **205**, 137-142.

Hucho, F. (1986) Eur. J. Biochem. **158**, 211-226.

Imoto, K., Methfessel, C., Sakmann, B., Mishina, M., Mori, Y.,Konno, T., Fukuda, K., Kurasaki, M., Bujo, H., Fujita, Y. and Numa, S. (1986), Nature **324**, 670-674.

Kao, P., Dwork, A., Kaldany, R., Silver, M., Wideman, J., Stein, S. and Karlin, A; (1984) J. Biol. Chem. **259**, 11662-11665.

Kieffer, B., Goeldner, M.P. and Hirth, C.G. (1981) J.C.S. Chem. Comm. 398-399.

Knecht, R., Seegmüller, U., Liersch, M., Fritz, H., Braun, D.G. and Chang, J.Y. (1983) Anal. Biochem. **130**, 65-71.

Kotzyba-Hibert, F., Langenbuch-Cachat, J., Jaganathen, A., Goeldner, M.P. and Hirth, C.G. (1985) FEBS Lett. **182**, 297-301.

Langenbuch-Cachat, J., Bon, C., Mulle, C., Goeldner, M., Hirth, C. and Changeux, J.P. (1988) Biochemistry **27**, 2337-2345.

Lazure, C., Seidah, N.G., Chrétien, M., Lallier, R. and St-Pierre, S. (1983) Can. J. Biochem.Cell.Biol. **61**, 287-292.

104

Lindstrom, J. (1986) TINS, 401-407.

Luyten, W. (1986) J. Neurosci. Res. **16**, 51-73.

Mautner, H.G. and Bartels, E. (1970) Proc. Natl. Acad. Sci. USA **67**, 74-78.

Merlie, J.P., Sebbane, R., Gardner, S. and Linstrom, J. (1983) Proc. Natl. Acad. Sci. USA **80**, 3845-3849.

Nathanson, N.M. and Hall, Z.W. (1980) J. Biol. Chem. **255**, 1698-1703.

Neubig, R.R., Boyd, N.D. and Cohen, J.B. (1982) Biochemistry **21**, 3460-3464.

Neumann, D., Barchan, D., Safran, A., Gershoni, J.M. and Fuchs, S. (1986) Proc. Natl. Acad. Sci. USA **83**, 3008-3011.

Noda, M., Takahashi, H., Tanabe, T., Toyosato, M., Furutani, Y., Hirose, T., Asai, M., Inayama, S., Miyata, T., Numa, S. (1982) Nature (London) **299** , 793-797.

Noda, M., Takahashi, H., Tanabe, T., Toyosato, M., Kikyotani, S., Hirose, T., Asai, M., Takashima, H., Inayama, S., Miyata, T. and Numa, S. (1983a) Nature **301**, 251-255.

Noda, M., Takahashi, H., Tanabe, T., Toyosato, M., Kikyotani, S., Furutani, Y., Hirobe, T., Takashima, H., Inayama, S., Miyata, T. and Numa, S. (1983b) Nature (London) **302** , 528-532.

Oswald, R. and Changeux, J.P. (1982) FEBS lett. **139**, 225-229.

Rosenfield, R.E., Jr. and Murray-Rust, P. (1982) J. Am. Chem. Soc. **104**, 5427-5430.

Saitoh, T., Oswald, R., Wennogle, L. and Changeux, J.P. (1980) FEBS Lett. **116**, 30-36.

Schmidt, J. and Raftery, M.A. (1973) Anal. Biochem. **52**, 349-354.

Schoffeniels, E. and Nachmanson, D. (1957) Biochim. Biophys. Acta **26**, 1-15.

Silman, H.I. and Karlin, A. (1969) Science **164**, 1420-1421.

Smith, M.M., Lindstrom, J. and Merlie J.P. (1987) J. Biol. Chem. **262**, 4367-4376.

Sumikawa, K., Houghton, M., Smith, J.C., Bell, L., Richards, B.M. and Barnard, E.A. (1982) Nucleic Acid. Res. **10**, 5809-5822.

Tzartos, S.J. and Changeux, J.P. (1984) J. Biol. Chem. **259**, 11512-11519.

Weber, M. and Changeux, J.P. (1974) Mol. Pharmacol. **10**, 1-14.

Weber, M. and Changeux, J.P. (1974) Mol. Pharmacol. **10**, 15-34.

Wieland, G., Frisman, D. and Taylor, P. (1979) Mol. Pharmacol. **15**, 213-226.

Witzeman, W., Muchmore, D. and Raftery, M.A. (1979) Biochemistry **18**, 5511-5518.

PHOTOSUICIDE LABELLING

L. EHRET-SABATIER, B. KIEFFER, M.P. GOELDNER and C.G. HIRTH
Laboratoire de Chimie Bio-organique UA CNRS 31
Universite Louis Pasteur Strasbourg, Faculte de Pharmacie
67400 Illkirch-Graffenstaden
France

ABSTRACT. The photochemical properties of aryldiazonium salts and 4-diazocyclohexa-2,5-dienones are described with regards to their use as photoaffinity probes of biological receptors. The high reactivity of the photogenerated species, the corresponding aryl cation and the cyclohexadienonylidene carbene are emphasized. The short lifetime of the photogenerated intermediate is an important factor for the specificity of alkylation at the target binding site during the photoaffinity labelling process. We designed the photosuicide concept (suicide inactivation using photolabile probes) to increase this specificity of labelling by selective photoactivation of the ligand molecule that is complexed at the binding site. Descriptions of three selective photoactivations are given, each induced by different physico-chemical properties of a receptor site. The use of the above mentionned highly reactive species in photosuicide labelling experiments opens diffent possibilities including the topographical mapping of a receptor site.

1. Introduction

The aim of photosuicide labelling (1) is to achieve an efficient and specific irreversible labelling of a receptor binding site through selective photoactivation of the ligand bound to the recognition site. Such labelling combines the specificity of suicide inhibitors with the reactivity of the photogenerated species in photoaffinity labelling (2). The term specificity refers to the labelling that occurs exclusively at the amino acid residues of the target binding site as opposed to all other portions of the polypeptide or *a fortiori* to another proteic structure present in the incubation mixture.

The suicide inhibitors (kcat inhibitors)(3) are enzyme substrates that are converted through the catalytic reaction to reactive species which ideally inactivate the enzyme through an alkylation reaction occuring at the enzyme active site. However, virtually all of the enzymatically generated reactive species are electrophiles with moderate chemical reactivity which absolutely require the presence of reactive (nucleophilic) amino-acid residues to allow such an inactivation.

Thus, the questions we asked were first how to increase the reactivity of those species and secondly how to transpose the suicide inactivation concept to the irreversible labelling of a hormone or neuromediator receptor binding site which lacks catalytic properties.

The use of light allows the generation of short lived chemical species (carbenes, nitrenes, radicals and carbocations) which are not otherwise easily provided. However, substantial differences in chemical reactivity exist between these reactive species and also

107

P. E. Nielsen (ed.), Photochemical Probes in Biochemistry, 107–122.
© *1989 by Kluwer Academic Publishers.*

within a given class, depending on their structure. The present article emphasizes first the high reactivity of aryl cations and cyclohexadienonylidene carbenes derived respectively from aryldiazonium salts and diazocyclohexadienones, a new class of potential photoaffinity reagents (4). Secondly, three different methods of photosuicide labelling will be described emphasizing the specificity of the labelling. The combination of the suicide concept with a hyper-reactivity of the photogenerated probe opens the possibility of mapping the direct environment of the ligand within its receptor binding site.

2. Experimental

This section is limited to a general description of the photochemical experiments as well as a general procedure for the synthesis of diazonium salts, including small scale experiments which are adapted to the synthesis of radiolabelled ligands.

Irradiation experiments: The light source (either a 1000 watt Xe/Hg or a 500 watt Xe lamp; Hanovia) is beamed through a water filter, a diaphragm and finally connected to a grating monochromator (Kratos). Using a lens, the resulting light beam is focused on a quartz cell (1 cm path, 1 or 3 ml) to a spot of fixed dimensions for a given set of experiments (the entire light spot is in contact with the medium). The light intensities are measured by use of a thermopile (Kipp & Sohnen) coupled to a microvoltmeter. The modulation of the light intensity is monitored with the diaphragm, preventing any change in the monochromaticity or in the geometry of the light spot. A constant and slow magnetic stirring during the irradiation experiments is always effective.This equipment allows for reproducible irradiation experiments throughout a project independent of the experimenter.

Synthesis of aromatic diazonium salts: The synthesis of diazonium salts is achieved by a classic nitrosation reaction of the corresponding aniline derivative. Aqueous tetrafluoroboric acid (33 or 48%) proved to be particularly useful for the synthesis of crystalline derivatives (generally 10 mg and up):

General procedure: This experiment is carried out in the absence of light. A 10% excess of either solid $NaNO_2$ or a 1M aqueous $NaNO_2$ solution is added over a period of 30 min. to a well-stirred cooled (-10°C) solution or suspension of the aromatic amine (0.1-2M), in aqueous HBF_4. After 45min. of additional stirring at the same temperature, the diazonium salt is filtered off and dried under reduced pressure. The addition of diethyl ether (a minimum amount) to the medium may be necessarry to induce the precipitation of the salt.

Small scale synthesis (1-10 μmoles): The following changes are made to the preceding experimental procedure: The solution of the aromatic amine is 10-15 mM in a 1/1 mixture of TFA and conc. HCl and the aqueous $NaNO_2$ solution is 10^{-2} M. After the additional stirring, the acids are evaporated under vacuum at room temperature and the residue resuspended in water (0.5-2 ml) to be analysized and purified by HPLC. The typical chromatographic conditions use a Waters μ-Bondapack (3.9x300 mm) column eluted at a flow rate of 2 ml per min. with a gradient of 100% H_2O- 0.05% TFA to 100% CH_3CN in 40 min. The elution of the products is monitored by absorbance at 229 nm. We checked on diazonium salts which were obtained in a pure crystalline form that this synthesis and purification procedure gave satisfactory results in terms of purity and excellent yields (over 90%).

3. Results and discussion

3.1.PHOTOCHEMICAL REACTIVITY OF AROMATIC DIAZONIUM SALTS AND 4-DIAZOCYCLOHEXA-2,5-DIENONES

The aromatic diazonium salts and the diazocyclohexadienones are among the few chemicals known, which on illumination would loose nitrogen in a reversible fashion (5). This unique feature is indicative of the unusual reactivity of the corresponding photogenerated species, namely the aryl cation and the cyclohexadienone carbene. The following chapter summarizes the photochemical properties of these probes with regard to their use as photoaffinity labels of biological receptor sites.

3.1.1. *Aromatic diazonium salts*

The primary process in the photochemical decomposition of an aryldiazonium salt is the formation of the corresponding aryl cation. It has been demonstrated that the triplet cation is the primary photoproduct of para-dimethylamino (6) and 2,4,5-trimethoxy benzenediazonium (7) by ESR spectroscopy at 77K. This second transient species has been shown to be reduced in the photolytic conditions to the corresponding aryl radical. Electron donating substituents (particularly at the 2,4,6 positions of the aromatic ring) will stabilize the cationic species by resonance. On the other hand, electron attracting groups favor the electronic photoreduction process.

$$Ar-N_2^+ \rightleftharpoons Ar^+ + N_2 \xrightarrow{1\,e} Ar^{\cdot}$$

Chemical reactions

			Half-life(h) pH = 7.2	λ max	ε	Quantum yields
	1	R = 4-NMe$_2$	3.5	379	37500	0.59
	2	R = 4-OMe	>24	313	23500	0.46
	3	R = 3-OMe	0.17	273	7300	
				350	1500	
	4	R = 3-Br	0.08	268	7000	
				323	1400	
	5	R = 4-I	0.5	326	12500	
	6	R = 4-N(Me)COMe	>72	340	16600	
	7	R = 4-NHCOMe 2,6-Me$_2$	>72	338	22500	0.33

Table I : Physico-chemical properties of aromatic diazonium salts

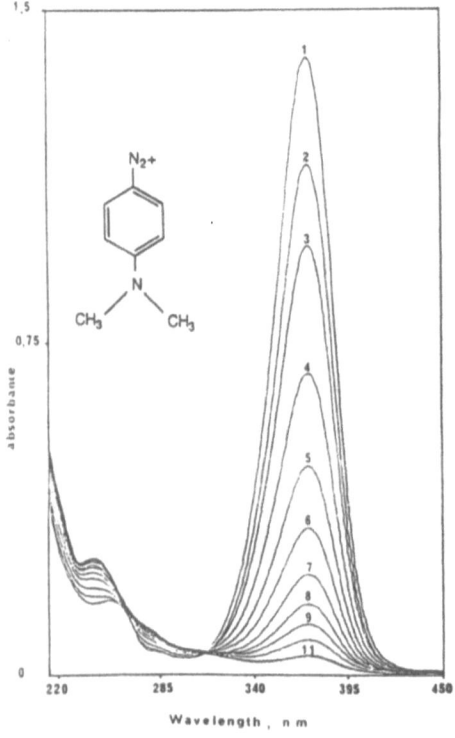

Wavelength , n m

Table I (8)(9), summarizes some physico
-chemical properties of substituted
diazonium salts. As expected, only the
diazonium salts substituted in the para
position with electron donating groups
show the required properties of potential
photo-affinity labels: sufficient chemical
stability in the absence of light in neutral
buffered medium and strong absorption at
wavelengths greater than 300nm which are
associated with high quantum yields.

Figure 1. Typical photodecomposition
kinetics of the diazonium salt 1 in water at
wavelengths > 300nm showing the total
disappearance of the high absorption band.

The aryl cation is one of the few reactive species (10) that displays some "universal"
reactivity, that is its capability to react with any chemical bond in its vicinity, including
reaction with a carbon-hydrogen bond. A direct study of the photochemical
decomposition of aromatic diazonium salts in hydrocarbons is difficult because of their
insolubility in these media. Nevertheless a few examples of reactions with C-H bonds are
described on particular substrates:

(11)

(12)

An interesting approach has been taken by the group of Speranza who have generated
the free unsolvated aryl cation by a radiolytic pathway from tritiated benzene. They were
able to study the chemical reactivity of this species with a series of reagents including
alcohols (13, 14), hydrocarbons (CH_4, C_2H_6)...(15) and several alkyl chlorides (16).

Several interesting features on the reactivity of the free aryl cation have been observed:

- efficient reactions with C-H bonds occur not only with hydrocarbons (15) but also with alcohols(13),

- no tritium scrambling occurs during the reaction with liquid CH_3OH for which a rate constant of 10^7-10^8 s^{-1} has been estimated(14),

- finally, the formation of over 50% indane during the reaction of the cation with cyclopropane (17), which involves an initial **C-C insertion** reaction, illustrates the hyper-reactivity of this cation.

The study of the photochemical reactivity of a class of compounds towards any biological binding site is tedious. Nevertheless two common features can be defined in all experiments, i.e., the reaction with water accounting for the surrounding medium and the reaction with a peptide bond assuming that the receptor of interest is a protein. The photochemical decomposition of a series of diazonium salts in water gave the expected corresponding phenol derivatives with nearly quantitative yields (8). The photochemical reaction with a peptide bond was tested using N-ethylacetamide (18). Over 90% of the reaction led to an O or N alkylation process which was expected from the solvatation of the charged moiety of the diazonium salt by the oxygen and nitrogen atoms (figure 2.).

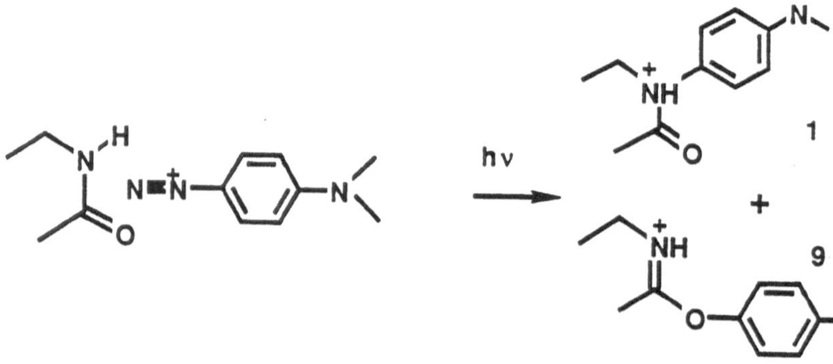

Figure 2. Photochemical reaction of diazonium 1 with an amide bond.

Finally, the finding that crown ethers complex diazonium salts (19), led to the design of a more realistic model of the reactivity of a diazonium salt at the receptor site (20). The photochemical reaction of the entraped diazonium salt (Figure 3) lacks the formation of the anisole and the halogenated derivatives which are the main photo-compounds formed in the corresponding medium (CH_2Cl_2,MeOH). Such an exclusive reaction with the neighboring crown groups observed on a fairly loose complex ($K_D = 10^{-3}M$) can only be explained by a faster alkylation process than the escape of the reactive species (the p.-dimethylamino benzene cation) for which no remaining affinity for the crown is expected. The identification of the reaction products on the crown ether is currently in progress.

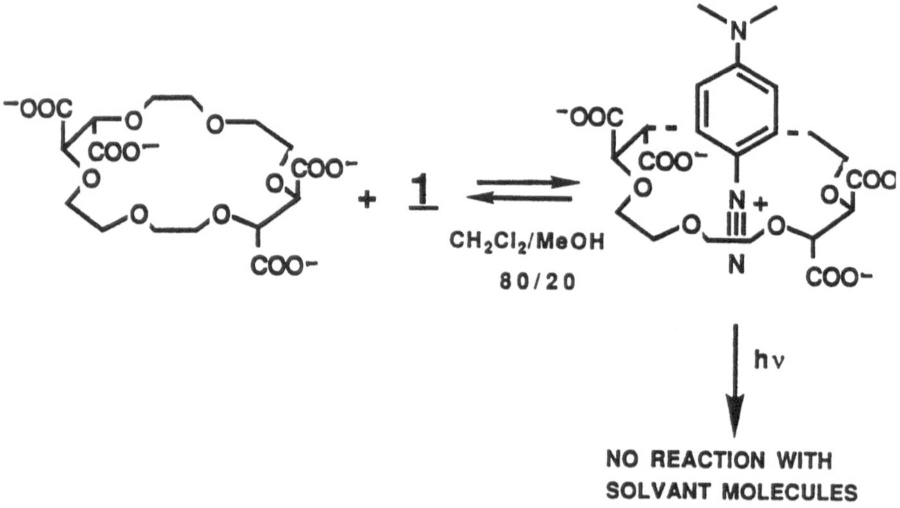

Figure 3. The crown-diazonium complex as a model of the reactivity of a species in a receptor site

3.1.2. *4-diazocyclohexa-2,5,dienones*

These chemicals derive typically from 4-hydroxy diazonium salts by deprotonation of the acidic phenol (pKa = 3.5). This acid-base equilibrium can be monitored by UV spectroscopy, the diazo species showing a red-shift with increased absorbance attributed to a quinonoid structure. For instance, 4-diazocyclohexa-2,5-dienone shows a λmax at 350 nm (ϵ = 44000) while the corresponding 4-hydroxy diazonium salt absorbs at λmax = 315 nm (ϵ = 23000) (21). Substitution on the ring induces only small changes in the UV absorption spectra and does not affect the high chemical stability of these compounds in aqueous buffer at neutral pH (half-life of several days).

The photochemical properties of this class of compounds has been reviewed (22). Although the formed cyclohexadienone carbenes have been described as ground-state triplets (23), their photochemical reactivity does not necessarily agree with a unique spin state. For example, 2,6-substituted derivatives show typical singlet carbene reactivity based on their stereoselective addition to double bonds (24). The resonance structures of the cyclohexadienone carbene (see below) illustrate the multiple reactions of this species (Figure 4).

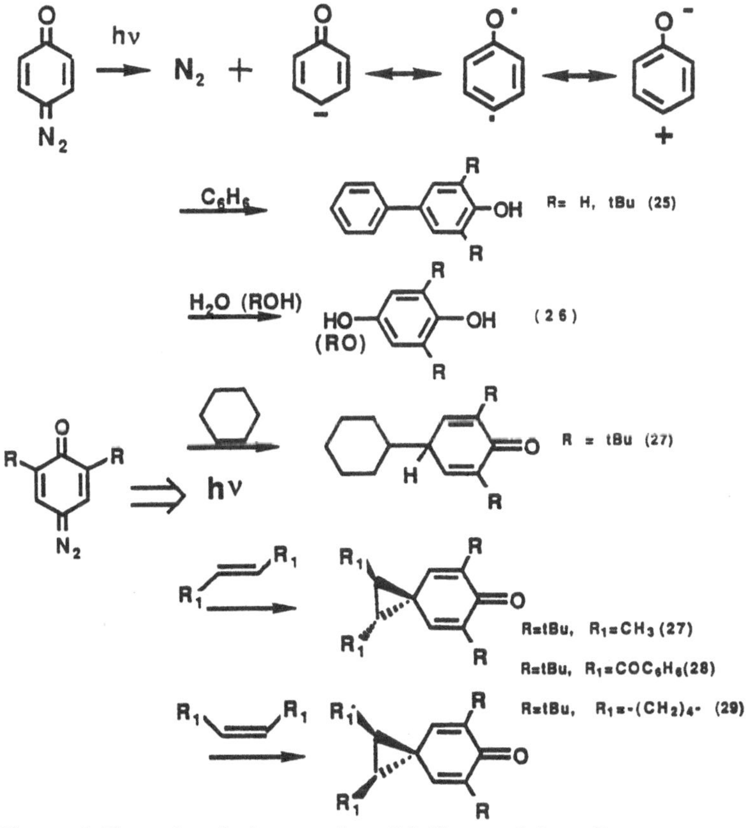

Figure 4. Photochemical properties of 4-diazocyclohexadienones

3.2. PHOTOSUICIDE LABELLING:

A photosuicide inactivator is an enzyme or receptor ligand analog, the photoactivation of which is selectively induced by the intrinsic physico-chemical properties of the binding site. The scheme below illustrates an ideal case where only the complexed probe is photoactivated as opposed to a classical photoaffinity labelling experiment where all the ligand molecules become activated by light.

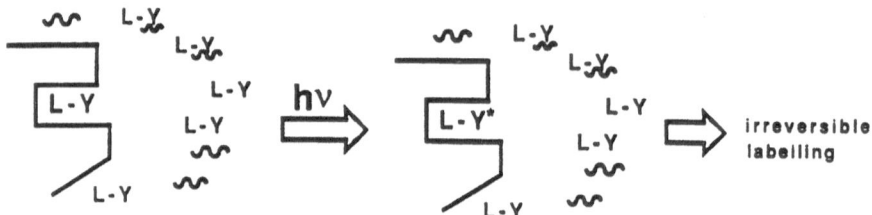

PHOTOSUICIDE CONCEPT:
⇨ SPECIFICITY AND EFFICENCY OF IRREVERSIBLE LABELLING

Two major consequences are expected from photosuicide labelling: selective excitation will increase the specificity (as defined in the Introduction) as well as the efficiency of labelling. The term efficency refers to the kinetics of inactivation for a given experiment. It is dependent on the fact that the ligand concentration remains (or should remain) constant during the irradiation process leading to a pseudo steady state situation.

Three examples of photosuicide labelling will be described making use of different physico-chemical characteristics of a receptor. First, the enzymatic reaction is used to generate a chromophore that can be specifically excited (chromophore transfer). Secondly, assuming that a binding site possesses hydrophobic character,the enhanced photosensitivity of some derivatives (nitrosamines) in hydrophobic media is used to achieve a selective photoactivation of the complexed ligand (phase transfer). Finally, the location of a tryptophan residue at, or nearby, the binding site allows for the photodecomposition of the complexed ligand molecule through an energy tranfer-mediated reaction (energy transfer).

3.2.1 *Chromophore transfer* (30)

Theoretically, there are two ways to use the enzymatic step to generate a photosensitive product that can be selectively photoactivated. In the first case the catalytic reaction would transform a non photosensitive substrate to a photosensitive enzyme product. In the second case, both the substrate and the enzyme product are photosensitive but at different wavelengths; this change in absorption is induced by the catalytic reaction. The use of an appropriate irradiation wavelength will allow the desired selectivity of photoactivation.

The neutral hydrolysis of para-phenyl diazonium esters leads to the quantitative formation of 4-diazocyclohexa-2,5-dienone derivatives. The half-time of hydrolysis of para-phenyl diazonium butyrate at different pHs is given in Table II and, as can be seen in Figure 5, the higher absorption band of the reaction product does not overlap with the starting diazonium salt.

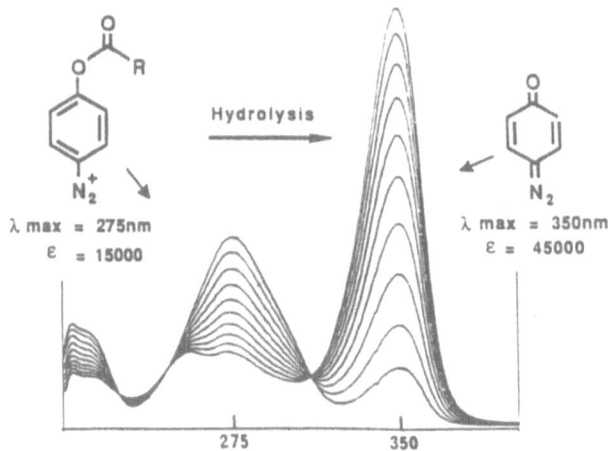

ESTER HYDROLYSIS (in the dark)

pH	chemical Half-life(h.)	enzymatic (pH 6.8 25°C)
6	12.5	AChE: $K_m = 6 \times 10^{-5}$ M
7	4.75	$V_{max} = 2.7 \times 10^{-4}$ µmoles/min/ml/U
8	0.75	
		BuChE $K_m = 1.3 \times 10^{-4}$ M
		$V_{max} = 0.67$ µmoles/min/ml/U

Table II : Chemical and enzymatic hydrolysis of para-phenyl diazonium butyrate

We first tested the possibility of using hydrolytic enzymes to carry out this reaction and then monitored the inactivation process by irradiating the incubation mixture at wavelength around 350nm which corresponds to the maximum absorption of the enzyme product. The phenyldiazonium butyrate we tested was a substrate for both acetylcholinesterase (Torpedo marmorata) and butyrylcholinesterase (human serum) as shown in Table II.

A substantial difference (2500 fold) in the rate of hydrolysis is observed for the two enzymes. The fact that this diazonium salt was an efficient affinity label of AChE (31), while no irreversible inactivation was noticeable for BuChE in the absence of light, might be related to this difference in the rates of enzyme hydrolysis. As a matter of fact, non stabilized aromatic diazonium salts have been described as affinity labels (32).

We tested the possibility of inactivating BuChE through a chromophore transfer-induced labelling using this para-benzene diazonium butyrate derivative as a substrate. In order to slow down the hydrolytic process (medium and enzyme), the experiments were done at low temperature (5°C) and at lower pH (Tris HCl 100mM, pH 5.5). The samples were irradiated at 360 nm, conditions in which both the starting diazonium salt and the enzyme activity were unaffected (Figure 6). It was also determined, as previously mentioned, that no inactivation occured in the absence of light (affinity labelling).

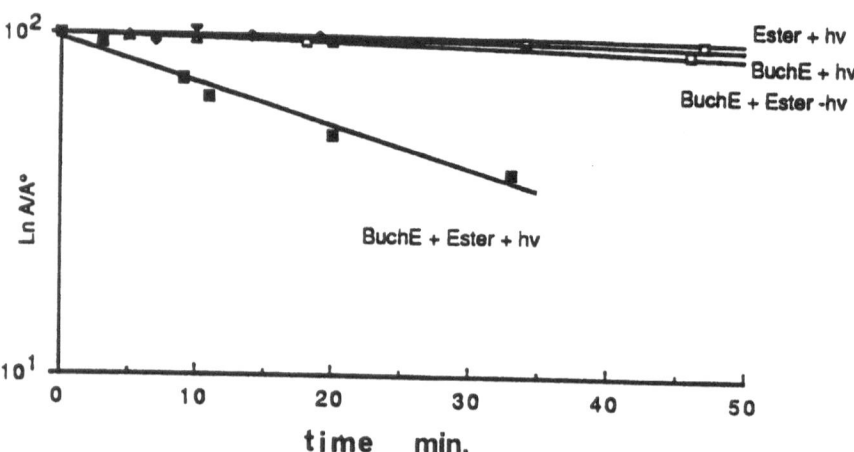

Figure 6. Chromophore transfer-induced labelling of BuChE using a para-benzene diazonium ester

Figure 6 shows kinetic data from chromophore transfer experiments. Using a substrate concentration of $9x10^{-5}$M, it was possible to inctivate up to 60% of the enzyme activity and this inactivation was protectable by tetraalkylammonium salts (not shown). Work is currently in progress to improve the efficiency of the reaction as well as to determine the specificity of the inactivation using a radiolabelled probe.

An interesting theoretical feature was noticed in this inactivation process. This particular enzymatic reaction generates within the active site a photosensitive species (the 4-diazo cyclohexa-2,5-dienone), which is transformed to the corresponding carbene. No remaining affinity for the binding site is expected from either of these species. This leads to an extreme situation where it will be possible (once the optimal reaction conditions are found) to compare the rate of alkylation of this probe with its rate of escape which should be close to the diffusion rate.

3.2.2. *Phase transfer* (33)

In this process we were looking for chemicals whose photosensitivity is increased in non-polar medium, assuming that a binding site possesses a hydrophobic nature. Indeed, a change in environment has been invoked as an important factor in enzyme catalysis (34). The photochemistry of nitrosamine derivatives has been extensively studied (35) and it was shown that these chemicals need acidic catalysis to become photosensitive. We have found, in addition to that catalysis, that the use of solvents of low dielectric constants enhanced this property considerably. This photochemical reaction induces a homolytic cleavage of the N-N bond leading to the formation of amminium and nitrosyl radicals (Figure below)

$$CH_3-\underset{\underset{O}{\overset{|}{N}}}{N}-OAc \quad \xrightarrow[h\upsilon]{H^+} \quad CH_3-\underset{H}{\overset{+\bullet}{N}}-OAc$$

$$\bullet \ N=O$$

AChE $h\upsilon$ } > 60% enzyme inactivation

$K_M = 10^{-2}M$ [ ] = 2.5 x 10^{-4}M   } Prot. by $^+NMe_4$

} ^{14}C Lab. $\longrightarrow$ Stoichiom. = 1

Even though N-methyl,N-acetoxymethyl nitrosamine is a very poor substrate of acetylcholinesterase ($K_m = 10^{-2}M$), we were able to completely inactivate this enzyme under photolytic conditions. For instance, using a concentration of nitrosamine of 2.5×10^{-4} M, that is far below the K_m value, it was possible to attain 60% enzyme inactivation. The use of tetraalkyl ammonium salts protected the enzyme from inactivation. As a general rule, the only way to demonstrate the specificity of a labelling experiment is to carry out the stoichiometry of inactivation. Using a ^{14}C labelled probe, we found that the incorporation of one molecule of label per molecule of enzyme was sufficient for complete enzyme inactivation.

The two features expected from a photosuicide labelling experiment namely specificity and efficiency are obtained. The former characteristic is related to the selective mode of photoactivation of the complexed ligand molecule. An important contribution to the understanding of this phase transfer mechanism was to show that the action spectrum of enzyme inactivation as a function of the irradiation wavelength fits the absorption spectrum of the nitrosamine derivative taken in heptane (or chloroform) in the presence of an acid (trifluoroacetic acid). The efficiency of the labelling was even more surprising in view of the relative limited reactivity of the reactive species involved. Effectively, the photochemically generated amminium radical or the derived immonium ion (loss of an H radical), constitute only moderate electrophiles. It is very likely that this inactivation of acetylcholinesterase involves the participation of an activated amino acid residue at the active site of the enzyme.

3.2.3. *Energy transfer* (36)

Most proteins emit light (fluorescence) when they are excited at the upper region of their absorption spectrum involving mainly the tryptophan residues. Using appropriate fluorophores, numerous examples of shifted fluorescence emission through an energy transfer reaction have been described in order to demonstrate the existence of a tryptophan residue within (or in proximity of) a binding site. The same principle of excitation can be applied to a photoactivatable molecule, which would result in a selective photodecomposition around the tryptophan(s) residue(s). Such an energy transfer reaction (singlet-singlet) is only efficient when the distances between the donor (tryptophan) and acceptor (labelling reagent) are short. It is therefore reasonable to expect this energy transfer reaction to become optimal at the level of the ligand-receptor complex assuming the existence of a tryptophan residue at or near by the binding site.

On the other hand, two conditions are required for the photoactivatable probe to be a good acceptor in such an energy transfer process; first the molecule should be photochemically stable at the tryptophane excitation wavelength (around 290nm) and secondly it should be as photosensitive as possible at the tryptophane emission wavelengths (320-350nm). In other words, the ideal chromophore would have a minimum of absorbance at 290 nm as well as a good overlap with the tryptophan emission spectrum. Among the different photoaffinity probes we have been using, only three types of molecules are likely to be good acceptors:

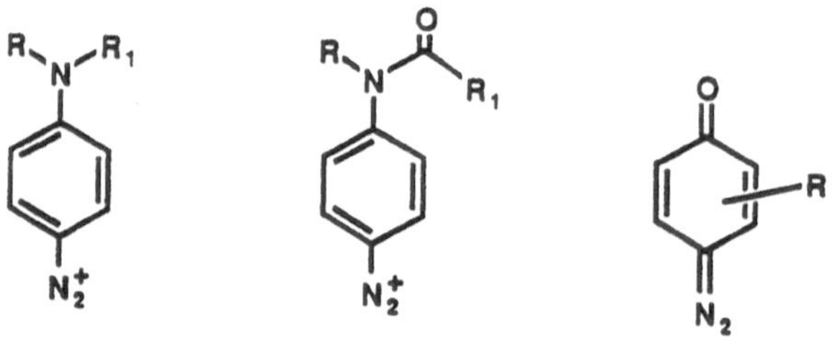

We demonstrated (36) that p-dimethylamino benzene diazonium 1 can be photo-decomposed by photoexcited tryptophan derivatives (N-acetyltryptophanamide) through an energy tranfer reaction (λexc. 295 nm). The same process was observed with acetylcholinesterase for which 1 was shown to be a competitive inhibitor (Ki = 2×10^{-5}M), leading to an irreversible inhibition of the enzyme. In this experiment it was possible to have a direct comparison with the classical photoaffinity labelling experiment simply by changing the wavelength of irradiation. To validate such a comparison, the simplest control was to assess an identical rate of decomposition of the probe at the two wavelengths of irradiation by modulating the light intensity. Table III summarizes some of this comparative study.

conditions	Energy transfer λ=295nm, int.=0.3mV.	Photoaffinity λ=410nm, int=0.048mV.
	Half-time (min.) of AChE inactivation	
control (Enz.)	>110	∞
Enz. + 1 [2×10^{-5}M]	6	48
Enz. + 1 ["] + $NR_4^{+}I^{-}$	15	>80
	Stoichiometries of covalent ^{3}H 1 incorporation	
1 [2×10^{-5}M]	1	1.2
1 [10^{-4}M]	2.4	3.9

TABLE III : Comparative study of AChE inactivation under energy transfer and classical photoaffinity labelling conditions.

This clearly demonstrates the expected advantages of the photosuicide process over the classical photoaffinity labelling experiment. This irreversible labelling of AChE protectable by ammonium salts led to a further study (18) giving some insight into the molecular structure of the ammonium binding site of this enzyme. The structural analogy of acetylcholine and our reagent is simply based on the existing charge on both molecules. Identification of the major radioactive peptide fragment labelled by our reagent (in energy transfer conditions) on purified electric eel acetylcholinesterase led, after purification, to the following sequence Gly-Ser-X-Phe. This peptide sequence is different from the decapeptide known to contain the reactive serine of electric eel AChE or active site sequences of other cholinesterases (see table IV).

Eel AChE (37)

Gly-Gly-Glu-<u>Ser</u>-Ser-Glu-Gly-Ala-Ala-Gly

Torpedo AChE (38)

Thr-Val-Thr-Ile-Phe-Gly-Glu-<u>Ser</u>-Ala-Gly-Gly-Ala-Ser-Val-Gly-Met-His-Ile-Leu-Pro-Gly-Ser-Arg

Human serum ChE (39)

Ser-Val-Thr-Leu-Phe-Gly-Glu-<u>Ser</u>-Ala-Gly-Ala-Ala-Ser-Val-Ser-Leu-His-Leu-Leu-Ser-Pro-Gly-Ser-His-Leu-Thr-Ser-Arg

TABLE IV : Active site region of cholinesterases

It is interesting to note that on the primary sequence of Torpedo californica (40) only one tetrapeptide, having the sequence Gly-Ser-Phe-Phe (positions 328-331), could match our found sequence. However, one has to keep in mind that major differences might exist between the esterases from different species. The accompanying article (F. Kotzyba-Hibert et al.) describes an identical approach using energy transfer-induced photoaffinity labelling of different binding sites existing on the acetylcholine receptor. Similarly, it could be demonstrated that the photosuicide labelling process was superior to the classical photoaffinity experiment for the irreversible labelling of the acetylcholine binding site (41).

Aryldiazonium salts are the ideal photosensitive probes for the study of the ammonium binding site of the cholinergic enzymes and receptor, by virtue of the positive charge existing in both structures. Nevertheless, their use is not restricted to the cholinergic system. For instance, we have shown that such an analogy can be extended to a protonated primary amine as found in the GABA molecule (21). On the other hand, an aromatic diazonium chromophore can be coupled to any ligand, we incorporated e.g. an p.N,N-dialkyl benzene diazonium into the rhamnose moiety of ouabain, an inhibitor of Na^+, K^+ ATPase. This new probe (42) led to an efficient irreversible photoinactivation (energy transfer process) of that enzyme which proved to be far superior in terms of specificity and efficiency than a similarly modified arylazido derivative (43).

Several other examples of energy transfer induced labelling experiments are currently in progress in our laboratory using the aryldiazonium (44) or the 4-diazo cyclohexa-2,5-dienone chromophores (21).

4. Conclusions

An important aspect in photoaffinity labelling is the specificity of alkylation at the target binding site during the irreversible labelling experiment. In fact, most of the failures in this area of research are related to a lack of specificity which is dependent upon the strength of the ligand receptor interaction and the nature of the photogenerated species. This phenomenon is generally observed with species of longer lifetimes which can diffuse into the medium. This paper describes presents two developments related to specificity in photolabelling experiments. First, new photoactivatable functions with high reactivity for the corresponding photogenerated species, and secondly an improvement of the method of labelling using the suicide concept are described.

The search for new chemical functions which can be used as potential irreversible photolabelling probes led to the discovery of two classes of chemicals, the aromatic diazonium salts and the diazo cyclohexadienones, which were both well known in the chemical litterature. When adequately substituted (see Chapter 3.), both of these reagents display several peculiar properties in addition to the classical chemical and photochemical properties required in photoaffinity labelling several peculiarities. When light-activated they are indeed converted to short live species (the aryl cation and the cyclohexadienonyl carbene) which are very likely more reactive than most of the reagents used in photoaffinity labelling, related in part to the fact that none of them rearrange to a less reactive intermediates. For instance, by using nanosecond laser photolysis techniques, it has not been possible to directly detect thie aryl cation (including the more stabilized species) which consequently does not allow an accurate lifetime measurement of these species. However, an estimate of the lifetime of the phenylcation in water (<500 ps) has been described (45) using laser flash techniques. This value was obtained by analyzing, in the presence and absence of inorganic anions, the bleaching of chromate ions induced by the proton release in the reaction of the cation with water. This extreme short lifetime explains why this species reacts not only with a carbon hydrogen bond, but also with gaseous nitrogen (46) (the reverse of the photochemical reaction) which can be considered as an ultimate criterion of reactivity. The same reaction is demonstrated for the cyclohexadienonylidene carbene (5) for which an identical degree of reactivity is expected. Both of these classes of chemicals approach "universal rectivity" which will allow a non-descriminative alkylation reaction with any chemical function.

In addition, we showed that it was possible to transpose the suicide concept to photolabelling experiments. This implies that, through the action of light, it is possible to selectively photoactivate the ligand analogue that is complexed at the binding site leading to the ideal "configuration" for specific labelling. Of the three different examples of photosuicide labelling we described (Chapter 3.2), we have most extensively developed the energy transfer process which uses the fluorescence properties of tryptophanyl residues to induce the photoactivation reaction. Both of the above mentionned chemical structures (some aryldiazonium salts and 4-diazocyclohexa-2,5-dienone derivatives) have ideal chromophores for such an energy transfer reaction. As a consequence, it will be possible using this mode of photoactivation to generate selectively within the binding site a species which reacts at a diffusion rate with its direct environment.

The scheme below emphasizes the importance of the lifetime of the photogenerated species. It takes advantage of both the reactivity of the aryl cation and the selective mode of photoactivation of its precursor (diazonium salt) to achieve a topographical mapping of a receptor site (acetylcholine binding site from the nicotinic acetylcholine receptor)(47). This work is described in the accompanying article (F. Kotzyba-Hibert et al.).

Recently the primary amino acid sequence of a series of neuromediator receptors have been deduced from their cDNA sequences(48). This technology requires knowing the sequence of a peptide fragment belonging to the polypeptide, frequently obtained from a

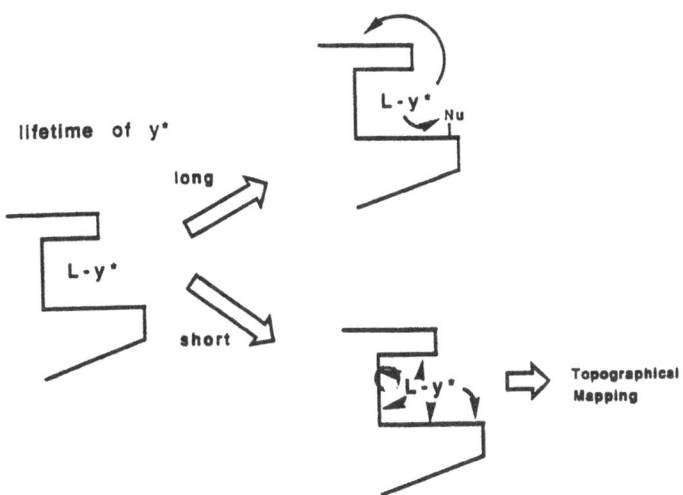

highly purified receptor. Photoaffinity labelling experiments could allow the determination of the protein primary structure without previous purification steps. However, the molecular characterization of a hormone or neuromediator receptor present in very low concentrations requires highly specific and extremely efficient labelling. We therefore propose our approach of photosuicide labelling using highly reactive alkylating species.

Acknowledgments: We thank Drs J. Hawkinson and M. Sanders for critical reading of the manuscript.

References
(1)   M.P. Goeldner, C.G. Hirth, B. Kieffer and G. Ourisson (1982) Trends Biochem. Sci. **7**, 310-312.
(2)   H. Bayley (1983) Photogenerated Reagents in Biochemistry and Molecular Biology in Laboratory Techniques in Biochemistry and Molecular Biology, Elsevier Amsterdam.
(3)   K. Bloch (1969) Acc. Chem. Res. **2**, 193-202.
(4)   P. Kessler and L. Ehret-Sabatier unpublished
(5)   D.M.A. Grieve, G.E. Lewis, M.D. Ravenscroft, P. Skrabal, T. Sonoda, I.Szele and H. Zollinger (1985) Helv. Chim. Acta **68**, 1427-1443.
(6)   A. Cox,T.J. Kemp, D.R. Payne, M.C.R. Symons and P. Pino de Moira (1978) J. Am. Chem. Soc. **100**, 4779-4783.
(7)   H.B. Ambroz and T.J. Kemp (1982) J. Chem. Soc., Chem. Commun.172-173.
(8)   B. Kieffer, M.P. Goeldner and C.G. Hirth (1981)J. Chem. Soc., Chem. Commun. 398-399.
(9)   J.L. Galzi (1987) These de l'Universite Louis Pasteur Strasbourg, France.
(10)   H.B. Ambroz and T.J. Kemp (1979) Chem. Soc. Rev. **8**, 353-365.
(11)   R.W. Stumpe (1980) Tetrahedron Lett. **21**, 4891-4892.
(12)   R. Huisgen and W.D. Zahler (1963) Chem. Ber. **96**, 736-746.
(13)   M. Speranza (1980) Tetrahedron Lett. **21**, 1983-1986.
(14)   G. Angelini, S. Fornarini and M. Speranza (1982) J. Am. Chem. Soc **104**, 4473-4480.
(15)   G. Angelini, C. Sparapani and M. Speranza (1984) Tetrahedron **40**,4865-4871.
(16)   Y. Keheyan and M. Speranza (1985) Helv. Chim. Acta **68**, 2381-2388.

122

(17) M. Colosimo, M. Speranza, F. Cacace and G. Ciranni (1984) Tetrahedron **40**, 4873-4883.
(18) B. Kieffer, M.P. Goeldner, C.G. Hirth, R. Aebersold and J.-Y. Chang (1986) FEBS Lett. **202**, 91-96.
(19) R.M. Izatt, J.D. Lamb, B.E. Rossiter, N.E. Izatt and J.J. Christensen (1978) J. Chem. Soc., Chem. Commun. 386-387.
(20) J.P. Behr and C.G. Hirth unpublished.
(21) M.J. Bouchet, A. Rendon, C.G. Wermuth, M.P. Goeldner and C.G. Hirth (1987) J. Med. Chem. **30**, 2222-2227.
(22) H. Dürr (1975) Photochemie in Methoden der Organischen Chemie, Houben-Weyl, (E. Müller Ed.) Bd. IV/5b teil II, pp.1210-1237, Georg Thieme Verlag Stuttgart.
(23) E. Wasserman (1964) J. Am. Chem. Soc. **84**, 4203-4204.
(24) M.L. Kaplan and H.D. Roth (1972) J. Chem. Soc., Chem. Commun.970-971.
(25) H. Dürr and H. Kober (1972) Tetrahedron lett. 1259-1262.
(26) O. Süs, K. Möller and H. Heiss (1956) Ann. **598**, 123-138.
(27) W.H. Pirkle and G.F. Koser. (1968) Tetrahedron Lett. 3959-3962.
(28) J.C. Fleming and H. Shechter (1969) J. Org. Chem. **34**, 3962-3969.
(29) G.F. Koser and W.H. Pirkle (1967) J. Org. Chem. **32**, 1992-1994.
(30) L. Ehret-Sabatier unpublished
(31) L. Ehret-Sabatier unpublished
(32) J.P. Changeux, T.R. Podleski and L. Wofsy (1967) Proc. Natl. Acad. Sci.USA **58**, 2063-2070.
      H.G. Mautner and E. Bartels (1970) Proc. Natl. Acad. Sci. USA **67**, 74-78.
(33) P. Eid, M.P. Goeldner, C.G. Hirth and P. Jost (1981) Biochemistry **20**, 2256-2260.
(34) W.P. Jencks (1975), Advances in Enzymology, (Meister, A. Ed.) Vol.43 pp.219-410. John Wiley & sons Ltd., New-York.
(35) Y.L. Chow (1973) Acc. Chem. Res. **6**, 354-360.
(36) M.P. Goeldner and C.G. Hirth (1980) Proc. Natl. Acad. Sci. USA **77**, 6439-6442.
(37) N.K. Schaffer, H.O. Michel and A.F. Bridge (1973) Biochemistry **12**, 2946-2950.
(38) K. MacPhee-Quigley,P. Taylor and S. Taylor (1985) J. Biol. Chem. **260**, 12185-12189.
(39) O. Lockridge (1984) in: Cholinesterases, Fundamental and Applied Aspects (Brzin, M. et al. eds.) pp5-11, De Gruyter, Berlin
(40) M. Schumacher, S. Camp, Y. Maulet, M. Newton, K MacPhee-Quigley, S. Taylor, T. Friedmann and P. Taylor (1986) Nature **319**, 407-409.
(41) J. Langenbuch-Cachat, C. Bon, C. Mulle, M. Goeldner, C. Hirth and J.P. Changeux (1988) Biochemistry **27**, 2237-2245.
(42) M.P. Goeldner, C.G. Hirth, B. Rossi, G. Ponzio and M. Lazdunski (1983) Biochemistry **22**, 4685-4690.
(43) T.B. Rodgers and M. Lazdunski (1979) Biochemistry **18**, 135-140.
(44) J.L. Galzi, A. Mejean, B. Illien, M. Goeldner and C. Hirth submitted
(45) J.C. Scaiano and NGuyen Kim-Thuan (1983) J. Photochem. **23**, 269-276.
(46) M.D. Ravenscroft and H. Zollinger (1988) Helv. Chim. Acta **71**, 507-514, and references cited therein.
(47) M. Dennis, J. Giraudat, F. Kotzyba-Hibert, M. Goeldner, C. Hirth, J.Y. Chang, C. Lazure, M. Chretien and J.P. Changeux (1988) **27**, 2346-2357.
(48) P.R. Schofield, M.G. Darlison, N. Fujita, D.R. Burt, F.A. Stephenson, H. Rodriguez, L.M. Rhee, J. Ramachandran, V. Reale, T.A. Glencorse, P.H. Seeburg and E.A. Barnard (1987) Nature **328**, 221-227.

ANTIBIOTIC PHOTOAFFINITY LABELING PROBES OF <u>ESCHERICHIA COLI</u> RIBOSOMAL STRUCTURE AND FUNCTION

BARRY S. COOPERMAN, MELISSA A. BUCK, CARMEN L. FERNANDEZ, CARL
J. WEITZMANN, and BARBARA F. D. GHRIST
Department of Chemistry
University of Pennsylvania
Philadelphia, PA 19104
USA

ABSTRACT. The use of antibiotic photoaffinity label probes to identify functionally important E. coli ribosomal components is presented, emphasizing in particular recent results obtained in the authors' laboratory with the antibiotics puromycin, chloramphenicol, and tetracycline. These results have led to the identification of 50S proteins and bases within 23S rRNA at the peptidyl transfer center as well as of proteins involved in tRNA binding to the 30S subunit. Particular stress is placed on the methodologies used in component identification and on the different approaches that are available for demonstrating the functional significance of photoaffinity labeling results.

1. Introduction

1.1 RIBOSOME STRUCTURE AND FUNCTION

The ribosome is a complex ribonucleoprotein particle, in reality a very large enzyme, that catalyzes the overall process of protein synthesis (excellent contemporary discussions of various aspects of ribosome structure and function may be found in Hardesty and Kramer, 1986 and Moldave and Noller, 1988). The E. coli ribosome is the one best understood, and this article will concern itself with this ribosome exclusively. However, the reader should be aware of photoaffinity labeling studies on ribosomes isolated from other sources, both procaryotic and eucaryotic, as recently reviewed elsewhere (Cooperman, 1988).

The E. coli ribosome has a molecular weight of some 2.5×10^6 daltons and is designated as a 70S particle, following its Svedberg number. It is made up of two dissociable subunits, a 30S subunit (MW $\sim 0.8 \times 10^6$) and a 50S subunit (MW $\sim 1.7 \times 10^6$). The 30S subunit is a complex of one large RNA strand (16S rRNA, 1542 bases long) and 21 proteins (denoted S1 -S21), and has approximate dimensions (in Å) 50-80 X 100 X 200. The 50S subunit is a complex of two RNA strands (23S rRNA, 2904 bases long; 5S rRNA, 120 bases long) and at least 33 proteins (denoted L1-L34; there is no L8), although recent work suggests the presence of an additional four proteins (Wada, 1986a,b).

123

P. E. Nielsen (ed.), Photochemical Probes in Biochemistry, 123–139.
© *1989 by Kluwer Academic Publishers.*

Just as the structure of the ribosome is complex, so too is the detailed mechanism by which it carries out protein synthesis. As with any polymerization process, protein synthesis can be divided into three steps, initiation, elongation and termination. The initiation process (Maitra et al., 1982) takes place via the binding of mRNA and fMet-tRNA$_f^{Met}$ to isolated 30S subunits to form the initiation complex. This process requires three separate protein cofactors (IF-1, IF-2, IF-3), each of which binds directly to the ribosome. Initiation also involves specific recognition by the ribosome, within mRNA, of both the initiation codon AUG and a purine-rich sequence of bases (the Shine-Dalgarno sequence) upstream from the AUG. At the close of initiation, the 50S subunit binds to the 30S initiation complex to form the 70S ribosome, placing fMet-tRNA$_f^{Met}$ in the peptidyl-tRNA binding site (or P-site) of the ribosome. The elongation process (Nierhaus, 1984; Spirin, 1985) begins with the binding of the appropriate aminoacyl-tRNA to the aminoacyl-tRNA binding site (or A-site) of the ribosome in response to the triplet codon adjacent to the initiation codon AUG. The first peptide bond is made via nucleophilic attack of the α-amino group of aminoacyl-tRNA bound in the A- site on the ester group of fMet-tRNA$_f^{Met}$ bound in the P-site. Bond formation occurs within the peptidyl transferase center of the ribosome, which has been shown to be localized on the 50S subunit. It is followed by a translocation process, in which the discharged tRNA^{Met} is transferred from the P-site to the exit or E-site, while the tRNA charged with the newly-formed dipeptide, still bound to its mRNA codon, is transferred to the P-site. The resulting open A-site is then available for binding by the aminoacyl-tRNA corresponding to the third codon triplet. Such binding is accompanied by the release of discharged tRNA$_f^{Met}$ from the E-site into solution, setting the stage for formation of the second peptide bond. The elongation process also requires three protein cofactors, EF-Ts, EF-Tu, and EF-G, the latter two of which bind directly to the ribosome. It continues in the cyclical manner described above until any one of three termination codons (UAA, UAG, or UGA) is bound in the A-site. These codons induce the binding of proteins called release factors into the A-site, leading to termination of protein synthesis via hydrolysis of peptidyl-tRNA in the P-site. Polypeptide is then released into solution. As is clear from this brief description, ribosome-catalysis of peptide synthesis proceeds via a large number of highly-ordered steps and requires the binding of a multitude of ligands including mRNA, the tRNAs, and an imposing array of protein factors.

1.2 ANTIBIOTIC PROBES OF RIBOSOMAL STRUCTURE AND FUNCTION

Antibiotic inhibitors of ribosome function constitute another very large class of ribosomal ligands. These molecules, whether formed biologically by various strains of Streptomyces or by total or partial synthesis, collectively inhibit virtually every step of protein biosynthesis. Antibiotics are, for the most part, rather small molecules (< 500 daltons), remarkably diverse in structure, that generally exert their inhibitory effects by binding to the ribosome at specific sites (Gale et al., 1981). Antibiotics and their chemically reactive derivatives thus have great potential as photoaffinity label probes of ribosomal structure and function, since identifying the site of interaction with the ribosome of an antibiotic of known inhibitory function provides strong evidence for the identification of the functional site.

Below we present some recent photoaffinity labeling results obtained in our laboratory with four different antibiotics, puromycin, p-azidopuromycin, chloramphenicol, and tetracycline (Figure 1). The first three of these inhibit peptidyl transferase, while the fourth inhibits aminoacyl-tRNA binding to the A-site of the ribosome. A comprehensive review of antibiotic photoaffinity labeling of ribosomes has been presented elsewhere (Cooperman, 1987). The reader is also directed to some more recent studies of Arevalo et al. (1988) using photolabile erythromycin derivatives.

1.3 PHOTOAFFINITY LABELING STUDIES ON RIBOSOMES

In carrying out photoaffinity labeling studies on ribosomes we seek to answer three principal questions: 1) What is the identity of the ribosomal component (or components) into which photoincorporation occurs? 2) What is the evidence that photoincorporation occurs into a functionally important site of the ribosome? 3) Where does photoincorporation occur in a three-dimensional sense within the ribosome? In virtually all cases, photoaffinity labeling is monitored through use of radioactive photoaffinity labels.

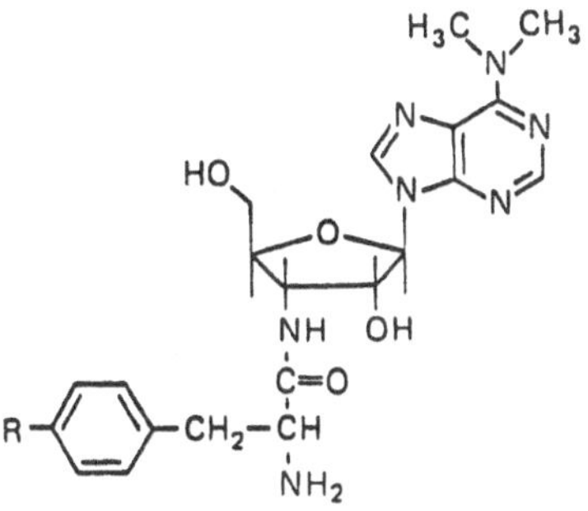

Puromycin: R = OCH$_3$
p-Azidopuromycin: R = N$_3$

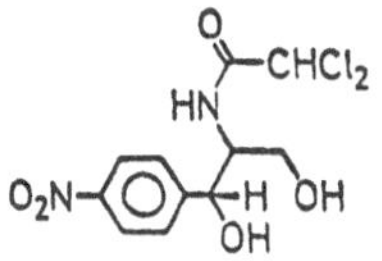

Chloramphenicol

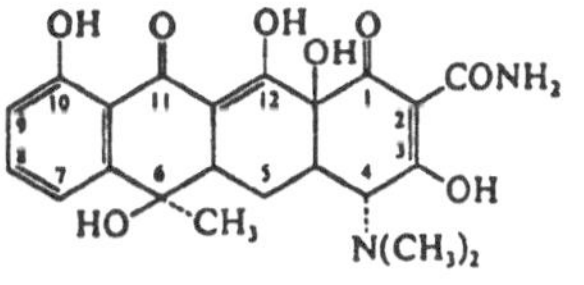

Tetracycline

Figure 1. Antibiotic photoaffinity labels

1.3.1a. Identification of labeled protein. We and other workers in the field have generally considered the individual protein as the unit of ribosome structure and have not attempted to identify labeled amino acid residues within the labeled protein. This is because the low-resolution model for ribosome structure currently available makes it questionable whether the effort to identify labeled amino acid residues is justified at this time. We have used three different methods to identify photoaffinity-labeled proteins. These are: polyacrylamide gel electrophoresis (PAGE), both one-and two-dimensional; specific immunoprecipitation with antisera raised against purified ribosomal proteins, and, most recently, high-performance liquid chromatography (HPLC), both reverse phase (RP-HPLC) and ion-exchange (IE-HPLC) (Kerlavage et al., 1983a,b; Weitzmann and Cooperman, 1985; Fernandez and Cooperman, 1989). In our view, the identification of affinity-labeled proteins by HPLC is the clear method of choice. The immunological approach suffers from the extreme tediousness of preparing more than 50 protein antisera in pure (i.e., non-cross-reacting) form, although it is true that once such antisera are available analysis is straightforward. With respect to PAGE, HPLC analysis is easier to perform, more rapid, more precise and affords higher yields and superior reproducibility, all important advantages given both the large number of control experiments that a successful photoaffinity labeling experiment usually requires and the need to quantitate the extent of labeling as a function of experimental protocol. Further, although the resolution afforded by HPLC is typically lower than that available with two-dimensional PAGE, it is usually possible to manipulate elution conditions to obtain very high resolution in a particular region of the chromatogram. We published the first standardized RP-HPLC chromatograms of 30S proteins in 1982 (Kerlavage et al., 1982) and of 50S proteins in 1983 (Kerlavage et al., 1983a). These were already highly resolved separations, yielding 15-17 peaks for the twenty one 30S proteins and a total of 22 peaks for the thirty three or so 50S proteins, and were based on the use of simple linear gradients. More recently we have shown that the behavior of ribosomal proteins on RP-HPLC, with respect both to elution time and peak-to-peak separation, can be predicted (Ghrist et al., 1988, 1989) and have used this result to design complex gradients affording still higher resolution. Some recent examples of this work are shown in Figure 2.

Use of either PAGE or HPLC analysis to identify labeled proteins is subject to the potential problem that a labeled protein may migrate or elute with sufficient difference compared to unmodified protein to make unambiguous identification difficult, particularly in crowded regions of the gel or chromatogram. There are now two straightforward ways around this problem. First, if the labeled protein can be prepared in sufficient quantity and purity, then it can be identified by classic protein identification methods, such as tryptic fingerprints or partial N-terminal sequence analysis (Kerlavage and Cooperman, 1986). This approach has the additional advantage that it can be extended to identify labeled amino acid residues, when such information is useful (vide supra). Second, one can apply an additional high-resolution method to the analysis of proteins from photoaffinity labeled ribosomes, the underlying reasoning being that the cohort of proteins close to which a labeled protein elutes (or migrates) in one method will, in general, be different from the corresponding cohort using the second method. Suitable pairs of methods are RP-HPLC and IE-HPLC, for both of which standardized chromatograms are now available (Cooperman et al., 1988; Capel et al., 1988), and RP-HPLC and PAGE.

1.3.1b. Identification of labeling sites within rRNA. A two-step methodology has recently been introduced to first localize a site of rRNA photoaffinity labeling to a limited region of rRNA and second, to identify the exact labeled base (Barta et al., 1984; Hall et al., 1985, 1988; Gravel et al., 1987; Kuechler et al., 1989). In the first step, which is totally general in its application, rDNA restriction fragments are used as hybridization probes that protect labeled rRNA from single-stranded nuclease digestion. The use of several restriction enzymes having overlapping restriction sites usually results in fairly rapid localization of the labeled rRNA to 50-200 bases. Because this first step allows quantitation of the extent of labeling, it is particularly useful for the analysis of the results of control experiments (Hall et al., 1988). In the second step photoaffinity labeled rRNA is hybridized with a single-stranded oligoDNA that is complementary to a region of rRNA to the 3'-side of the region of rRNA found to be labeled in the first step. This heteroduplex is then used as a substrate for reverse transcriptase, and the exact position (or positions) of labeling is (are) identified, using a classical sequencing gel, as the position(s) following that (those) for which a halt or pause is observed (Youvan and Hearst, 1979, 1981). The second step, despite its power and elegance, is subject to significant ambiguities: it fails to distinguish between interference with reverse transcriptase arising from photoincorporation or other photochemically-induced effects and it is unable to detect photoaffinity label interaction at those sites that are preceded by a significant halt or pause in primer extension when control rRNA is the template.

1.3.2. Evidence that photoaffinity labeling takes place at a functionally important site. Given the underlying rationale for studying the photoaffinity labeling of ribosomes by antibiotic photoaffinity labels, it is important to demonstrate that such labeling takes place at a site related to the inhibitory action of the antibiotic. In doing so, we have asked the following questions, recognizing that it may be feasible to answer only some of these questions for a given photoaffinity labeling experiment: 1) Does the photoaffinity label have comparable antibiotic activity to the native antibiotic? 2) Does photoaffinity labeling proceed from a saturable site on the ribosome? 3) Is the dependence on photoaffinity label concentration of photoincorporation from this site consistent with the apparent affinity of the photoaffinity label for the inhibitory site? 4) Does the structural specificity of the photoaffinity labeling reaction parallel that seen for effects on ribosomal function? 5) Is the effect of photoaffinity labeling on ribosomal function consistent with the known mode of action of the antibiotic? 6) In the case of photoaffinity labeling of a protein, does omission of that protein in a reconstitution experiment lead to a corresponding loss in antibiotic binding and, as relevant, ribosomal function? 7) Is there a correspondence between sites of mutation conferring antibiotic resistance and sites of photoaffinity labeling?

1.3.3. Three-dimensional localization of sites of photoaffinity labeling. We (Olson et al. 1980, 1982, 1985; Grant et al., 1983) have used immunoelectron microscopy to obtain approximate three-dimensional localization of sites of photoaffinity labeling, by determining the structures of complexes formed between ribosomes photoaffinity-labeled with antibiotics and antibodies to the antibiotics. Here, the site of antibiotic photoincorporation is indicated by the position of antibody binding to the ribosome.

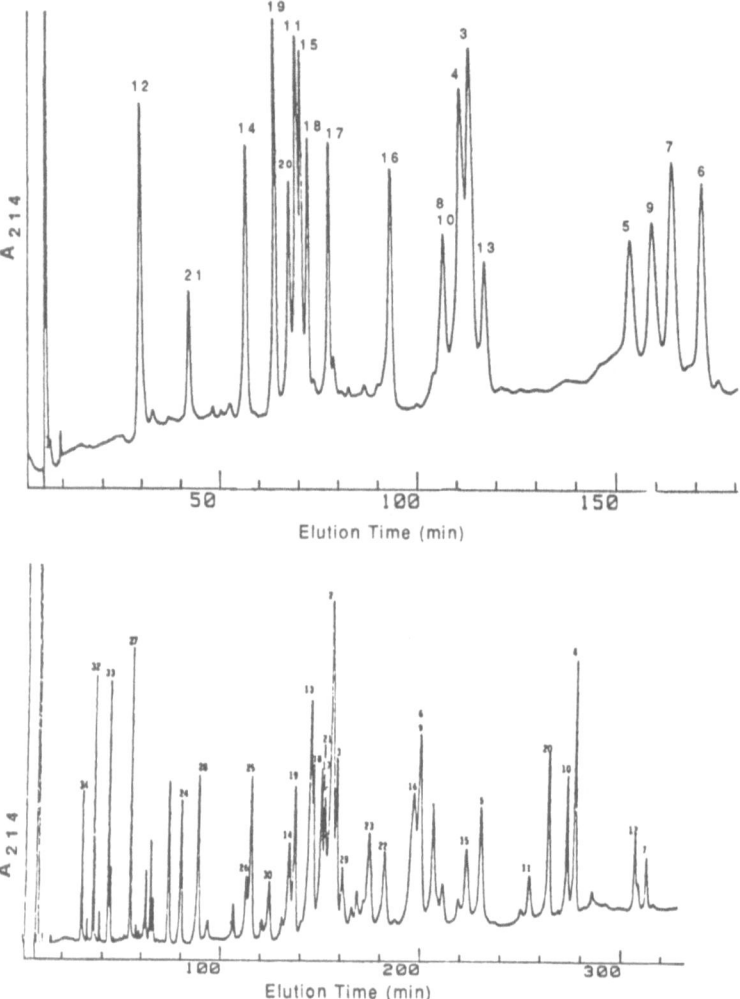

Figure 2. High resolution Reversed Phase HPLC separations of 30S and 50S proteins using multistep gradients. Upper panel: TP30 (total protein from 30S subunit). Note that at least partial resolution is obtained for all 19 proteins S3-S21. S1 and S2 elute considerably later and are well resolved from each other and from other TP30 proteins. Lower panel: TP50 (total protein from 50S subunit). Note that all 50S proteins are at least partially resolved with the exception of the pair L6,L9. For details see Ghrist et al. (1988,1989).

2. Photoaffinity Labeling By Puromycin and p-Azidopuromycin

Puromycin is an inhibitor of peptidyl transferase that is both a structural and functional analogue of the 3'-end of Tyr-tRNA[Tyr]. It has two inhibitory modes of action, acting as a substrate in accepting, at its α-amino position, a peptidyl group from P-site bound peptidyl-tRNA to form an aborted peptide chain (Allen and Zamecnik,1962; Nathans, 1964) and competing with aminoacyl-tRNA for binding to the A-site (Kuechler et al., 1989). We have used both native puromycin and p-azidopuromycin (Figure 1), a functionally competent derivative in which a p-azidophenyl group replaces the p-methoxy group in puromycin, as photoaffinity labels.

Although puromycin photoincorporates into both protein and rRNA, only protein labeling has been examined in detail. Using relatively high light doses because of the limited photoreactivity of puromycin, two proteins have been shown to be labeled from relatively high affinity sites, L23 and S14, with L23 labeling proceeding in considerably higher yield (Jaynes et al., 1978; Weitzmann and Cooperman, 1985). These proteins are labeled from two different sites on the ribosome, having somewhat different specificities toward the puromycin structure, as shown by the following observations. First, when isolated 50S and 30S subunits are used as the targets, puromycin photoincorporates into L23 and S14, respectively, with approximately the same quantum yields as when 70S ribosomes are the target. Second, although both non-radioactive puromycin and non-radioactive puromycin aminonucleoside (which lacks the O-methyl tyrosine moiety of puromycin, see Figure 1) inhibit photoincorporation of [^{3}H]-puromycin into both L23 and S14, non-radioactive N^6,N^6-dimethyladenosine (in which the 3'-amino group of puromycin aminonucleoside is replaced by a 3'-hydroxyl group) only inhibits photoincorporation into S14. Third, while [^{3}H]-puromycin aminonucleoside shows almost the same photoincorporation pattern into ribosomes as does [^{3}H]-puromycin, thus demonstrating that puromycin photoincorporation proceeds through its N^6,N^6-dimethyladenosyl moiety, [^{3}H]-N^6,N^6-dimethyladenosine photoincorporates into S14 but not into L23.

Since puromycin aminonucleoside but not N^6,N^6-dimethyladenosine is also a competitive inhibitor for puromycin as a substrate in peptidyl transferase (Nicholson et al., 1982b; Weitzmann and Cooperman, 1985), these results are also consistent with L23 being labeled from the A'-site of the ribosome, which is defined as the site of binding of the 3'-end of aminoacyl-tRNA within the peptidyl transferase center. Supporting this interpretation is the finding that the 3'-O-phenylalanyl ester of the dinucleoside phosphate CpA, which is a very good ligand for the A'-site, is also a strong inhibitor of puromycin photoincorporation into L23 (Jaynes et al., 1978).

Recently, an alternative approach has been utilized to investigate the functional consequences of photoaffinity labeling of L23 and of S14 by puromycin. In this work RP-HPLC was used to prepare puromycin-L23 (i.e, L23 that has been photoaffinity labeled with puromycin in intact 70S ribosomes) free of L23 and puromycin-S14 (S14 that has been photoaffinity labeled with puromycin in intact 70S ribosomes) free of S14. These derivatized proteins were then used in place of the native proteins in reconstituting 50S and 30S subunits, giving subunits in which

L23 or S14 were the only puromycin-modified components (the rRNA and other proteins were prepared from unmodified ribosomes). Although these proteins were stoichiometrically labeled with puromycin, in practice it proved impossible to achieve a full stoichiometry of 1.0 puromycin-L23/50S particle, largely due to contamination of protein pools with unmodified L23. The functional properties of these subunits, along with those of various control subunits, were then measured, as summarized in Tables I and II. The results with puromycin-S14 are particularly easy to interpret (Kerlavage and Cooperman, 1986). Native 30 subunits bind Phe-tRNA^{Phe} in a poly(U)-dependent manner. Protein S14 is required for such binding, as shown by the poor binding of 30S subunits reconstituted in the absence of S14. When S14 is added to the reconstitution mix binding is restored, but addition of puromycin-S14, though it results in puromycin-S14 uptake into the reconstituted 30S subunit, does not restore Phe-tRNA^{Phe} binding. We conclude that the puromycin group in puromycin-S14 directly interferes with tRNA binding, possibly by occupying a site normally filled by the 3'-end of aminoacyl-tRNA. Support for this idea comes from the photoaffinity labeling results of Girshovich et al. (1974a,b) showing that S14 is photoaffinity labeled by an N^{α}-arylazide derivative of Phe-tRNA^{Phe}. The results with puromycin-L23 (Weitzmann and Cooperman, 1989) are less straightforward, both because of the limited uptake of puromycin-L23 into reconstituted 50S particles and because L23 is not required for either of the two activities measured, i.e., mRNA-independent peptidyl transferase or 50S stimulation of poly(U)-dependent Phe-tRNA^{Phe} binding. Nevertheless, the results obtained do support the simple conclusions that puromycin-L23 has no effect on peptidyl transferase but does prevent 50S stimulation of Phe-tRNA^{Phe} binding.

Table I.   Activities of Reconstituted 30S Subunits

protein in reconstitution mix	puromycin-S14 30S	relative Phe-tRNA^{Phe} binding activity
TP30 - S14 + S14	---	1.00
TP30 - S14	---	0.01 ± .01
TP30 - S14 + puromycin-S14	0.85	0.05
TP30 - S14 + hv - S14	---	0.85 + .06

Table I. TP30 refers to total protein from 30S subunit. Protein hv-S14 refers to S14 isolated from ribosomes irradiated in the absence of puromycin.

Table II. Activities of Reconstituted 50S Subunits

protein in reconstitution mix	puromycin-L23 50S	relative stimulation of Phe-tRNA^{Phe} binding	relative peptidyl transferase
TP50 - L23 + L23	- - -	1.27 ± .04	1.10 ± .03
TP50 - L23	- - -	1.00	1.00
TP50 - L23 + puromycin-L23	0.3	0.64 ± .07	0.97 ± .23
TP50 - L23 + hv - L23	- - -	1.24 ± .09	1.07 ± .13

Table II. TP50 refers to total protein from 50S subunit. Protein hv-L23 refers to L23 isolated from ribosomes irradiated in the absence of puromycin.

Earlier (Jaynes et al., 1978) we had shown that puromycin photoincorporated into ribosomes (including , of course, into L23) does not serve as an acceptor substrate in peptidyl transferase. Thus our overall conclusion is that while L23 is not directly involved in peptidyl transferase, it is close enough to the peptidyl transferase center that puromycin-L23 can interfere with tRNA binding.

As with puromycin, photoincorporation of p-azidopuromycin also takes place into both proteins and rRNA (Nicholson et al., 1982a,b). When the reaction is carried out with relatively low light doses and in the presence of 2-mercaptoethanol, added to reduce, by scavenging, labeling by photoexcited p-azidopuromycin coming from solution or via a slowly reacting photogenerated intermediate, the protein labeled to the greatest extent is again L23. Such labeling proceeds via an azide-dependent photochemistry not available to puromycin and is additional evidence for the site-specificity of L23 labeling. The other major labeled proteins are L18/22 and L15. Using the RNA site identification methodology described above, p-azidopuromycin has also been shown to photoincorporate into bases U-2504 and G-2502 in 23S rRNA (Hall et al., 1988). As added puromycin inhibits the photoincorporation of p-azidopuromycin into proteins L23, L18/22 and L15 to the same extent as into a small fragment of rRNA containing bases U-2504 and G-2502, it is at least plausible that these ribosomal components are all being labeled by p-azidopuromycin bound at the same site. Bases U-2504 and G-2502 fall into a region of the secondary structure of 23S rRNA known as the central loop of Domain V, for which there is a considerable body of evidence that it forms part of the peptidyl transferase center (Hall et al., 1988; Kuechler et al., 1989). Particularly pertinent examples of such evidence are first, that mutation of U-2504 to a C yields a ribosome that is resistant to chloramphenicol, a known peptidyl transferase inhibitor (vide infra), and second, that a photoreactive derivative of Phe-tRNA^{Phe} also photolabels U-2504.

The results cited above suggest that proteins L23, L18/22, and L15 as well as the central loop of Domain V form a ribonucleoprotein neighborhood within the 50S ribosome that is located at the peptidyl transferase center. Immunoelectron microscopy studies on 50S subunits provide a strong indication of the three-dimensional location of this center. Antibodies that recognize the N^6,N^6 -dimethyl adenine moiety of puromycin bind to 50S subunits that have been photoaffinity-labeled with either puromycin or p-azidopuromycin in the location shown in Figure 3 (Olson et al., 1982, 1985). Antibodies to chloramphenicol bind in essentially the same location to 50S subunits that have been affinity-labeled with an electrophilic chloramphenicol derivative (Lührmann et al., 1981). Furthermore, antibodies to three proteins that have been implicated as being at the peptidyl transferase center by both our own and other affinity and photoaffinity labeling experiments, L15, L18, and L27 (vide infra), either overlap or border this location (Stöffler and Stöffler-Meilicke, 1984). Lastly, antibodies to haptens incorporated synthetically into 5S RNA at both the 3' and 5' termini bind nearby on the central protuberance of the 50S subunit (Vasiliev and Shatsky, 1984), and 5S RNA is an essential 50S component for peptidyl transferase. A fuller discussion of the evidence placing the peptidyl transfer center at the location shown in Figure 3 is presented in Cooperman (1987).

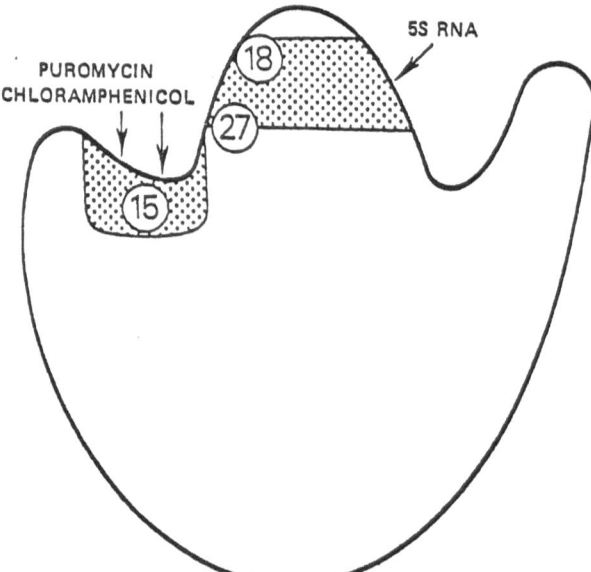

Figure 3. Immunoelectron microscopy determinations of the locations of 50S components and sites of antibiotic binding implicated as being at or near the peptidyl transferase center. The circles indicate the locations of 50S proteins.

3. Photoaffinity Labeling By Chloramphenicol

Chloramphenicol inhibits peptidyl transferase. Inhibition is thought to result from its ability to interfere with the correct binding of the 3'-end of aminoacyl-tRNA to the A'-site, but the exact nature of such inhibition is unclear. A number of photoaffinity labeling experiments with chloramphenicol and chloramphenicol derivatives have been reported (Nielsen et al., 1978; Le Goffic et al., 1980; Bouthier de la Tour et al., 1983), but no clearly significant results were obtained. Recently, using a modification of the method of Le Goffic et al. (1980) we synthesized [^{3}H]-chloramphenicol at very high specific radioactivity. Through the use of RP-HPLC, IE-HPLC, and one-dimensional PAGE analyses, we were able to exploit the very low yield photoincorporation of native chloramphenicol into ribosomes to identify protein L27 as the major specific site of chloramphenicol photoaffinity labeling (Fernandez and Cooperman, 1989).

As may be seen in Figure 1, chloramphenicol has two optically active centers. Of the four stereoisomers, only one, the D-threo form, has high antibiotic activity. This same sterochemical preference is displayed by the single high affinity ($K_D \sim 2 \times 10^{-6}$M) site for chloramphenicol binding to the ribosome, which is localized to the 50S subunit. This and other evidence suggests that the high affinity site is located at or near the peptidyl transferase center. In our work we found, in agreement with others (LeGoffic et al., 1980), that native [^{3}H]-chloramphenicol photoincorporated into a large number of ribosomal proteins with very low photochemical yield. However, at low concentrations ($\leq 10\mu$M) and using limited light doses, protein L27 was labeled to the greatest extent of any ribosomal protein. More importantly, only photoincorporation into protein L27 was site-specific.

Evidence for site-specificity is typically sought by determining whether a non-radioactive ligand will compete for the binding of the corresponding radioactive ligand, as manifested by a reduction in the photoincorporation of radioactivity in the presence of the non-radioactive ligand. In the case of chloramphenicol this approach is complicated by the photodynamic action of the nitrophenyl moiety within the chloramphenicol molecule, which gives rise to non-specific increases in photoincorporation of [^{3}H]-chloramphenicol. To address this problem we took advantage of the stereochemical specificity of chloramphenicol stereoisomer binding to the ribosome, by subtracting the labeling pattern (using RP-HPLC analysis) obtained for photoincorporation of [^{3}H]-D-threo-chloramphenicol in the presence of excess non-radioactive D-threo-chloramphenicol from that obtained in the presence of excess non-radioactive L-erythro-chloramphenicol. Here the reasoning was that the non-specific effects of the p-nitrophenyl group should be the same in both experiments and give rise to no differential labeling. By contrast, a positive peak of labeling would only result from specific site labeling , since the D-threo isomer but not the L-erythro isomer should inhibit such labeling. These expectations were well born out by the experimental results, as shown in Figure 4. The positive peak of radioactivity shown elutes slightly behind L27 in chromatograms of either 70S or 50S proteins. Its identity as labeled L27 was confirmed by additional analyses by IE-HPLC and by one-dimensional PAGE. A similar exploitation of the stereospecificity of chloramphenicol binding was provided

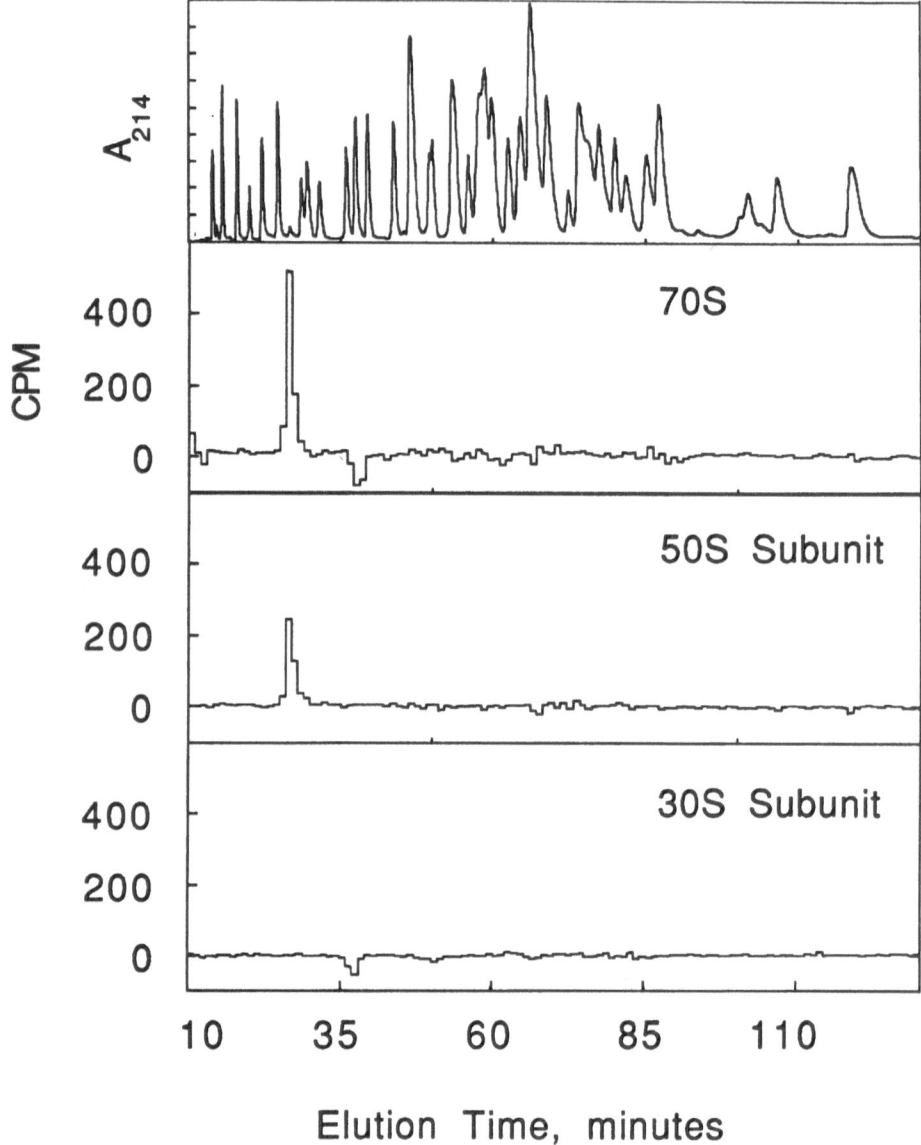

Figure 4. Reversed phase HPLC analysis of [³H]-chloramphenicol photoaffinity labeling of 70S ribosomes. The top panel shows an RP-HPLC pattern of 70S proteins. The bottom three panels show the radioactivity due to covalently attached [³H]-chloramphenicol, expressed as the difference in labeling occurring in the presence of excess non-radioactive L-erythro-chloramphenicol minus that occurring in the presence of non-radioactive D-threo-chloramphenicol, in TP70, TP50, and TP30.

by showing that the photoincorporation of [^{3}H]-D-erythro-chloramphenicol, which we obtain as a byproduct of our radioactive synthesis of [^{3}H]-D-threo-chloramphenicol, did not proceed selectively into L27.

4. Photoaffinity Labeling By Tetracycline

Tetracycline binds to a single high affinity site localized to the 30S subunit within the 70S ribosome ($K_D \sim 2\mu M$), as well as to a large number of low affinity sites. There is good evidence that the major inhibitory effect of tetracycline on protein synthesis, the blocking of aminoacyl-tRNA binding to the A site, is a direct consequence of its binding to the high affinity site. We have taken advantage of the photolability of tetracycline to use the unmodified, [^{3}H]-labeled molecule as a photoaffinity label, and our discussion below focuses on results obtained with this material. Some promising initial results have also been obtained using both the 7-azido and 9-azido derivatives of sancycline (a tetracycline analogue) as photoaffinity labels (Hasan et al., 1985).

The observed labeling pattern using tetracycline is complex due to contributions both from light-dependent incorporation of native tetracycline and from light-dependent and light-independent incorporation of tetracycline photoproduct. However, by using carefully controlled conditions (with respect to light dose and tetracycline concentration) and separately determining both the rate of photoproduct formation and the labeling pattern due to photoproduct, we were able to determine the labeling pattern due to photoincorporation of native tetracycline alone and show that the major protein labeled is S7, identified by both PAGE and immunoprecipitation analyses (Goldman et al., 1983).

Several lines of evidence lead to the conclusion that labeling of S7 takes place from a high affinity site related to the inhibitory effect of tetracycline on tRNA binding. First, radioactive labeling of S7 is decreased in the presence of non-radioactive tetracycline. Second, the labeling of S7 is insensitive to scavenging by 2-mercaptoethanol and thus is unlikely to proceed from solution or from a low affinity site. Third, photoaffinity labeling of S7 proceeds to a much lower extent when [^{3}H]-4-epi-tetracycline is used in place of tetracycline (see Figure 1). The logic of this experiment is similar to that presented above in connection with [^{3}H]-D-erythro-chloramphenicol, since 4-epi-tetracycline is both a weaker inhibitor of aminoacyl-tRNA binding and binds more weakly to subunits as compared with tetracycline. Fourth, the results of single-protein omission reconstitution experiments (Buck and Cooperman, 1989) demonstrate that S7 is one of only three 30S proteins (the others being S14 and S19) that are absolutely crucial for tetracycline binding. The most direct interpretation of these latter results and the related results of others (Wiener and Brimacombe, 1987) is that the three proteins S7, S14 and S19 , when added to 16S RNA, create a structural domain or fold within the 30S particle, that the presence of this fold is necessary for high affinity tetracycline binding, and that within this fold tetracycline binds directly to S7. Fifth, again as shown by single-protein omission reconstitution experiments, each of these same three proteins is necessary for poly(U)-dependent Phe-tRNA[Phe] binding to 30S subunits (Nomura et al., 1969; Buck and Cooperman, 1989).

136

5. Summary and Conclusions

The results summarized above illustrate the use of photoaffinity labeling studies in identifying specific ribosomal components at the binding sites for different ribosomal ligands, and show how a variety of control experiments can be used to obtain evidence that such components are present in areas of functional importance for the ribosome. These results provide direct evidence for the presence of proteins L23, L27, L18/22 and L15 as well as 23S rRNA bases U-2504 and G-2502 as being at or near the peptidyl transferase center, show where this center is within the 50S subunit, and provide indirect evidence for the importance of proteins S7 and S14 in tRNA binding to the 30S subunit and the 70S ribosome. Clearly the approaches outlined above can be applied to study the interaction of other antibiotics with the ribosome, and such studies are underway in several laboratories, including our own. The information gained from these experiments has been and will continue to be of major importance in the construction of detailed functional models of the ribosome.

ა. Acknowledgement

This work was supported by grants AI-16806 and GM-32769 from the National Institutes of Health.

7. References

Allen, D.W. and Zamecnik, P.C. (1962) The effect of puromycin on rabbit reticulocyte ribosomes. *Biochim. Biophys. Acta* 55: 865-874

Arevalo, M. A., Tejedor, F., Polo, F., and Ballesta, J. P. G. (1988) Protein components of the erythromycin binding site in bacterial ribosomes. *J. Biol. Chem.* 263: 58-63

Barta, A., Steiner, G., Brosius, J., Noller, H.F. and Kuechler, E. (1984) Identification of a site on 23S ribosomal RNA located at the peptidyl transferase center. *Proc. Natl. Acad. Sci. U.S.A.* 81: 3607-3611

Bouthier DeLa Tour, C., Moreau, B., Baillargé, M., Blazejewski, J.-C., Capmau, M.-L. and LeGoffic, F. (1983) Synthesis of affinity label structural analogues of chloramphenicol. *Eur. J. Med. Chem.-Chim. Ther.* 18: 315-318

Buck, M. A. and Cooperman, B. S. (1989) manuscript in preparation

Capel, M. S., Datta, D. B., Nierras, C. R., and Craven, G. R. (1988) Ion-exchange high-performance liquid chromatographic separation of ribosomal proteins. *Methods Enzymol.* 164: 532-541

Cooperman, B. S. (1987) Photoaffinity labeling of Escherichia coli ribosomes. *Pharmac. Ther.* 34: 271-302

Cooperman, B. S. (1988) Affinity labeling of ribosomes. *Methods Enzymol.* 164: 341-361

Cooperman, B. S., Weitzmann, C. J., and Buck, M. A. (1988) Reversed-phase high-performance liquid chromatography of ribosomal proteins. *Methods Enzymol.* 164: 523-532

Fernandez, C. L. and Cooperman, B. S. (1989) manuscript in preparation

Gale, E. F., Cundliffe, E., Reynolds, P. E., Richmond, M. H., and Waring, M. J. (1981) *The Molecular Basis of Antibiotic Action* , 2nd edition, Wiley, New York

Ghrist, B. F. D., Cooperman, B. S., and Snyder, L. R. (1988) The design of optimized gradients for the separation of either small or large molecules. I. Minimizing errors in computer simulation. *J. Chromatogr.*, in press

Ghrist, B. F. D., Snyder, L. R., and Cooperman, B. S. (1989) High-performance liquid chromatography of ribosomal proteins. High resolution separation of the E. coli 30S ribosomal proteins by reversed-phase HPLC, using computer simulation to explore different gradient conditions. In: *HPLC of Biological Macromolecules: Methods and Applications*, Gooding, K. M. and Regnier, F. (eds.) Dekker, New York, in press

Girshovich, A.S., Bochkareva, E.S., Kramarov, V.A. and Ovchinnikov, Yu. A. (1974a) E. coli 30S and 50S ribosomal subparticle components in the localization region of the tRNA acceptor terminus. *FEBS Lett.* 42: 213-217

Girshovich, A., S., Bochkareva, E.S. and Pozdnyakov, V.A. (1974b) Affinity labelling of functional centres of Escherichia coli ribosomes. *Acta Biol. Med. Germ.* 33: 639-648

Goldman, R.A., Hasan, T., Hall, C.C., Strycharz, W.A. and Cooperman, B.S. (1983) Photoincorporation of tetracycline into Escherichia coli ribosomes. Identification of the major proteins photolabeled by native tetracycline and tetracycline photoproducts and implications for the inhibitory action of tetracycline on protein synthesis. *Biochemistry* 22: 359-368

Grant, P. G., Olson, H.M., Glitz, D. and Cooperman, B.S. (1983) Puromycin binding to the small subunit of Escherichia coli ribosomes. Localization of the antibiotic in subunits reconstituted with puromycin-modified components. *J. Biol. Chem.* 258: 11305-11312

Gravel, M., Melançon, P., and Brakier-Gingras, L. (1987) Cross-linking of streptomycin to the 16S ribosomal RNA of Escherichia coli. *Biochemistry* 26: 6227-6232

Hall, C.C., Smith, J.E. and Cooperman, B.S. (1985) Mapping labeled sites in Escherichia coli ribosomal RNA: distribution of methyl groups and identification of a photoaffinity-labeled RNA region putatively at the peptidyltransferase center. *Biochemistry* 24: 5702-5711

Hall, C. C., Johnson, D., and Cooperman, B. S. (1988) [^{3}H]-p-Azidopuromycin photoaffinity labeling of Escherichia coli ribosomes: Evidence for site-specific interaction at U-2504 and G-2502 in domain V of 23S ribosomal RNA *Biochemistry* 27: 3983 - 3990

Hardesty, B. and Kramer, G. , eds., (1986) *Structure, Function, and Genetics of Ribosomes* , Springer, New York

Hasan, T., Goldman, R.A. and Cooperman, B.S. (1985) Photoaffinity labeling of the tetracycline binding site of the Escherichia coli ribosome. The uses of a high intensity light source and of radioactive sancycline derivatives. *Biochem. Pharmacol.* 34: 1065-1071

Jaynes, E.N., Jr., Grant, P.G., Giangrande, G., Wieder, R. and Cooperman, B.S. (1978) Photoinduced affinity labeling of the Escherichia coli ribosome puromycin site. *Biochemistry* 17: 561-569

Kerlavage, A.R., Kahan, L. and Cooperman, B. S. (1982) Reverse-phase high-performance liquid chromatography of Escherichia coli ribosomal small subunit proteins. *Anal. Biochem.* **123**: 342-348

Kerlavage, A.R., Hasan, T. and Cooperman, B.S. (1983a) Reverse phase high performance liquid chromatography of Escherichia coli ribosomal proteins: standardization of 70S, 50S and 30S protein chromatograms. *J. Biol. Chem.* **258**: 6313-6318

Kerlavage, A.R., Weitzmann, C.J., Hasan, T and Cooperman, B.S. (1983b) Reversed-phase high-performance liquid chromatography of Escherichia coli ribosomal proteins: characteristics of the separation of a complex protein mixture. *J. Chromatogr.* **266**: 225-237

Kerlavage, A.R. and Cooperman, B.S. (1986) Reconstitution of Escherichia coli ribosomes containing puromycin-modified S14: functional effects of the photoaffinity labeling of a protein essential for tRNA binding. *Biochemistry* **25**: 8002-8010

Kuechler, E.., Steiner, G., and Barta, A. (1989) Dissection of the ribosomal peptidyl transferase center by photoaffinity labeling: P- and A- sites are located at different positions in domain V of 23S RNA. In: This volume, pp. xxx- xxx

LeGoffic, F., Capmau, M.-L., Chausson, L. and Bonnet, D. (1980) Photo-induced affinity labeling of *Escherichia coli* ribosomes by chloramphenicol. *Eur. J. Biochem.* **106**: 667-674

Lührmann, R., Bald, R. Stöffler-Meilicke, M. and Stöffler, G. (1981) Localization of the puromycin binding site on the large ribosomal subunit of Escherichia coli by immunoelectron microscopy. *Proc. Natl. Acad. Sci. U.S.A.* **78**: 7276-7280

Maitra, U., Stringer, E.A. and Chaudhuri, A. (1982) Initiation factors in protein biosynthesis. *Ann. Rev. Biochem.* **51**: 869-900

Moldave, K. and Noller, H. F., eds., (1988) *Ribosomes. Methods Enzymol.* **164**

Nathans, D. (1964) Puromycin inhibition of protein synthesis: incorporation of puromycin into peptide chains. *Proc. Natl. Acad. Sci. U.S.A.* **51**: 585-592

Nicholson, A.W., Hall, C.C., Strycharz, W.A. and Cooperman, B.S. (1982a) Photoaffinity labeling of Escherichia coli ribosomes by an aryl azide analogue of puromycin. On the identification of the major covalently labeled ribosomal proteins and mechanisms of photoincorporation. *Biochemistry* **21**: 3797-3808

Nicholson, A.W., Hall, C.C., Strycharz, W.A. and Cooperman, B.S. (1982b) Photoaffinity labeling of Escherichia coli ribosomes by an aryl azide analogue of puromycin. Evidence for the functional site specificity of labeling. *Biochemistry* **21**: 3809-3817

Nielsen, P.E., Leick, V. and Buchardt, O. (1978) On photoaffinity labeling of Escherichia coli ribosomes using an azidochloramphenicol analogue. *FEBS Lett.* **94**: 287-290

Nierhaus, K.H. (1984) New aspects of the ribosomal elongation cycle. *Molec. Cell. Biol.* **61**: 63-81

Nomura,M., Mizushima, S., Ozaki, M., Traub, P. and Lowry, C.V. (1969) Structure and function of ribosomes and their molecular components. *CSH Symp. Quant. Biol.* **34**: 49-61

Olson, H.M., Grant, P.G., Glitz, D.G. and Cooperman, B.S. (1980) Localization by immunoelectron microscopy of the site of photoinduced affinity labeling of the small ribosomal subunit with puromycin. *Proc. Natl. Acad. Sci. U.S.A.* **77**: 890-894

Olson, H.M., Grant, P.G., Cooperman, B.S. and Glitz, D.G. (1982) Immunoelectron microscopic localization of puromycin binding on the large subunit of the Escherichia coli ribosome. *J. Biol. Chem.* **257**: 2649-2656

Olson, H.M., Nicholson, A.W., Cooperman, B.S. and Glitz, D.G. (1985) Localization of sites of photoaffinity labeling of the large subunit of Escherichia coli ribosomes by an arylazide derivative of puromycin. *J. Biol. Chem.* **260**: 10326-10331

Spirin, A.S. (1985) Ribosomal translocation: facts and models. *Prog. Nucleic Acid Res. Molec. Biol.* **32**: 75-113

Stöffler, G. and Stöffler-Meilicke, M. (1984) Immunoelectron microscopy of ribosomes. *Ann. Rev. Biophys. Bioeng.* **13**: 303-330

Vasiliev, V.D. and Shatsky, I.N. (1984) Structural organization of the Escherichia coli ribosome and its functional centers. *Sov. Sci. Rev. D. Physiochem. Biol.* **5**: 141-178

Wada, A. (1986a) Analysis of Escherichia coli ribosomal proteins by an improved two dimensional gel electrophoresis. I. Detection of four new proteins. *J. Biochem.* **100**: 1583-1594

Wada, A. (1986b) Analysis of Escherichia coli ribosomal proteins by an improved two dimensional gel electrophoresis. II. Characterization of four new proteins. *J. Biochem.* **100**: 1583-1594

Weitzmann, C.J. and Cooperman, B.S. (1985) On the structural specificity of puromycin binding to *Escherichia coli* ribosomes. *Biochemistry* **24**: 2268-2274

Weitzmann, C. J. and Cooperman, B. S. (1989) manuscript in preparation

Wiener, L. and Brimacombe, R. (1987) Protein binding sites on Escherichia coli 16S RNA; RNA regions that are protected by proteins S7, S14 and S19 in the presence or absence of protein S9. *Nucleic Acids Res.* **15**: 3563-3670

Youvan, D.C. and Hearst, J.E. (1979) Reverse transcriptase pauses at N2-methylguanine during *in vitro* transcription of E. coli 16S ribosomal RNA. *Proc. Natl. Acad. Sci. U.S.A.* **76**: 3751-3754

Youvan, D.C. and Hearst, J.E. (1981) A sequence from Drosophila melanogaster 18S rRNA bearing the conserved hypermodified nucleoside amΨ: analysis by reverse transcription and high-performance liquid chromatography. *Nucleic Acids Res.* **9**: 1723-1741

DISSECTION OF THE RIBOSOMAL PEPTIDYL TRANSFERASE CENTER BY PHOTOAFFINITY LABELING: P- AND A-SITES ARE LOCATED AT DIFFERENT POSITIONS IN DOMAIN V OF 23 S RNA

E. KUECHLER, G. STEINER and A. BARTA
Institute of Biochemistry
University of Vienna
Währinger Strasse 17
A-1090 Vienna
Austria

ABSTRACT. Photoreactive 3-(4′-benzoylphenyl)propionyl-Phe-tRNA was bound specifically to the P-site or the A-site of the *E. coli* ribosome. Photoreaction led to high yield crosslinking between the aminoacyl-terminus of Phe-tRNA and 23S RNA. The site of reaction was identified by hybridization and by taking advantage of the fact that reverse transcriptase stops one base before a modified nucleotide. The nucleotides labeled at the P-site were identified as A-2451 and C-2452. A-site specific labeling occurred at U-2584 and U-2585. Although distant in the primary sequence, these residues lie in close proximity within the central loop of domain V according to the secondary structure model of 23S RNA. Only antibiotics known to act at the peptidyl transferase site strongly inhibit the photoaffinity reaction, thus providing further evidence for the specificity of the labeling. The demonstration of specific, high yield crosslinks at the level of 23S RNA strongly supports the involvement of an RNA-catalyzed activity in the peptidyl transferase reaction.

1. Introduction

1.1. THE SEARCH FOR THE PEPTIDYL TRANSFERASE CENTER

Despite recent progress, many of the basic mechanisms underlying protein biosynthesis are still poorly understood. In particular, the role of the various ribosomal components and their interplay in the process of peptide bond formation is still unclear. Peptidyl transferase - the quintessential ribosomal enzymatic activity - has been known for a long time to be located on the large subunit (Monro, 1967; Maden *et al.*, 1968). Since enzymatic catalysis was then considered to be exclusively a property of proteins, the search for the peptidyl transferase center concentrated on the ribosomal proteins from this subunit (for reviews see Kuechler, 1978; Cantor, 1979; Kuechler and Ofengand, 1979; Cooperman, 1980; Ofengand, 1980; Cooperman *et al.*, 1986; Ofengand *et al.*, 1986). Indeed, several proteins (e.g. L2, L15, L16, L27) were detected by a variety of thermal affinity labels designed to react specifically with residues at the peptidyl transferase site. The affinity labels employed were derivatives of aminoacyl-tRNAs as well as of antibiotics. Subsequently, these

141

P. E. Nielsen (ed.), Photochemical Probes in Biochemistry, 141–156.
© *1989 by Kluwer Academic Publishers.*

proteins were shown by immunoelectron microscopy to be located on the side of the central protuberance opposite the L7/L12 stalk of the *E. coli* 50S subunit, allowing for the first time a topographical description of the peptidyl transferase center on the ribosome (Oakes *et al.*, 1986; Stoeffler & Stoeffler-Meilicke, 1986). Furthermore, some of these proteins were also shown to be essential or strongly stimulatory in the reconstitution of peptidyl transferase activity (Nierhaus, 1982). Nevertheless, no consistent picture which could serve as the basis for a molecular model of the process of transpeptidation could be deduced from these data. Moreover, the efficiencies of labeling were usually low and the results varied depending on the particular affinity reagents employed. There were also practical considerations for pursuing the analysis of proteins. At that time powerful techniques were available for separating ribosomal proteins by two-dimensional gel electrophoresis (Kaltschmidt & Wittmann, 1970). This and the availability of specific antibodies (Tischendorf *et al.*, 1974) made the identification of affinity labeled proteins a rather easy and straightforward undertaking. Attempts to analyse labeled sites on RNA, however, met with insuperable technical difficulties. For these reasons, affinity and photoaffinity labels introduced into 23S RNA (Bochkareva *et al.*, 1971; Barta *et al.*, 1975; Sonenberg *et al.*, 1975; Breitmeyer & Noller, 1976) could not be localized with any reasonable degree of accuracy at that time.

1.2. A BOOST FROM GENE TECHNOLOGY

With the advent of recombinant DNA technology, the situation changed dramatically. The sequences of rRNAs from *E. coli* and other sources were determined and secondary structure models were deduced by sequence comparison (Noller, 1984). In addition, a technique was developed which allowed the identification of modified nucleotides within an RNA without the necessity of isolating the particular modified nucleotide directly (Barta *et al.*, 1984). This method was based on the observation made by Youvan and Hearst (1979, 1981) that reverse transcriptase slows down or halts chain elongation one base before encountering a modified nucleotide. Employing this chain elongation technique, a suitable set of oligonucleotide primers can be used to scan an RNA for sites of modification.

At about the same time, RNAs were discovered which exhibit catalytic functions (Guerrier-Takada *et al.*, 1983; Kruger *et al.*, 1982). This discovery greatly stimulated the search for potential catalytic sites on rRNA. Unfortunately, all attempts to reconstitute an active peptidyl transferase system based solely on rRNAs failed. Therefore the direct proof of a catalytic involvement of the rRNA in ribosomal transpeptidation could not be achieved. However, it was consistently found that thermal affinity labels of high reactivity reacted predominantly with 23S RNA. This was especially true for photoaffinity labels. This strengthened the idea that the RNA does not simply provide a framework to hold functionally active proteins in the proper position.

1.3. A HIGH YIELD PHOTOAFFINITY LABEL

One of the problems with photoaffinity labels in general is that many of the highly reactive ones exhibit rather low yields of reaction. This problem has been circumvented in many cases by using radioactive reagents of high specific activity which allow the incorporation of sufficient radioactivity to permit identification of the labeled proteins. In the chain termination assay employing reverse transcriptase, however, a high reaction yield is important, because reverse transcriptase also terminates at other sites, such as sites of strong secondary and/or tertiary structure. Therefore bands brought about by base modification are expected to appear above a pattern of bands due to spontaneous chain termination events.

The photoaffinity reagents developed by our group were derived from aromatic ketones. The carbonyl group of aromatic ketones can be photoactivated with UV light of 320 nm (Turro, 1967). The activation leads to an $n \rightarrow \pi^*$ transition of a lone electron pair on the carbonyl oxygen. This is followed by a rapid spin inversion resulting in a triplet state, with the triplet state energy amounting to about 290 kJ. The activated molecule is capable of reacting with covalent bonds of similar energy. Activated benzophenone was shown to undergo reaction with model compounds such as N-acetylglycine methyl ester in a photoaddition to the α-C-H bond of the glycine molecule (Galardy et al., 1973) and also with nucleotides (Barta & Kuechler, 1983). Since the energy of the O-H bond in a water molecule amounts to about 525 kJ, the energy of the activated triplet state is too low to result in a reaction with water. Thus, loss of photoreagent due to side reactions with water molecules does not occur. The reagent has been proven useful in the identification of the gastrin receptor (Galardy et al., 1974), in the study of the influenza virus cap-binding protein (Blaas et al., 1982 a, b) and the nuclear and the cytoplasmic cap-binding proteins (Patzelt et al., 1983). As aromatic ketone derivatives are activated by light at 320 nm (which does not destroy ribosomal activity) and are capable of reacting both with polypeptides and polynucleotides but not with water, they are ideally suited for use in the search for the peptidyl transferase center.

tRNA derivatives were chosen as substrates because they offer several advantages over other types of ligands such as antibiotics. The binding site is defined a priori and the specificity of the labeling can be checked and controlled both positively by the stimulation with messenger RNA and negatively by competition with non-derivatized aminoacyl-tRNA. The poly (U) - Phe-tRNA system was employed because of the high efficiency binding; only "tight couple" E. coli ribosomes were used.

In order to attach benzophenone on to the aminoacyl-moiety of the Phe-tRNA, a propionic acid residue was introduced in the para position of the aromatic ring. The three carbon atom bridge was found to be optimal in terms of the binding properties of the Phe-tRNA derivative. Obviously it was desirable to keep the length of the bridge to a minimum. Moreover, on building a space-filling model of the aminoacyl-terminus it was seen that at least a three carbon bridge was required to allow maximal flexibility of the benzophenone residue.

2. Materials and Methods

2.1. MATERIALS

Synthesis of 3-(4'benzoylphenyl)propionyl-N-hydroxy-succinimide ester, charging of tRNA[Phe] (Boehringer Mannheim, FRG) with [³H]Phe (specific activity 70-110 Ci/mmol from the

Radiochemical Center Amersham, UK) and the synthesis of 3-(4'benzoylphenyl) propionyl-[³H]Phe-tRNA (BP-[³H]Phe-tRNA) have been described (Kuechler and Barta, 1977). Non-derivatized tRNA was removed by adsorption of BP-Phe-tRNA to BD-cellulose (Gillam and Tener, 1971). BP-Phe-tRNA was eluted from the column with a buffer containing 1M NaCl, 10 mM sodium acetate, pH5, 10 mM MgCl₂, 30 % (v/v) ethanol. Purity of the BP-[³H]Phe-tRNA obtained was checked after hydrolysis by thin layer chromatography. The preparation obtained was free of uncharged tRNA as calculated from the specific activity obtained. *E. coli* MRE 600 ribosomes ("tight couples") were obtained as described (Kuechler and Barta, 1977).

2.2. SPECIFIC BINDING OF BP-[³H]PHE-tRNA TO RIBOSOMES

For specific binding to the P-site, 250 pmol *E. coli* ribosomes, 0.1 mg poly (U) and 100 pmol BP-[³H]Phe-tRNA were incubated in a buffer containing 100 mM KCl, 20 mM Tris-HCl, pH 7.4, 6 mM MgCl₂, 0.2 mM EDTA, 1 mM dithiothreitol (final volume 1 ml) at 25°C for 10 min. For A-site binding, 250 pmol *E. coli* ribosomes, 0.1 mg poly (U) and 750 pmol uncharged tRNA^{Phe} were incubated in 1 ml of the same buffer except that the MgCl₂ concentration was raised to 10 mM. Incubation was for 3 min at 37°C. 100 pmol BP-[³H]Phe-tRNA was then added and the incubation continued for 10 min at 25°C. The amount of ribosomal bound BP-[³H]Phe-tRNA was determined by chromatography on Sepharose 6B columns (80 x 5 mm) (Kuechler *et al.*, 1988). For translocation, 0.1 mg/ml EF-G (kind gift of K. Nierhaus, MPI für Molekulare Genetik, Berlin, FRG) and 1 mM GTP were added to A-site complexes. Binding to P- and A-site and the extent of translocation were checked by adding 1 mM puromycin and subsequent incubation for 20 min at 25°C.

2.3. LOCALIZATION OF THE SITE OF THE PHOTOAFFINITY LABELING ON 23S RNA

2.3.1. Photoreaction. The apparatus employed for irradiation has been described before in detail (Kuechler and Barta, 1977). Briefly, it consists of a super-high-pressure mercury lamp, Philips Type SP500 W; light below 300 nm is removed by use of a WG 320 cut off filter, thickness 2 mm (Schott & Gen., Mainz, FRG). Samples were irradiated in conical quartz tubes at 0-4°C. In order to protect the eyes high-quality safety goggles must be worn when manipulating the samples. Photocrosslinking of BP-[³H]Phe-tRNA to 23S RNA was measured after phenol extraction and chromatography on a Sepharose 6B column. Incorporation was shown to occur exclusively into 23S RNA (Barta and Kuechler, 1983).

2.3.2. Localization of the photoaffinity labeled region on 23S RNA. In order to determine the region of photoaffinity labeling use was made of plasmid pKK123 which contains part of the 23S RNA gene (between nucleotide 843 and the 3'terminus) integrated between an *Eco*RI and a *Bam*HI site (Brosius *et al.*, 1981). The *Eco*RI-*Bam*HI fragment was digested further by various restriction enzymes (*Hin*fI, *Hae*III, *Hpa*II and *Cfo*I) to produce suitable fragments for Southern blot hybridization. The DNA fragments were separated on a 2% agarose gel and the DNA bands were transferred to a nitrocellulose filter (Southern, 1975). The filter was incubated in 50% formamide,

5 x SSC (1 x SSC : 150 mM NaCl, 15 mM sodium citrate), 50 mM sodium phosphate, pH 6.5, 0.2% sodium dodecyl-sulfate (SDS), 5 x Denhardt's solution, 0.1 mg/ml of tRNA and 0.05 mg/ml of denatured salmon sperm DNA with 80 μg (1.25 x 10^6 cpm) of photoaffinity [^{3}H]Phe-labeled 23S RNA (heated briefly to 96°) for 2 days at 42°. The filter was washed with 2 x SSC and with 0.1 x SSC and subsequently incubated for 30 min at 37° with 100 μg/ml ribonuclease A in 2 x SSC to release unhybridized RNA. Subsequently, the filter was washed with 0.1 x SSC, 0.1% SDS, then with 0.1 x SSC, 25% ethanol and dried (Kuechler *et al.*, 1988). [^{3}H]Phe-labeled bands were visualized by placing the filter in 1M sodium salicylate, drying and exposing at -70° to a preflashed Kodak XAR-5 film.

2.3.3. Primer extension using reverse transcriptase. A 56 bp *Ava*II fragment corresponding to nucleotides 2607-2663 of 23S RNA was prepared from plasmid pKK 123 and purified by electrophoresis as described (Kuechler *et al.*, 1988). The fragment was 5'-labeled using polynucleotide kinase and [γ-^{32}P]ATP (specific activity 5000 Ci/mmol). The labeled fragment was further purified on a Bio Gel P-30 column. Strand separation was achieved by dissolving the 5'-labeled 56 bp *Ava*II fragment in 30 μl of a solution containing 30% dimethyl sulfoxide, 1 mM EDTA and heating for 1 min at 96°. The mixture was cooled and subjected to electrophoresis on a 30 cm long 16% polyacrylamide gel (acrylamide/bisacrylamide = 59/1) at 150 V in 100 mM Tris-borate, pH 8.3, 1 mM EDTA for 15 h. After exposing the gel to an X-ray film for a few minutes, the bands were cut out and the DNA was extracted by diffusion. The strands were identified by their ability to hybridize to 23S RNA.

For the primer extension assay 0.6 μg of modified 23S RNA, photoaffinity labeled either under P-site or A-site binding conditions was incubated with an equimolar amount of the 5'-[^{32}P]-labeled single-stranded primer in 10 μl of 40 mM KCl, 25 mM Tris-HCl, pH 8.3, 5 mM MgCl$_2$ for 15 min at 65°. After cooling to room temperature, 10 μl of the same buffer containing 1 mM each of dATP, dCTP, dGTP and dTTP and 10 units of avian myeloblastosis virus (AMV) reverse transcriptase (Life Sciences, St. Petersburg, FL) was added and the mixture incubated for 1 h at 42°. Subsequently 20 μl of 0.2 M NaOH and 0.025 M EDTA were added and the incubation continued for 5 min at 96 °. After cooling the solution was neutralized by addition of 0.2 M HCl. The sample was precipitated after addition of 2 volumes of ethanol for 30 min at -70° and the pellet was washed with 70% ethanol and dried. After dissolving in 5 μl of sequencing buffer (100 mM Tris-borate, pH 8.3, 1 mM EDTA, 80% formamide) containing 0.05% each of xylene cyanol and bromophenol blue, the sample was heated for 2 min at 96° and cooled to 0°. 3 μl were layered on to a 7% denaturing sequencing gel and electrophoresis was carried out at 50 W for 2 h 20 min.

Bands were identified by running a dideoxynucleotide sequencing gel alongside. A similar reaction mixture was prepared using the 5'-[^{32}P]-labeled primer but with unmodified 23S RNA as a template. Reverse transcriptase and the four dNTPs were added in the cold and the solution was divided into five aliquots. To four aliquots 0.5 mM of ddATP, ddGTP and ddTTP and 0.25 mM ddCTP were added respectively at a final volume of 20 μl, the fifth served as a control. Incubation conditions and gel electrophoresis were as described above. Gels were fixed in 10% methanol, 10% acetic acid for 30 min, transferred to Whatman 3MM paper, dried at 80° on a gel drier, and exposed to Kodak XAR-5 film.

3. Results

3.1. SPECIFIC LABELING AT THE P- AND A-SITE

Well defined conditions for binding of the BP-Phe-tRNA^{Phe} were the prerequisite for obtaining specific photoaffinity labeling at the P- or at the A-site. It was observed that in addition to the use of a low Mg^{++} - concentration for P-site binding and a high Mg^{++} - concentration for A-site binding the presence or absence of uncharged tRNA was essential for specificity. Uncharged tRNA is known to have a high affinity for the P-site. For that reason BP-Phe-tRNA^{Phe} was purified by chromatography on BD-cellulose in order to completely remove uncharged tRNA^{Phe}. Furthermore, a 2.5 fold molar excess of "tight couple" *E. coli* ribosomes over BP-Phe-tRNA^{Phe} was employed. These ribosomal preparations routinely contained more than 50% active particles.

A 6 mM Mg^{++} - concentration was found to result in optimal P-site binding of BP-[^{3}H]Phe-tRNA under these conditions. For A-site binding ribosomes were preincubated with a threefold molar excess of uncharged tRNA^{Phe} at 10 mM Mg^{++} concentration. Under these conditions the uncharged tRNA^{Phe} binds to the P-site thereby blocking this site. After preincubation of this mixture, BP-[^{3}H]Phe-tRNA was added and the incubation continued. Translocation from the A- to the P-site was achieved by addition of EF-G and GTP. The puromycin reaction was used as a test for specific P-site binding. Effective translocation was checked by comparing puromycin reactivity before and after incubation with EF-G and GTP. Since BP-[^{3}H]Phe-tRNA adsorbs to nitrocellulose filters, binding to ribosomes and incorporation into RNA had to be checked by chromatography on small Sepharose 6B-columns.

Data on specific binding to ribosomes and photocrosslinking to 23S RNA both under P- and A-site conditions are presented in Table I. It can be seen that both under P- and A-site binding conditions a large fraction of the bound BP-[^{3}H]Phe-tRNA becomes photocrosslinked to 23S RNA. The specificity of the crosslink is indicated by the dependence on poly(U) and on irradiation. Preincubation of the P-site complex with puromycin strongly reduced the amount of radioactivity bound and virtually abolished photocrosslinking as expected. Under A-site conditions binding was slightly reduced by puromycin to about 80% indicating that the preparation indeed contained BP-[^{3}H]Phe-tRNA predominantly bound to the A-site. However, much to our surprise puromycin also inhibited photocrosslinking under A-site conditions to a considerable extent. Only 28% of the radioactivity was recovered in 23S RNA. Since puromycin closely resembles the 3'-terminus of a Tyr-tRNA it is conceivable that the drug behaves as a competitive inhibitor and prevents proper arrangement of the 3'-terminus of BP-[^{3}H]Phe-tRNA within the ribosomal A-site. This interpretation was substantiated by purifying ribosomal A-site complexes following incubation with puromycin. Removal of puromycin before irradiation resulted in restoration of photocrosslinking to the control value (data not shown). This experiment indicates that inhibition by puromycin alone is not necessarily a reliable criterium for discriminating between P- and A-site labeling.

Functionally active A-site complexes are best characterized by their capability to translocate peptidyl-tRNA to the P-site. For this reason complexes containing BP-[^{3}H]Phe-tRNA bound under A-site conditions were incubated with EF-G and GTP and the effect of puromycin was

Table I

	% Binding	% Photocrosslinking
	(196 000 cpm = 100%)	(152 000 cpm = 100%)
"P-site"		
− Poly(U), − UV	10.7	n.d.
+ Poly(U), − UV	93.4	8.6
+ Poly(U), + UV	100.0	100.0
+ Poly(U), + puromycin, + UV	17.9	3.3
	(168 000 cpm = 100%)	(150 000 cpm = 100%)
"A-site"		
− Poly(U), − UV	31.0	n.d.
+ Poly(U), − UV	88.7	6.0
+ Poly(U), + UV	100.0	100.0
+ Poly(U), + puromycin, + UV	80.4	28.0

Ribosomal binding of BP-[³H]Phe-tRNA and photocrosslinking to 23S RNA. 10 pmol of ribosomes or 23S RNA were used respectively. n.d. = not determined.

determined. As seen in Table II, 74% of BP-[³H]Phe-tRNA bound under A-site conditions was refractory to treatment with puromycin. This value fell to 18% upon addition of EF-G and GTP indicating that a large portion of the BP-[³H]Phe-tRNA was indeed bound to functionally active ribosomal A-site complexes.

3.2. THE SITE OF PHOTOAFFINITY LABELING ON 23S RNA

Early attempts to localize the site of the photocrosslink by isolation of an oligonucleotide containing the site of the labeling met with considerable difficulty. By digestion with ribonuclease under well defined conditions it was, however, possible to demonstrate that photocrosslinking occurred at a site located within an 11S fragment which comprises 1100 nucleotides from the 3′terminus of 23S RNA (Barta & Kuechler, 1983). Based on this result, a strategy was developed to narrow down the region by Southern blot hybridization with a set of restriction fragments derived from the region of the 23S RNA gene between nucleotide 843 and the 3′-terminus. In this way, it was possible to locate the site of photoaffinity labeling to a region of 183 nucleotides between bases 2442 and 2625 (Barta *et al.*, 1984).

To localize the site more precisely, two primers were employed and the lengths of the reverse transcripts determined. The experiment is depicted schematically in Fig. 1. One primer was a 124 nucleotide long *Hpa*II fragment corresponding to the region from nucleotide 2715 to 2839. The second primer was obtained from a 56 base pair *Ava*II fragment corresponding to nucleotides 2607 to 2663 of 23S RNA. Both oligonucleotides were 5′-labeled using polynucleotide kinase and

Table II

	A-site	+ EF-G/GTP
	(69 000 cpm = 100%)	(46 000 cpm = 100%)
− Poly(U)	12.3	n.d.
+ Poly(U)	100.0	100.0
+ Poly(U), + puromycin	73.9	18.5

Translocation of BP-[³H]Phe-tRNA bound to A-site by EF-G and GTP. n.d. = not determined.

[γ-³²P]ATP and were employed to prime reverse transcription. The same results were obtained with both primers; the discussion is confined to the primer corresponding to nucleotides 2607 to 2663 as the cDNA bands were more clearly resolved. 23S rRNA labeled either under P-site or A-site conditions was used as a template for reverse transcription, with unlabeled 23S RNA serving as a control. The transcripts obtained were then separated on a gel with dideoxynucleotide sequencing reactions run in parallel. The results are presented in Fig. 2. The positions of the bands can be read from the sequencing gel. It can be seen that under P-site conditions (lane: P-site) two new bands corresponding to C-2452 and A-2453 appeared above the general background. These bands were not present either in the sample containing 23S RNA from non-irradiated ribosomes (lane: control

23S rRNA

Photoaffinity-labeled 23S rRNA

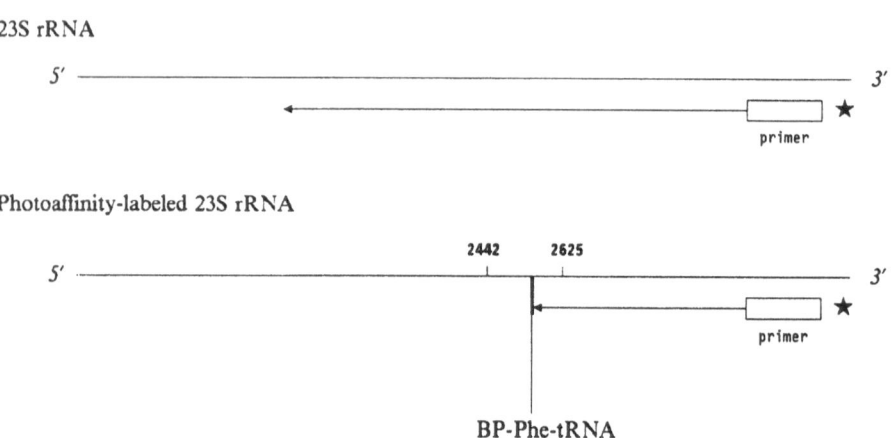

Fig. 1. Scheme of the primer extension experiment employing reverse transcriptase to determine the site of photoaffinity labeling on 23S rRNA. The region between nucleotides 2442 and 2625 identified previously by Southern blot hybridization is indicated. 5′-[³²P] label of the primer is indicated by an asterisk (★). The arrow shows the direction of reverse transcription.

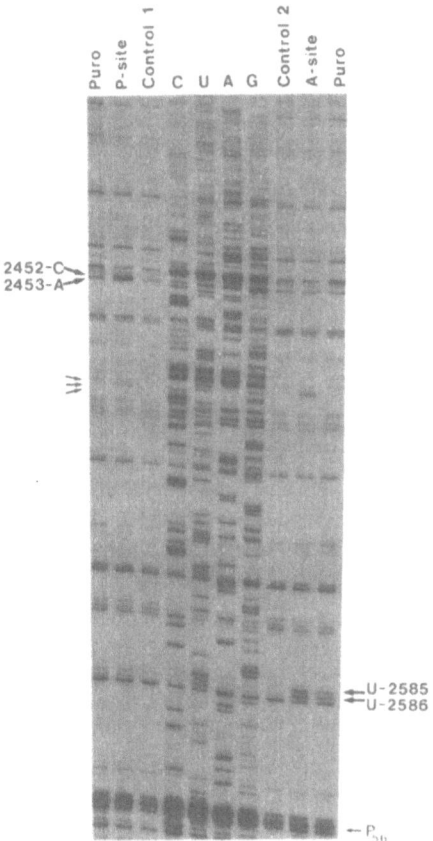

Fig. 2. Autoradiograph of the gel pattern of reverse transcripts obtained in the primer extension experiment using 23S RNA photoaffinity labeled at the P-site and at the A-site. Specific binding and photoaffinity labeling was performed as described in the Methods Section. Labeled RNAs were employed as templates for reverse transcription. The primer used was a single stranded 56 nucleotides long *Ava*II DNA fragment [^{32}P]-labeled at the 5'-end corresponding to positions 2607 - 2663 of 23S RNA. Lane *P-site*: Complete photoaffinity reaction system with BP-Phe-tRNA bound at the P-site. Lane *A-site*: Complete photoaffinity reaction system with BP-Phe-tRNA bound at the A-site. Stops at the designated nucleotides are considered to be due to photoaffinity labeling of the ensuing nucleotide. Little arrows on both sides represent sites of minor crosslinks. Lanes *Puro*: Samples incubated with 1 mM puromycin after binding of BP-Phe-tRNA but before irradiation, respectively. Lane *control 1*: rRNA from non-irradiated ribosomes used as template; lane *control 2*: rRNA from ribosomes irradiated in the absence of BP-Phe-tRNA used as template. Lanes *C, U, A, G*: indicate the respective dideoxynucleotide sequencing lanes used as reference.

1) or from ribosomes irradiated in the absence of BP-[^{3}H]Phe-tRNA (lane: control 2). In addition to these two strong bands, three weak bands corresponding to cDNAs terminating at G-2505, U-2506 and C-2507 were seen lower down on the gel. Assuming that the observation of Youvan & Hearst (1979, 1981) also applies to our system (that reverse transcriptase pauses or halts one base before the modified nucleotide), the bands obtained should be expected to be due to P-site photocrosslinking at A-2451 and at C-2452, respectively. Likewise the three minor bands should be due to photoreactions at U-2504, G-2505 and U-2506, respectively. Consistent with this interpretation, it was found that the bands disappeared when the samples were preincubated with puromycin before irradiation (lane: Puro, left side).

When the same primer extension experiment was carried out using 23S RNA photoaffinity labeled under A-site binding conditions, a different pattern of bands appeared above the same general background. Strong bands were found corresponding to U-2585 and U-2586, which should result from photocrosslinking at U-2584 and U-2585. In addition, two weak bands were seen resulting from reactions at A-2503 and U-2504 (small arrows at right side). When the same experiment was carried out in the presence of puromycin a partial inhibition was observed in agreement with crosslinking data shown in Table I. However, when puromycin was removed by purifying ribosomal complexes on Sepharose - 6B columns before irradiation, the strong bands reappeared indicating that the inhibition was due to competition rather than to a removal of the photoaffinity label by the puromycin reaction (data not shown). The results obtained are summarized in Fig. 3.

4. Discussion

The data obtained in this photoaffinity labeling study demonstrate that different sites on 23S RNA are labeled at the P-site (A-2451, C-2452) and at the A-site (U-2584, U-2585) by BP-Phe-tRNA (Fig. 3). Even though it is difficult to predict the reactivity of a photoaffinity label at a particular binding site, the fact that BP-Phe-tRNA reacts exclusively with RNA with a better than 70% yield indicates that the labeled sites in the central loop of domain V are integral components of the peptidyl transferase . The two strong labeling sites occur at different locations according to the primary sequence. The sites are, however, brought together by the secondary structure folding of the molecule. The discovery of this bipartite structural arrangement is of great significance for any molecular model of the peptidyl transferase center. An intimate association of this region with the peptidyl transferase has also been suggested by experiments in which mutations conferring resistance to antibiotics have been localized on ribosomes (Fig. 3, closed symbols). A-2451 and C-2452 appear to be involved in chloramphenicol binding, an antibiotic known to inhibit peptidyl transferase. Other antibiotics used in the studies of resistance mutants were erythromycin and vernamycin B which affect translocation. Although the mechanism of action is different, erythromycin and vernamycin B have been shown to compete with chloramphenicol for binding to the 50 S subunit (Celma et al., 1970; Fernandez-Munoz et al., 1971). These antibiotics also inhibit chemical modifications of 23S RNA in ribosomes by kethoxal and/or by dimethylsulfate at sites indicated in Fig. 3 (open symbols). Photoaffinity labeling with

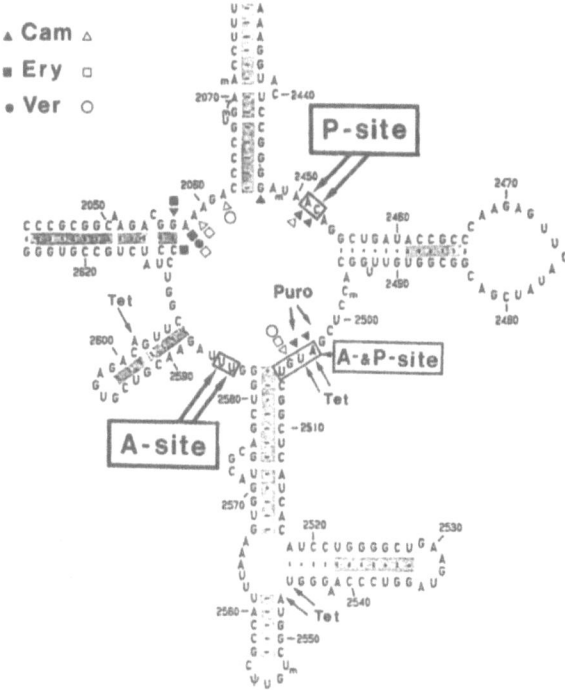

Fig. 3. Schematic Diagram of the Secondary Structure of the Central Loop Region of Domain V of 23 S rRNA.

The nucleotides specifically affinity labeled by P- and A-site bound BP-Phe-tRNA identified in this paper are boxed and designated. *Arrows* labeled *Tet* point to nucleotides modified by irradiating ribosomes with tetracycline alone (Steiner *et al.*, 1988). *Arrows* labeled *Puro* designate nucleotides photoaffinity labeled by p-azidopuromycin (Hall *et al.*, 1988). *Cam,* chloramphenicol; *Ery,* erythromycin; *Ver,* vernamycin B. Filled in symbols indicate nucleotides the mutation of which confers resistance to the respective antibiotic (reviewed in Noller, 1984; Douthwaite *et al.*, 1985; Sor and Fukuhara, 1984; Ettayebi *et al.*, 1985); open symbols designate antibiotics whose binding to the ribosome cause an alteration of reactivity of the respective nucleotide towards chemical modification (Moazed and Noller, 1987).

BP-Phe-tRNA has revealed a third region of minor reactivity (2503-A-U-G-U-2506) which is labeled at both P- and A-sites. Even though the bands are weak, they can be clearly distinguished due to the fact that they are well resolved from bands present in the non-irradiated control. Furthermore, nucleotides involved in binding of chloramphenicol (Ettayebi et al., 1985), erythromycin (Sor & Fukuhara, 1984), vernamycin B (Moazed & Noller, 1987), puromycin (Hall et al., 1988) and tetracycline (Steiner et al., 1988) are also located in this region (Fig. 3).

In other experiments the effects of a variety of antibiotics on the photoaffinity labeling of 23 S RNA have been investigated (Steiner et al., 1988). As expected, known inhibitors of the peptidyl transferase reaction such as chloramphenicol, tiamulin and sparsomycin strongly inhibited photoaffinity labeling at the P- and at the A-site. Vernamycin B and erythromycin which are considered to be inhibitors of translocation (Menninger & Otto, 1982) also reduced the photoaffinity reaction. Inhibitors of enzymatic aminoacyl-tRNA binding such as fusidic acid, gentamycin, kanamycin, streptomycin and thiostrepton showed no or only very weak effects. Tetracycline on the other hand was photocrosslinked owing to its intrinsic photoreactivity close to the site of BP-Phe-tRNA labeling (Steiner et al., 1988).

One caveat, intrinsic to the method of primer extension employed in this study remains, however. The identification of the residues photoaffinity labeled with BP-Phe-tRNA is an indirect one and is based on the observation that reverse transcriptase pauses or halts one base before a modified nucleotide as has been found for nucleotides methylated in ring positions and for hypermodified nucleotides (Youvan & Hearst, 1979; 1981). It should be stressed, however, that this is only a presumption since the nature of the photoproduct of the reaction between BP-Phe-tRNA and 23S RNA is unknown. We can therefore not exclude that the reason why we find doublets is simply that the photoproducts generated affect the enzyme in such a way that it pauses or halts either one or two bases before the modified nucleotide. By the same token it cannot be excluded that photoreaction at a unique nucleotide results in two or more products which affect the reverse transcription process differently. Due to the small amount of the modified nucleotide generated in the photoreaction this problem would be difficult to solve to any degree of satisfaction at the present time.

The sites labeled by BP-Phe-tRNA are all located in regions which are highly conserved in evolution (Barta et al., 1984). Furthermore, these areas are depicted as single strands in secondary structure models of 23S RNA (Noller, 1984). Even though these regions are probably involved in tertiary structure folding of 23 S RNA, they are nevertheless more likely to be able to react with -CCA termini than sequences in regions considered to be helical within the secondary structure map.

What might be the function of 23S RNA in the peptidyl transfer reaction? Since there is no requirement for additional energy, peptide bond formation might be brought about by arranging the two -CCA termini of the tRNAs in the proper position to allow transfer of the growing peptide chain. One could therefore speculate that 23S RNA might provide the framework for such an arrangement and in this way act as a catalytical center promoting the peptidyl transfer. Investigating the molecular mechanism of this reaction remains a challenging subject for future research.

5. Acknowledgement

This work was supported by a grant from the "Anton Dreher-Gedächtnisschenkung für Medizinische Forschung". We thank Tim Skern for critical reading of the manuscript and Klaus Hartmuth for stimulating discussions and help with the layout. The incisive and penetrating reflections of Z. Rattler were essential for this work.

6. References

Barta, A. & Kuechler, E. (1983) 'Part of the 23 S RNA Located in the 11 S RNA Fragment is a Constituent of the Ribosomal Peptidyltransferase Centre.' FEBS Lett. 163, 319 - 323.

Barta, A., Kuechler, E., Branlant, C., Sri Widada J., Krol, A. & Ebel, J.-P. (1975) 'Photoaffinity Labeling of 23S RNA at the Donor-Site of the Escherichia coli Ribosome.' FEBS Lett. 56, 170 - 174.

Barta, A., Steiner, G., Brosius, J., Noller, H.F. & Kuechler, E. (1984) 'Identification of a Site on 23 S Ribosomal RNA Located at the Peptidyltransferase Center.' Proc. Natl. Acad. Sci. USA 81, 3607 - 3611.

Blaas, D., Patzelt, E. & Kuechler, E. (1982) 'Cap-Recognizing Protein of Influenza Virus.' Virology 116, 339 - 348.

Blaas, D., Patzelt, E. & Kuechler, E. (1982) 'Identification of the Cap Binding Protein of Influenza Virus.' Nucleic Acids Res. 10, 4803 - 4812.

Bochkareva, E.S., Budker, V.G., Girshovich, A.S., Knorre, D.G. & Teplova, N.M. (1971) 'An Approach to Specific Labeling of Ribosome in the Region of Peptidyl-Transferase Center using N-Acylaminoacyl-tRNA with an Active Alkylating Grouping.' FEBS Lett. 19, 121 - 124.

Breitmeyer, J.B. & Noller, H.F. (1976) 'Affinity Labeling of Specific Regions of 23 S RNA by Reaction of N-Bromoacetyl-phenylalanyl-transfer RNA with Escherichia coli Ribosomes.' J. Mol. Biol. 101, 297 - 306.

Brosius, J., Dull, T.J. & Noller, H.F. (1980) 'Complete Nucleotide Sequence of a 23S Ribosomal RNA gene from Escherichia coli.' Proc. Natl. Acad. Sci. USA 77, 201 - 204.

Cantor, C.R. (1979) 'tRNA - Ribosome Interactions.' In: Transfer RNA: Structure, Properties and Recognition, eds. Schimmel, P.R., Soell, D. & Abelson, J.N. (Cold Spring Harbor Laboratory, Cold Spring Harbor, N.Y.), pp. 363 - 392.

Celma, M.L., Monro, R.E. & Vazquez, D. (1970) 'Substrate and Antibiotic Binding Sites at the Peptidyl Transferase Centre of E. coli Ribosomes.' FEBS Lett. 6, 273 - 277.

Cooperman, B.S. (1980) 'Functional Sites on the E. coli Ribosome as Defined by Affinity Labeling.' In: Ribosomes - Structure, Function and Genetics, eds. Chambliss, G., Craven, G.R., Davies, J., Davis, K., Kahan, L. & Nomura, M. (University Park Press, Baltimore), pp. 531 - 554.

Cooperman, B.S., Hall, C.C., Kerlavage, A.R., Weitzmann, C.J., Smith, J., Hasan, T. & Friedlander, J.D. (1986) 'Photoaffinity Labeling of *Escherichia coli* Ribosomes: New Approaches and Results. In: *Structure, Function, and Genetics of Ribosomes (Springer Series in Molecular Biology)*, eds. Hardesty, B. & Kramer, G. (Springer Verlag, N.Y.), pp. 362 - 378.

Douthwaite, St., Prince, J.B. & Noller, H.F. (1985) 'Evidence for Functional Interaction between Domains II and V of 23S Ribosomal RNA from an Erythromycin-Resistant Mutant.' *Proc. Natl. Acad. Sci. USA* **82**, 8330 - 8334.

Ettayebi, M., Pasad, S.M. & Morgan, E.A. (1985) 'Chloramphenicol - Erythromycin Resistance Mutation in a 23S rRNA Gene of *Escherichia coli*.' *J. Bacteriol.* **162**, 551 - 557.

Fernandez-Munoz, R., Monro, R.E., Torres-Pinedo, R. & Vazquez, D. (1971) 'Substrate- and Antibiotic -Binding Sites at the Peptidyl- Transferase Centre of *Escherichia coli* Ribosomes: Studies on the Chloramphenicol, Lincomycin and Erythromycin Sites.' *Eur. J. Biochem.* **23**, 185 - 193.

Galardy, R.E., Craig, L.C., Jamieson, J.D. & Printz, M.P. (1974) 'Photoaffinity Labeling of Peptide Hormone Binding Sites.' *J. Biol. Chem.* **249**, 3510 - 3518.

Galardy, R.E., Craig, L.C. & Printz, M.P. (1973) 'Benzophenone Triplet: A New Photochemical Probe of Biological Ligand-Receptor Interactions.' *Nature (New Biol.)* **242**, 127 - 128.

Gillam,. I.C. & Tener, G.M. (1971) 'The Use of BD-Cellulose in Separating Transfer RNAs.' *Meth. Enzymol.* **20**, 55 - 70.

Guerrier-Takada, C., Gardiner, K., Marsh, T., Pace, N. & Altman, S. (1983) 'The RNA Moiety of Ribonuclease P is the Catalytic Subunit of the Enzyme.' *Cell* **35**, 849 - 857.

Hall, C.C., Johnson, D. & Cooperman, B.S. (1988) ' [^{3}H]-p-Azidopuromycin Photoaffinity Labeling of *Escherichia coli* Ribosomes: Evidence for Site-Specific Interaction at U-2504 and G-2502 in Domain V of 23S Ribosomal RNA.' *Biochemistry* **27**, 3983 - 3989.

Kaltschmidt, E. & Wittmann, H.G. (1970) 'Ribosomal Proteins. XII: Number of Proteins in Small and Large Ribosomal Subunits of *Escherichia coli* as Determined by Two-Dimensional Gel Electrophoresis.' *Proc. Natl. Acad. Sci. USA* **67**, 1276 - 1282.

Kruger, K., Grabowski, P.J., Zaug, A.J., Sands, J., Gottschling, D.E. & Cech, T.R. (1982) 'Self-Splicing RNA: Autoexcision and Autocyclization of the Ribosomal RNA Intervening Sequence of *Tetrahymena*.' *Cell* **31**, 147 - 157.

Kuechler, E. (1978) 'Affinity Labels for tRNA and mRNA Binding Sites on Ribosomes.' In: *Theory and Practice in Affinity Techniques*, eds. Sundaram, P.V. & Eckstein, F. (Academic Press Inc. Ltd., London), pp. 151 - 168.

Kuechler, E. & Barta, A. (1977) 'Aromatic Ketone Derivatives of Aminoacyl-tRNA as Photoaffinity Labels for Ribosomes.' *Meth. Enzymol.* **46**, 676 - 683.

Kuechler, E. & Ofengand, J. (1979) 'Affinity Labeling of tRNA Binding Sites on Ribosomes.' In: *Transfer RNA: Structure, Properties and Recognition,* eds. Schimmel, P.R., Soell, D. & Abelson, J.N. (Cold Spring Harbor Laboratory, Cold Spring Harbor, N.Y.), pp. 413 - 444.

Kuechler, E., Steiner, G. & Barta, A. (1988) 'Photoaffinity Labeling of Peptidyltransferase.' *Meth. Enzymol.* **164,** 361 - 372.

Maden, B.E.H., Traut, R.R. & Monro, R.E. (1968) 'Ribosome Catalyzed Peptidyltransfer: The Polyphenylalanine System.' *J. Mol. Biol.* **35,** 333 - 345.

Menninger, J.R. & Otto, D.P. (1982) 'Erythromycin, Carbomycin, and Spiramycin Inhibit Protein Synthesis by Stimulating Dissociation of Peptidyl-tRNA from Ribosomes.' *Antimicrob. Agents Chemother.* **21,** 811 - 818.

Moazed, D. & Noller, H.F. (1987) 'Chloramphenicol, Erythromycin, Carbomycin, and Vernamycin B Protect Overlapping Sites in the Peptidyl Transferase Region of 23S Ribosomal RNA.' *Biochimie* **69,** 879 - 884.

Monro, R.E. (1967) 'Catalysis of Peptide Bond Formation by 50 S Ribosomal Subunits from *Escherichia coli.*' *J. Mol. Biol.* **26,**147 - 151.

Nierhaus, K.H. (1982) 'Structure, Assembly and Function of Ribosomes.' *Current Topics in Microbiol. and Immun.* **97,** 82 - 155.

Noller, H.F. (1984) 'Structure of Ribosomal RNA.' *Ann. Rev. Biochem.* **53,** 119 - 162.

Oakes, M., Henderson, E., Scheinman, A., Clark, M. & Lake, J.A. (1986) 'Ribosome Structure, Function, and Evolution: Mapping Ribosomal RNA, Proteins, and Functional Sites in Three Dimensions'. In: *Structure, Function, and Genetics of Ribosomes (Springer Series in Molecular Biology),* eds. Hardesty, B. & Kramer, G. (Springer Verlag, N.Y.), pp. 47 - 67.

Ofengand, J. (1980) 'The Topography of tRNA Binding Sites on the Ribosome.' In: *Ribosomes - Structure, Function, and Genetics,* eds. Chambliss, G., Craven, G.R., Davies, J., Davis, K., Kahan, L. & Nomura, M. (University Park Press, Baltimore), pp. 497 - 529.

Ofengand, J., Ciesiolka, J., Denman, R. & Nurse, K. (1986) 'Structural and Functional Interactions of the tRNA - Ribosome Complex.' In: *Structure, Function, and Genetics of Ribosomes (Springer Series in Molecular Biology),* eds. Hardesty, B. & Kramer, G. (Springer Verlag, N.Y.), pp. 473 - 494.

Patzelt, E., Blaas, D. & Kuechler, E. (1983) 'Cap Binding Proteins Associated with the Nucleus.' *Nucleic Acids Res.* **11,** 5821 - 5835.

Sonenberg, N., Wilchek, M. & Zamir, A. (1975) 'Identification of a Region in 23S rRNA Located at the Peptidyl Transferase Center.' *Proc. Natl. Acad. Sci. USA* **72,** 4332 - 4336.

Sor, F. & Fukuhara, M. (1984) 'Erythromycin and Spiramycin Resistance Mutations of Yeast Mitochondria: Nature of the *rib 2* Locus in the Large Ribosomal RNA Gene.' *Nucleic Acids Res.* **12,** 8313 - 8318.

Southern, E.M. (1975) 'Detection of Specific Sequences among DNA Fragments Separated by Gel Electrophoresis.' *J. Mol. Biol.* **58**, 503 - 517.

Steiner, G., Kuechler, E. & Barta, A. (1988) 'Photoaffinity Labeling at the Peptidyl Transferase Center Reveals Two Different Positions for the A- and P-Sites in Domain V of 23S rRNA.' *EMBO J.*, in press.

Stoeffler, G. & Stoeffler-Meilicke, M. (1986) 'Immuno Electron Microscopy on *Escherichia coli* Ribosomes.' In: *Structure, Function and Genetics of Ribosomes (Springer Series in Molecular Biology)*, eds. Hardesty, B. & Kramer, G. (Springer Verlag, N.Y.), pp. 28 - 46.

Tischendorf, G.W., Zeichardt, M. & Stoeffler, G. (1974) 'Determination of the Location of Proteins L14, L17, L18, L19, L22, and L23 on the Surface of the 50S Ribosomal Subunit of *Escherichia coli* by Immune Electron Microscopy.' *Mol. Gen. Genet.* **134**, 187 - 208.

Turro, N.J. (1967) *Molecular Photochemistry* (Benjamin, N.Y.) p. 44 - 67.

Youvan, D.C. & Hearst, J.E. (1979) 'Reverse Transcriptase Pauses at N2-Methylguanine during *in vitro* Transcription of *E. coli* 16 S Ribosomal RNA.' *Proc. Natl. Acad. Sci. USA* **76**, 3751 - 3754.

Youvan, D.C. & Hearst, J.E. (1981) 'A Sequence from *Drosophila melanogaster* 18S rRNA Bearing the Conserved Hypermodified Nucleoside amΨ: Analysis by Reverse Transcription and High-Performance Liquid Chromatography.' *Nucleic Acids Res.* **9**, 1723 - 1741.

STUDYING THE CYTOSKELETON BY LABEL TRANSFER CROSSLINKING:
USES AND LIMITATIONS

Martin Alexander Schwartz
Dept Cellular and Molecular Physiology
Harvard Medical School
Boston MA USA

ABSTRACT. HAHS is a new type of photoactivated crosslinking reagent
that can transfer a radiolabel from an initially labeled protein to
its neighbors, due to the presence of a radiolabeled phenyl azide
distal to a cleavable ester bond. HAHS has identified several
protein interactions in the red blood cell cytoskeleton. However,
when used to study how actin binds to integral plasma membrane
proteins from Dictyostelium discoideum, only actin-actin crosslinks
formed; no membrane protein was identified. Several nonradioactive
phenyl azide crosslinkers gave similar results. Nonspecific labeling
with a water-soluble precursor of HAHS showed that proteins varied by
a factor of 20 in their susceptibility to reaction with nitrenes,
with actin being the most reactive protein tested. Correspondingly,
amino acids vary by a factor of $>10^5$ in their reactivity, with
cysteine and the aromatic residues being the most reactive. Thus,
chemical reactivity of proteins, probably because of differences in
exposed amino acids, may significantly bias the outcome of
crosslinking experiments.

1. Introduction

     Understanding in molecular detail how cytoskeletal proteins
interact with the plasma membrane, and how these interactions
regulate membrane events has been a continuing area of interest in
cell biology (reviewed by Geiger, 1986). Solving this problem
amounts to working out the complex and sometimes dynamic binding
interactions among a large number of proteins. This sort of problem
is made more difficult by the shortage of techniques that can provide
high resolution information about complex protein interactions.
     In order to study systems of this sort, we have developed a new
chemical crosslinking technique called label transfer crosslinking.
This technique uses reagents that can transfer a radiolabeled group
from an initially labeled protein to nearby molecules. The first such
reagent (Chong and Hodges, 1982,Schwartz et al, 1982) contained
disulfide bonds, which prevented their use with cytoplasmic proteins

P. E. Nielsen (ed.), Photochemical Probes in Biochemistry, 157–168.
© 1989 by Kluwer Academic Publishers.

that generally have many free sulfhydryl groups and require a reducing environment. This paper will describe studies carried out with a second reagent (Schwartz, 1985) that uses a new cleavable group that is resistant to disulfide interchange. Our results indicate that label transfer crosslinking is a useful alternative to other photoactivated crosslinking approaches. However, they also show that the usefulness of phenyl azide crosslinking reagents in general is limited by their high chemical specificity for reactive sites on target proteins. Although phenyl nitrenes are generally considered to be comparatively nonspecific (Ji, 1979, Staros, 1980), we show that they react preferentially with particular amino acids and with particular proteins.

2. Materials and Methods

2.1 Chemicals

HAHS (1-(N-(2-hydroxy-5-azidobenzoyl)-2-aminoethyl)-4-(N-hydroxysuccinimidyl)-succinate), and its precursor HAHS amide (ethanolamine 5-azido-2-hydroxy benzoic acid), were synthesized as described (Schwartz, 1985). The nonradioacive crosslinking reagents ANB-NOS (N-5-azidobenzoyloxysuccinimide); EADB (ethyl-4-azidophenyl 1,4-dithiobutyrimidate); MABI (methyl-4-azidobenzimidate); sulfo-SADP (sulfosuccinimidyl-4-azidophenyl-dithioproprionate); MBS (m-maleimido benzoylsuccinimide); and BSS (bis-sulfosuccinmidyl suberate) were obtained from Pierce Chemicals, Rockville IL, USA. EDC (1-ethyl-3-(3-dimethylaminopropyl) carbodiimide) was from Aldrich Chemicals, Milwaukee WI, USA.

2.2 Proteins and Biochemicals

Rabbit muscle actin, purified by gel filtration, was prepared as described (Schwartz and Luna, 1986). Plasma membranes from Dictyostelium discoideum were a generous gift from Dr. E.J. Luna. Gelsolin was a generous gift from Dr. H.L. Yin. Protein A, rabbit IgG, BSA, ovalbumin, cytochrome c, ribonuclease A and amino acids were purchased from Sigma Chemicals, St. Louis MO, USA. Fibronectin was purified from human plasma as described (Engvall and Ruoslatti, 1977), and collagen was purified from rat tail tendons as described (Lee et al, 1984).

2.3 Label Transfer with HAHS-Protein A.

Experiments were carried out as described in Schwartz, 1985. Briefly, freshly iodinated HAHS was reacted with 1 mg/ml protein A in PBS, at a 1:1 mole ratio, and the protein gel filtered to remove unreacted reagent. It was incubated for 30 min in PBS with equimolar IgG, and photolyzed for 30s with a 100W long wave UV lamp. In some cases samples were treated with 20 mM NaOH from a 1M stock, and neutralized with 1M HCl. Samples were analyzed by SDS-PAGE and autoradiography. For experiments with amino acids, the indicated amino acid was added to concentrations of 1×10^{-6} to 1 M before photolysis. After SDS-PAGE, the gels were stained, the IgG heavy chain bands cut out and the radioactivity determined by gamma counting. The concentration that gave half maximal inhibition of label transfer was determined graphically.

2.4 Label Transfer with HAHS-Actin

Actin in 1 mM pH 7.0 phosphate/0.2 mM ATP was reacted with HAHS at a 1:0.2 mole ratio of protein:HAHS, and gel filtered to remove unreacted HAHS. Actin was bound to dictyostelium membranes in polymerizing buffer (50mM KCl, 2mM $MgCl_2$, 50 uM $CaCl_2$, 10 mM PIPES pH 7.0, 200 uM PMSF, 8 ug/ml leupeptin) lacking DTT, essentially as described (Schwartz and Luna, 1986). Samples were photolyzed and analyzed as before.

2.5 Crosslinking with Nonradioactive Crosslinkers

Actin that was radiolabeled with Bolton and Hunter reagent (Schwartz and Luna, 1986) was labeled with the indicated nonradioactive crosslinking reagents under conditions identical to those for HAHS. An initial mole ratio of 1:1 was used, which gives a final ratio of approximately 0.5 reagent per protein. Actin at 50 ug/ml was bound to dictyostelium membranes as before, and photolyzed. In some cases, the photolyzed samples were treated with 0.1 M NaOH and the membranes sedimented to remove peripheral proteins and actin that was not covalently bound. Samples were analyzed by SDS-PAGE on a 6% gel, and autoradiography. For samples without membranes, actin at 0.2 mg/ml was polymerized in the same buffer for 30 min, then photolyzed and analyzed as before.

3. Results and Discussion

3.1 Lable Transfer Crosslinking

The scheme by which HAHS transfers a radiolabel from an initial protein to nearby components is shown in figure 1. First, purified protein 1 is incubated with HAHS in the dark, where the succinimide ester group of HAHS reacts with amino groups on the protein to form an amide bond (step 1). If, at this point, the derivative is exposed to pH 12 or higher for 1 min (step 2), the internal ester bond is hydrolyzed (step 3). This cleavage is about 100x faster than the base-catalyzed hydrolysis of a normal aliphatic ester, suggesting that hydrolysis may be catalyzed by the deprotonated phenol anion. The HAHS ester shows normal stability at pH 7.

Usually, the protein derivative is reconstituted into a biological system where it binds to other components such as protein 2 (step 4). Upon exposure to UV light, the azido group is photolyzed to form a nitrene (step 5), which reacts rapidly with nearby molecules. In favorable cases it will react with protein 2 to form a crosslink (step 6). If the crosslink is cleaved with base as before, the radiolabeled phenyl group remains associated with protein 2 (step 7). Thus, the radiolabel has been transferred from protein 1 to protein 2.

160

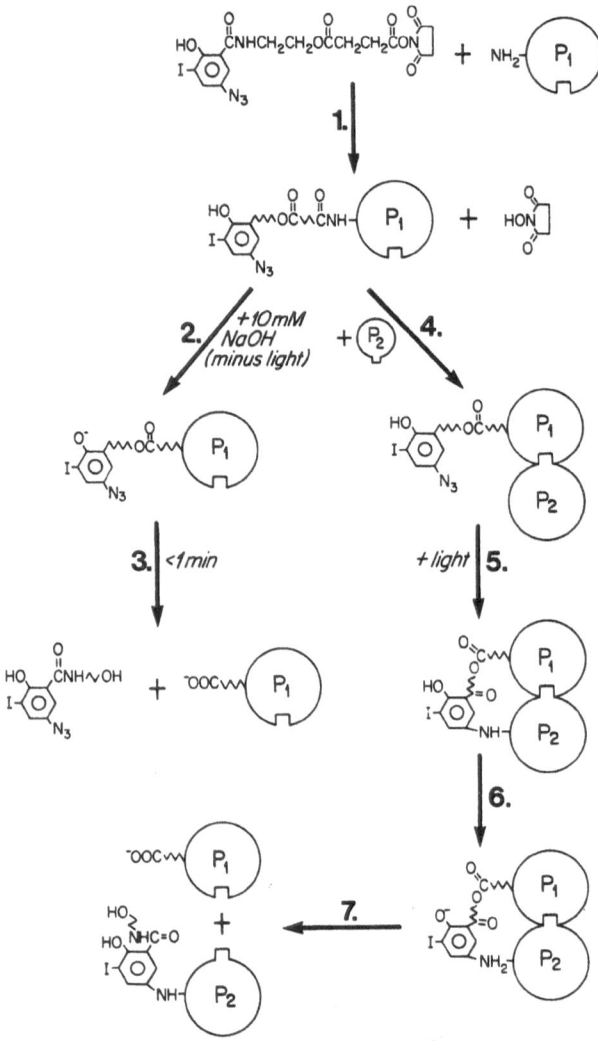

Figure 1.   Scheme for Label Transfer.
HAHS reacts with a purified protein P_1 to form a covalent derivative
(step 1).   If at this stage it is treated with base (step 2), the
phenol is deprotonated and the ester bond is rapidly hydrolyzed (step
3).   Normally, the protein is reconstituted into a biological system
where it will bind to other components such as P_2 (step 4).
Photolysis leads to formation of a crosslink to P_2 (step 5).   The
phenol can be deprotonated either by treatment with base or at
neutral pH due to the change in the substituent on the phenyl ring
(step 6), leading to hydrolysis of the ester bond (step 7), and
leaving the radiolabeled group attached to P_2.

After photolysis, hydrolysis of the ester group also occurs at
neutral pH. Thus, label transfer can take place without the need for
an additional step, though base treatment improves the efficiency of
cleavage. We hypothesize that hydrolysis at neutral pH is most
likely due to structural changes that follow photolysis. The nitrene
is most commonly converted to an amine, which, when *para* to the
phenolic hydroxyl, substantially lowers the pK of the phenol.
Deprotonation of the phenol at neutral pH leads to catalytic
hydrolysis of the ester. Hydrolysis at neutral pH after photolysis
takes on the order of several hours at 4°, and can still be
accelerated by base.

3.2 Testing HAHS

A demonstration of this scheme is shown in figure 2, using
protein A from staph aureus as protein 1 and rabbit IgG as protein 2.
Protein A was reacted with HAHS in the dark, separated from unreacted
HAHS, and mixed with IgG. Some samples were photolyzed, some were
treated with 20 mM NaOH, and all were analyzed by SDS-PAGE and
autoradiography.
Figure 2 shows that initially only the protein A was radiolabeled.
Before photolysis, most of the radiolabel was removed by base. After
photolysis, higher molecular weight complexes formed, and after
cleavage with base, most of the radiolabel was found on the IgG heavy
chain. In this case, 27% of the starting radioactivity was found on
the IgG heavy chain. Also note that in lane 2, after photolysis but
without treatment with base, a substantial amount of the
radioactivity is found in the IgG heavy chain, illustrating the slow
hydrolysis at neutral pH.
To test the specificity of label transfer, HAHS-protein A bound
to IgG was photolyzed in the presence of a 100x excess of unlabeled
protein A. Label transfer was completely blocked. Photolysis was
also carried out in the presence of a 20-fold mass excess of
ribonuclease A, ovalbumin, or bovine serum albumin. Label transfer
to these noninteracting proteins occurred to a low but detectable
extent. Correcting for the mass ratio of noninteracting protein to
IgG heavy chain, it was calculated that IgG is preferred by a factor
of 200:1 over ovalbumin and ribonuclease, and by 30:1 over BSA.
When incubated with cytochrome c, however, label was transferred
in a light-independent manner. Note that a 100-fold excess of
cytochrome c was used in order to demonstrate this point more
clearly. The mechanism for this transfer is unknown, though it
presumably involves the catalytic site of the enzyme. No other
proteins have been observed to exhibit this behavior.
We have verified the usefulness of label transfer crosslinking
with the red blood cell cytoskeleton as a model system. It has been
shown that actin filaments bind to the red blood cell membrane via an
indirect linkage: actin binds to spectrin and band 4.1 to form a
ternary complex; spectrin binds to ankyrin which in turn binds to the
integral membrane protein band 3 (reviewed by Branton et al, 1981).
We found that HAHS-spectrin reconstituted onto red cell membrane
vesicles transferred the label to ankyrin and band 3, and transfer

was specific since it was abolished by adding excess unlabeled spectrin or by carrying out the incubations in a low salt buffer that inhibits binding (Schwartz, 1985).

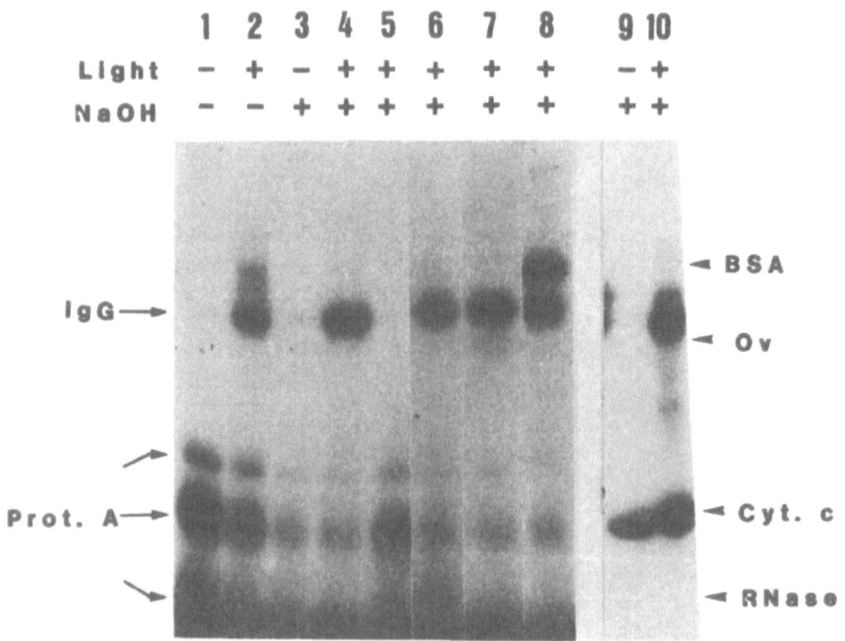

Figure 2. Label Transfer with HAHS-Protein A.
Exposure to UV light or treatment with 20 mM NaOH are indicated by a + or - at the top of the gel. HAHS-Proein A was bound to nonspecific rabbit IgG. Lane 5 contained a 100-fold molar excess of unlabeled protein A. Lane 6-8 contained 20-fold mass excesses of RNase, ovalbumin or BSA, respectively. Lanes 9 and 10 contain a 100-fold mass excess of cytochrome c.

HAHS has also been used to map the domain of spectrin to which band 4.1 binds. When HAHS-4.1 bound to spectrin was photolyzed and cleaved, label was transferred to the spectrin, which was then isolated, subject to partial proteolytic cleavage and the fragments analyzed on 2-dimensional peptide maps. A small number of unique peptides from near the N-terminus of the spectrin B-chain became labeled (Becker and Lux, 1984).

3.3 Actin-Membrane Interactions
More recently, we have attempted to identify actin binding proteins in isolated plasma membranes from dictyostelium Discoideum. In this system, integral membrane proteins bind actin directly (Luna et al, 1981). Binding studies with purified membranes and

radiolabeled actin indicate that binding is cooperative, and half maximal binding occurs at 3×10^{-7} M actin. Only filamentous actin binds with measurable affinity, (Schwartz and Luna, 1988).

We therefore attempted to identify the actin-binding integral membrane protein(s) by label transfer crosslinking with HAHS. Purified actin was labeled with HAHS and incubated with dictyostelium membranes under binding conditions. Negative controls were included in which the HAHS-actin was photolyzed before mixing with membranes and in which membranes were heat denatured to inhibit binding. As a positive control, actin was incubated with gelsolin, a 90 kD actin-capping protein that binds actin with very high affinity in a Ca^{++} dependent manner.

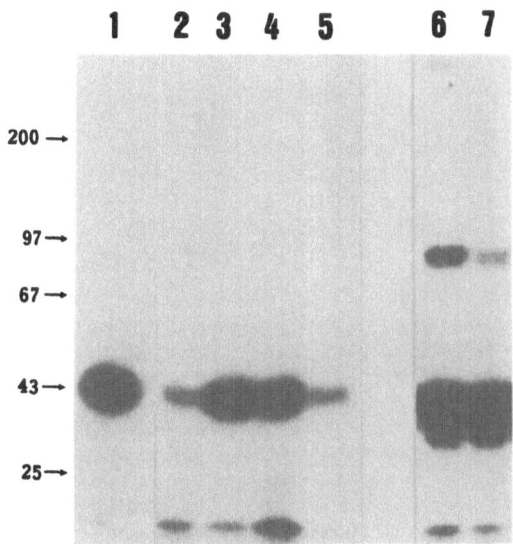

Figure 3.   Label Transfer with HAHS-Actin
Samples contained 15 ug/ml HAHS-actin, and where indicated, 100 ug/ml dictyostelium membranes or a 2-fold molar excess of gelsolin.   All samples were photolyzed and treated with NaOH, except where otherwise indicated.   Lane 1, HAHS-actin alone.   Lane 2, +membranes, -light. Lane 3, +membranes.   Lane 4, HAHS-actin photolyzed before adding membranes.   Lane 5, +heat-denatured membranes. Lane 6, +gelsolin and 50 uM calcium.   Lane 7, +gelsolin and 4 mM EGTA.

Figure 3 shows the results of such an experiment. After photolysis and cleavage of HAHS-actin bound to dictyostelium membranes, actin itself was heavily labeled but no additional bands are present that are not also seen in the negative controls. In the presence of gelsolin, there was some Ca^{++} dependent labeling of a 90 kD band, but actin itself was still very heavily labeled.

A number of attempts were made to improve the crosslinking of actin to membrane proteins. Actin was used at various concentrations in the hope that at low concentrations where assembly into filaments would be minimized, or that at high concentrations where binding sites were saturated, some labeling of membrane proteins would occur. In some experiments, the membrane-bound actin was dissociated after photolysis and the membranes washed, in the hope that removing the heavily labeled actin would reduce the background and allow labeled membrane proteins to be detected. In no case was a membrane protein specifically labeled by HAHS-actin (not shown).

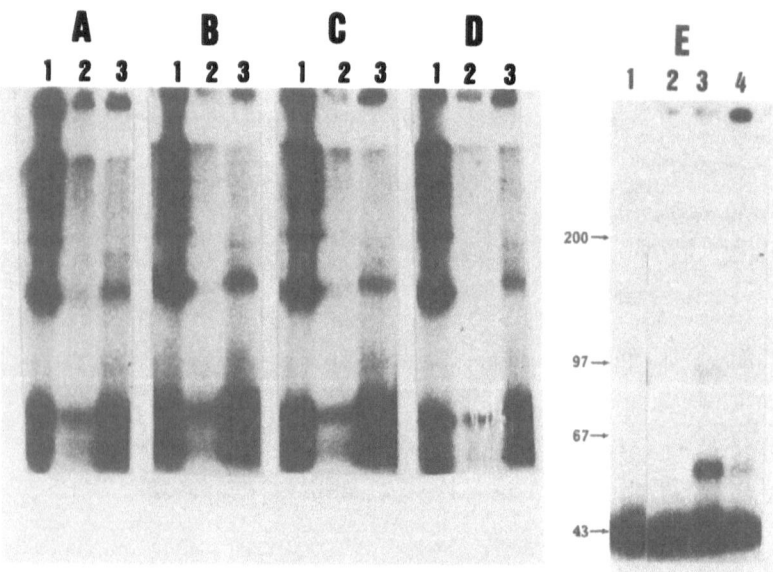

Figure 4. Nonradioactive Crosslinkers.
A-D: ^{125}I-actin, labeled with the indicated photoactivated crosslinker, was incubated for 30 min alone at 200 ug/ml, or at 50 ug/ml with dictyostelium membranes at 200 ug/ml. Samples were photolyzed and analyzed by SDS-PAGE and autoradiography without cleavage of the crosslinks.
Lane 1, actin-xl alone. Lane 2, actin-xl bound to membranes, photolyzed, then extracted with 0.1M NaOH. Lane 3, actin-xl bound to membranes. A, ANB-NOS. B, EADB. C, MABI. D, Sulfo-SADP.
E: ^{125}I-actin, alone or bound to membranes as above, was treated with 10 mM EDC for 30 min, except where indicated otherwise. Lane 1, actin alone. Lane 2, +membranes, without EDC. Lane 3, +membranes. Lane 4, +membranes that had been stripped with 0.1M NAOH.

It was apparent from the heavy labeling of actin that internal crosslinking or actin-actin crosslinking must be highly favored. To see if other phenyl azide crosslinkers with slightly different structures behaved similarly, a number of commercially availble nonradioactive photosensitive crosslinking reagents were tested. Radioactive actin was labeled with four such crosslinking agents at about 0.5 reagent per actin, then was polymerized or bound to membranes and photolyzed. Figure 4, panels A-D, show the samples after SDS-PAGE and autoradiography. It is apparent that all four reagents crosslinked actin into multimers with very high efficiency. This result indicates that actin-actin crosslinks form very readily and that several different phenyl azides give identical results, independent of substituents on the phenyl ring. However, there were no bands other than actin multimers that could be detected when membrane-bound actin was photolyzed, even after most of the actin was extracted from the membranes with NaOH.

3.4 Chemical Specificity of the Nitrene

Nitrenes have generally been regarded to be nonspecific crosslinkers, capable of inserting into the peptide backbone and not requiring particular sites on the protein (Ji, 1979, Staros, 1980). However, to determine why actin crosslinks so readily to itself but not to membrane proteins, experiments designed to test the chemical specificity of the HAHS phenyl nitrene were carried out. First, we attempted to determine whether all proteins reacted equally well with phenyl nitrenes. A water-soluble precursor of HAHS (shown below), was iodinated and photolyzed in the presence of various proteins at 1 mg/ml. The proteins were then electrophoresed on an SDS polyacrylamide gel, stained, the bands cut out and counted for radioactivity.

HAHS-amide

Table I

Protein	% radiolabel on the protein	labeling relative to collagen
protein A	2.65	1.50
fibronectin	2.07	1.16
ovalbumin	6.18	3.5
actin	33.6	18.9
BSA	22.1	12.4
IgG	2.05	1.15
collagen	1.78	1.00

Table I shows the results of such an experiment. It is apparent that proteins vary widely in their reactivity. Actin is the most reactive of the proteins tested, being about 20 times more reactive than collagen. This difference is clearly large enough to bias the outcome of a crosslinking experiment.

To determine why some proteins are much more reactive than others, the reactivity of different amino acids was tested by measuring their ability to quench label transfer from protein A to IgG. HAHS-protein A bound to IgG was photolyzed in the presence of various amino acids over a range of concentrations. The samples were separated by SDS-PAGE, and the IgG heavy chain band cut out and counted. Table II shows the results of such an experiment, expressed as the concentration of amino acid that gave half-maximal inhibition of label transfer. It is apparent that amino acids differ in reactivity towards phenyl nitrenes by a factor of $>10^5$. It therefore seems reasonable that proteins could vary widely in their reactivity, depending on the number of exposed reactive amino acids. This idea is consistent with the fact that collagen, which was the least reactive protein, has an unusually low content of cysteine and aromatic amino acids, which are the most reactive residues.

Table II	amino acid	conc. half-max inhibition (M)
	cysteine	6×10^{-6}
	cystine	3×10^{-2}
	tyrosine	5×10^{-2}
	tryptophan	1×10^{-3}
	phenylalanine	5×10^{-2}
	histidine	1×10^{-1}
	lysine	1×10^{-1}
	glycine	$>1M$
	arginine	$>.1M$

This argument led us to hypothesize that actin failed to crosslink to the dictyostelium membrane binding sites because actin is much more reactive than the membrane protein. In effect, actin itself effectively competes for the nitrene. About this time, the major membrane binding site for actin was shown by affinity chromatography to be a 17 kD transmembrane protein (Wuestehube and Luna, 1987). Thus, its small size (which is even smaller when one considers that only the cytoplasmic domain should be available to the actin) probably contributes to its lack of reactivity.

Our hypothesis implies that a crosslinker that reacts poorly with actin itself should have a much better chance to crosslink actin to the membrane site. Therefore, several nonphotoactivated crosslinking reagents with different chemical specificities were tested for their ability to crosslink F-actin in solution into multimers. MBS, which crosslinks amino groups to sulfhydryl groups, BSS, which links amino groups to amino groups, and EDC, which links carboxyl groups to amino, hydroxyl or sulfhydryl groups were tested.

Only EDC, a water soluble carbodiimide, failed to efficiently generate actin multimers (not shown).

Next, actin bound to dictyostelium membranes was incubated with EDC, and the products analyzed by SDS-PAGE and autoradiography. As shown in figure 4E, a considerable amount of a single major crosslinked complex formed. The complex also formed using membranes that had been stripped of peripheral proteins with 0.1 M NaOH. It was inhibited by adding excess unlabeled actin or by heat denaturing the membranes(not shown). The apparent molecular weight of the complex was 60 kD, consistent with the crosslinking of 43 kD actin to a 17 kD membrane protein. Thus, efficient crosslinking to the membrane did occur once actin-actin crosslinking was reduced. Although more examples are needed to establish a firm correlation, the data is clearly consistent with the competition hypothesis.

3.5 Conclusions

Label transfer crosslinking was HAHS has several advantages over conventional photoactivated crosslinking techniques. It avoids the need to analyze or immune precipitate large crosslinked complexes. It allows the exact fate of the phenyl azide group to be followed, for example to distinguish between effective crosslinking to a nearby protein, internal crosslinking to the same protein, and ineffective reaction where crosslinking does not occur. And it provides for efficient detection by ^{125}I without the need to radiolabel the entire sample.

Label transfer crosslinking with HAHS has been useful in some systems, for example with the red blood cell cytoskeleton. It was also essential to the studies reported here, which could not have been carried out using conventional crosslinkers. However, it failed to identify an actin binding protein in dictyostelium membranes. Instead, only self-labeling of actin occurred. This result can be explained by the fact that actin is an unusually reactive protein, so that actin-actin crosslinks are favored over actin-membrane crosslinks. This property is shared by all other phenyl azide crosslinkers tested. Only when a crosslinking reagent with a different chemical specificity was used, one that did not generate actin-actin crosslinks, was actin crosslinked to a membrane protein.

These results have several implications for the use of phenyl azide crosslinkers in general. They imply that crosslinking experiments can be strongly biased by the reactivity of the proteins. A significant interaction may be missed if the protein is of low reactivity, and a minor or nonspecific interaction may be overrepresented if the protein is of high reactivity. Thus, crosslinkers with different chemical specificities should be tested when investigating a new system. Perhaps photochemical probes with different specificities could be developed.

Also, our data suggest that the yield of crosslinked products in any reaction is the result of a competition. Once generated, a nitrene will tend to crosslink to the most reactive moiety nearby, so that the proximity of a highly reactive group will diminish crosslinking to less reactive groups. Thus the outcome of an experiment may conceivably depend as much on the geometry and reactivity of the proteins, or on secondary interactions, as it does on the characteristics of the specific binding interaction.

In summary, these results suggest that considerable caution is required in the design and interpretation of photochemical experiments with proteins, and that progress in this field will depend on further research into the chemistry of the reactions with proteins.

Acknowledgements
    I would like to thank Imelda Daley for her skilled technical assistance. I am especially grateful to Dr. Elizabeth J. Luna for the gift of dictyostelium membranes, and for her advice and assistance during that portion of the work.

References
1. Becker, P., and Lux, S.E. 1984. J. Cell Biol 99, 113a (abstract).
2. Branton, D., Cohen, C.M., and Tyler, J. 1981. Cell 24, 24-32.
3. Chong, P.C.S. and Hodges. R.S. 1981. J. Biol. Chem. 256, 5064-5070.
4. Engvall, E. and Ruoslatti, E. 1977. Int. J. Cancer 20, 1-5.
5. Geiger, B. 1983. Biocchim. Biophys. Acta 737, 305-342.
6. Ji, T.H. 1979. Biochim. Biophys. Acta 559, 39-69.
7. Lee, Y., PArry, G. and Bissell, M. 1984. J. Cell Biol. 98, 146-155.
8. Luna, E.J., Fowler, V.M., Swanson, J., Branton, D. Taylor, D.L. 1981. J. Cell Biol. 88, 396-409.
9. Schwartz, M.A., Das, O.P. and Hynes, R.O. 1982. J. Biol. Chem. 257, 2343-2349.
10. Schwartz, M.A. 1985. Anal. Biochem. 149, 142-152.
11. Schwartz, M.A. and Luna, E.J. 1986. J. Cell Biol. 102, 2067-2075.
12. Schwartz, M.A. and Luna, E.J. 1988. J. Cell Biol. 107, 201-209.
13. Staros J.V. TIBS Dec. 1980. 320-322.
14. Wuestehube,L. and Luna, E.J. 1987. J. Cell Biol. 105, 1741-1752.

RecA-Directed Hybridization of Psoralen-Monoadducted DNA
      Oligonucleotides to Duplex Targets

Suzanne Cheng, Howard B. Gamper[¶], and John E. Hearst
Department of Chemistry, University of California and the
Division of Chemical Biodynamics, Lawrence Berkeley
Laboratory, Berkeley, CA 94720

ABSTRACT. RecA protein can be used to introduce a psoralen-
monoadducted DNA oligonucleotide into a duplex target
site-specifically. Irradiation of the three-stranded DNA complex
with long wavelength ultraviolet light results in formation of a
site-specific crosslink between the inserted oligomer and its
complement. Crosslinked complexes have been formed between pUC19
plasmid and modified oligomers ranging from 30 to 107 nucleotides in
length. The efficiency of this two-step process depends upon
several factors, including the nucleotide cofactor for RecA, the
length of the oligomeric substrate, and the position of the psoralen
adduct relative to the 5'-end of the oligomer. Absolute homology is
not required.

INTRODUCTION

Psoralens are photochemical crosslinking agents that have been used
extensively as probes of nucleic acid structure (reviewed by Cimino
et al., 1985). Adduct formation requires the psoralen molecule, a
furocoumarin with two photochemically reactive ends, to intercalate
between stacked base pairs. The preferential reactivity of
psoralens with thymidines at duplex 5'-TpA sites in B-form DNA
enables selective placement of crosslinks. Because all such sites
would potentially be available to free psoralen, site-specificity
and a high efficiency of crosslinkage per site are best achieved
using oligonucleotides that contain a unique psoralen-monoadducted
thymidine. Here we demonstrate the ability of the RecA protein from
Escherichia coli to direct the hybridization of psoralen-
monoadducted DNA oligonucleotides (fewer than 110 residues) to
target sequences in a duplex plasmid. The oligonucleotide and its

[*]This work was supported by NIH grant GM11180 and by DOE, Office of
Health and Environmental Research Contract DE-AC03-76F0093.

[¶]Present address: Microprobe Corporation, 1725 220th Street SE, No.
104, Bothell, WA 98021.

P. E. Nielsen (ed.), Photochemical Probes in Biochemistry, 169–177.
© 1989 by Kluwer Academic Publishers.

complement can then be covalently crosslinked through the psoralen by irradiating the sample with long wavelength ultraviolet light.

Hybridization between a single-stranded DNA molecule and its complementary sequence within a duplex DNA molecule is one of several reactions catalyzed by the RecA protein (reviewed by Radding, 1982; Cox and Lehman, 1987). This process of homologous pairing results in displacement of the duplex strand that is homologous to the single-stranded molecule. A three-stranded intermediate forms, comprised of the invading strand, its complement, and the displaced strand. If the inserted single strand is a psoralen-monoadducted oligomer, an irradiation to effect crosslink formation will yield site-specifically and covalently stabilized three-stranded DNA complexes.

We prepared a series of eight modified oligonucleotides to first examine the minimum length required for such a substrate in the RecA-mediated pairing reaction. One estimate was that between 30 and 150 base pairs of homology are needed (Gonda and Radding, 1983). To focus on the effect of varying the position of the monoadducted thymidine within the substrate, we then prepared another three oligomers. Using the RecA protein, we have formed crosslinked three-stranded complexes between each of these oligomeric substrates and duplex pUC19, and have identified factors which influence the efficiency of the overall process. This procedure has applications not only in the study of RecA and its interactions with DNA, but in DNA probe technology as well.

MATERIALS AND METHODS

Details have been presented elsewhere (Cheng et al., submitted).

RESULTS AND DISCUSSION

A total of eleven psoralen-monoadducted oligomers were prepared and tested as substrates for RecA-mediated pairing with pUC19. Each of these oligonucleotides (Figure 1) contained a specific monoadducted thymidine located within the KpnI recognition site. The process to form three-stranded complexes, diagrammed in Figure 2, was based upon the methods of Tsang et al. (1985). In brief, RecA is first allowed to polymerize along the single-stranded substrate, either at a low Mg^{2+} concentration or in the presence of a single-strand binding protein to minimize interference from secondary structure in the DNA. The pairing reaction is initiated by the addition of duplex pUC19. Crosslink formation is then induced by irradiating the reaction mixtures with long wavelength ultraviolet light.

Representative results of pairing reactions between the monoadducted oligomers and duplex pUC19 are shown in Figure 3. Since only the monoadducted oligonucleotide carries a ^{32}P label, formation of three-stranded complexes is indicated by the presence

(a) 5'-GCTCGGTACCCGG-3'

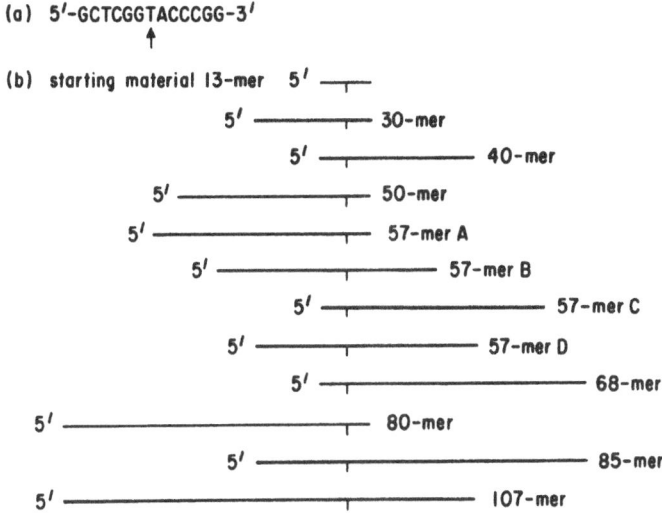

Figure 1. Monoadducted oligonucleotides. (a) Sequence of the
13-mer used to prepare site-specific psoralen monoadducts. The
thymidine within the unique KpnI recognition site (arrow) was
modified on the 3'-side with 4'-hydroxymethyl-4,5',8-trimethyl-
psoralen. Additional synthetic oligomers were ligated onto either
end of this 13-mer to generate the final substrates. (b) Schematic
of the final eleven substrates, aligned according to sequence. All
are homologous to the polylinker region in the plus strand of pUC19.
The position of the psoralen within each substrate is denoted by the
small vertical bar.

of ^{32}P label in the plasmid bands. The site-specificity of oligomer
insertion into the duplex was determined by restricting samples with
KpnI. The pUC19 plasmid contains only one KpnI recognition site,
and a psoralen crosslink within this site is known to interfere with
KpnI restriction (Gamper et al., 1984). Therefore, if the oligomer
has been crosslinked to its target, the complexes will show
resistance to KpnI. Nicking activity is presumably responsible for
the conversion of FI to FII that is observed.

Comparison of samples 5 and 9 in Figure 3 shows that the
nonhydrolyzable adenosine 5'-O-(3-thiotriphosphate) (ATPγS) is more
effective than ATP, the natural cofactor for RecA. Even with the
107-mer substrate, ATPγS was necessary for efficient formation of
three-stranded complexes. As Leahy and Radding (1986) have noted,

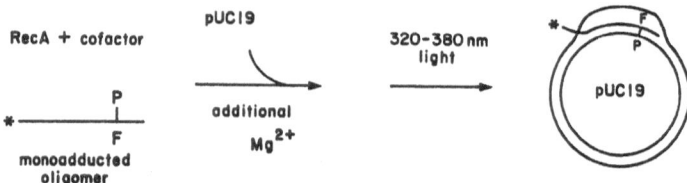

Figure 2. Formation of three-stranded DNA complexes.
Oligomers carrying a ^{32}P 5'-end-label (asterisk) were first
incubated with RecA at 2 mM Mg^{2+}. To initiate the pairing reaction,
the Mg^{2+} concentration was raised to 12 mM and the duplex substrate
was added. If present, single-strand binding protein was included
in the initial incubation mixture; in these instances, the Mg^{2+}
concentration was raised when this protein was added. Reaction
mixtures were then irradiated with 320-380 nm light to effect
crosslinkage. The pyrone (P) and furan (F) ends of the psoralen
molecule are indicated; only the furan-side monoadduct can be
driven to form a crosslink by irradiation with long wavelength
ultraviolet light. The exact structure of the final product has not
been determined.

excess RecA and this ATP analog are apparently needed to overcome
the weak affinity that RecA has for oligomeric substrates. The
effect of including a single-strand binding protein such as Gp32
from phage T4 was also studied. Helix-destabilizing proteins have
been found to facilitate the association of RecA with
single-stranded DNA (Muniyappa et al., 1984; reviewed by Cox and
Lehman, 1987). In general, single-strand binding protein from
either phage T4 (Figure 3, samples 6, 7, 10, and 11) or E. coli
(data not shown) does enhance the efficiency of complex formation if
ATP is the cofactor. Both proteins had a small effect on ATPγS
reactions, and the results varied among oligomers and experiments
(data not shown).
    Efficiency was estimated by quantitating the distribution of
^{32}P between the oligomer and plasmid bands excised from dried
agarose gels. As Figure 4 suggests, the extent of complex formation
does depend upon the length of the oligonucleotide. In general, the
longer the oligomer, the greater the efficiency of reaction. With
ATPγS as the cofactor for RecA, and in the absence of a
single-strand binding protein, typical values range from 22%

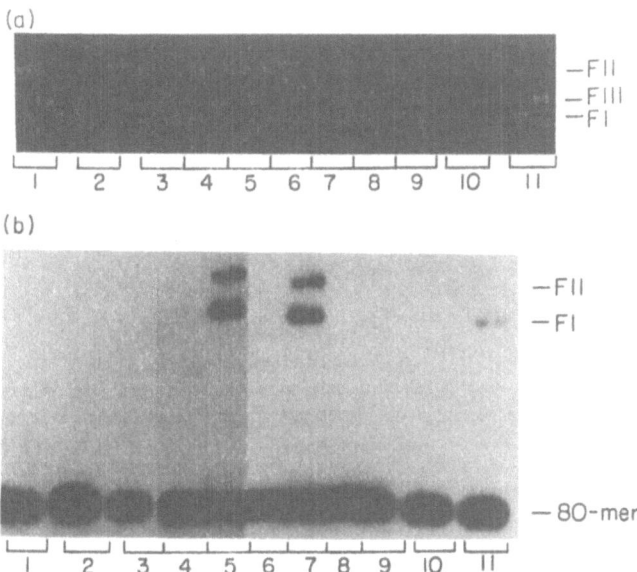

Samples	RecA	ATPγS	ATP	Gp32	320-380 nm hν
1	−	−	−	−	+
2	+	−	−	−	−
3	+	−	−	−	+
4	+	+	−	−	−
5	+	+	−	−	+
6	+	+	−	+	−
7	+	+	−	+	+
8	+	−	+	−	−
9	+	−	+	−	+
10	+	−	+	+	−
11	+	−	+	+	+

Figure 3. Deproteinized samples from RecA-mediated homologous pairing reactions between duplex pUC19 and the monoadducted 80-mer, analyzed by native agarose gel electrophoresis: (a) photograph of gel stained with ethidium bromide, and (b) autoradiogram of the dried gel. The right-hand lane of each pair is the sample following restriction by KpnI. A system to regenerate ATP was always included in reactions which depended upon ATP as the cofactor.

174

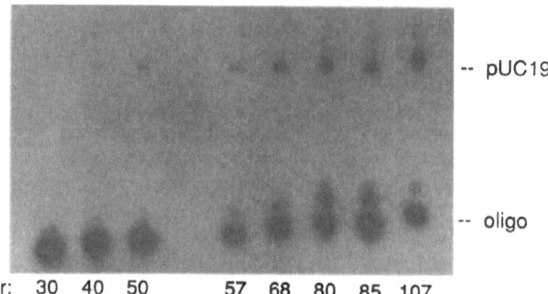

MA oligomer: 30 40 50   57 68 80 85 107

Figure 4.   Dependence of complex formation on oligomer length.
This autoradiogram shows deproteinized pUC19 complexes after agarose
gel electrophoresis.   These complexes were formed with different
oligomers in the presence of ATPγS and then crosslinked.   The length
of the monoadducted (MA) oligomer used in each sample is indicated
at the bottom.   57-mer D shown in Figure 1 was used here.   These
complexes were actually formed under conditions optimized for ABC
excinuclease studies, but the relative efficiences parallel those
observed under RecA reaction conditions.   The multiple bands of the
107-mer complex are attributed to species differing in the extent of
unwinding as a result of oligomer insertion.   The multiplets
observed in the free oligomer region are attributed to formation of
intra- and inter-oligomeric crosslinks.

incorporation for the 30-mer to 47% for the 50-mer and 84% for the
107-mer.   No incorporation of monoadducted 13-mer (Figure 1a) was
observed (data not shown).
    The position of the psoralen within the monoadducted substrate
also affects the yield of complexes.   The 40- and 68-mer substrates
are both incorporated with less than 20% efficiency, much less than
would be expected based upon their length.   As is indicated in
Figure 1, these two have the psoralen-adducted thymidine located
very near the 5'-end.   To confirm this observation, 57-mers A, B and
C were prepared, each having the psoralen at a different position
from the original 57-mer D (Figure 1).   The observed efficiencies of
reaction under ATPγS conditions were: 32% for 57-mer A, 16% for B,
5% for C, and 49% for 57-mer D.   Apparently, the adduct at the
5'-end interferes either with RecA association along the substrate
or with base pairing such that the monoadduct is unfavorably
oriented to form a crosslink.   Internally positioned adducts can be

stabilized by base pairing on both sides. The 5'-end is subject to displacement by the renaturing strands of pUC19; however, the 3'-end would presumably be equally likely to be displaced and yet 57-mer A shows a much greater level of complex formation than 57-mer C. The data may indicate a directional preference of RecA in the insertion and homologous alignment steps. RecA-mediated strand exchange (a unidirectional process) between longer substrates is not observed in the presence of ATPγS (Honigberg et al., 1985). The role of the sequences of these 57-mers in the stabilities of the three-stranded complexes formed has not been addressed.

Complexes can also be detected in ATPγS reactions which have not been irradiated to effect crosslinkage (longer exposure of the autoradiogram shown in Figure 3; additional data not shown). The three longest substrates may be incorporated at levels up to 50% of those observed in samples that have been irradiated. In some instances, then, complexes are of sufficient stability to survive phenol extraction and nondenaturing agarose gel electrophoresis despite the absence of a psoralen crosslink.

Denaturation of irradiated samples at 90°C revealed uncrosslinked complexes in these samples as well. This was most significant for the longest oligomers. For example, up to 35% of complexed 107-mer remains uncrosslinked. This incomplete photoreaction even at light doses in excess of that needed to effect 100% crosslinkage suggests that the newly formed DNA helix within the RecA-stabilized intermediate may have an altered B-form structure which is unfavorable to the formation of a psoralen crosslink. Alternatively, there may be more than one stable conformation, with differing capabilities of forming a crosslink.

As noted above, KpnI restrictions were used to test the specificity of oligomer insertion. To the extent that linearized plasmids carrying labeled oligomer could be detected, up to 5-10% of all complexes observed may have been formed at sites other than the target, leaving the KpnI site undisturbed. To further test the specificity of the RecA-mediated hybridization under ATPγS conditions, the 50- and 85-mer substrates were incubated with pBR322. Sequences from pBR322 constitute the bulk of pUC19 outside the polylinker region, the target for the oligomers. Complex formation with pBR322 was detected, if the samples had been irradiated to form crosslinks (data not shown). For the 50- and 85-mer, respectively, levels of incorporation were 10% and 30% of that usually observed into pUC19. Complexes would not be expected if absolute homology is required by RecA for hybridization of these oligomers.

In an effort to understand these observations, the sequences of the two oligomers were compared with that of pBR322 and of φX174, expected to have even less homology with pUC19. For both duplexes, the sites of the greatest homology with the 50- and 85-mers had 50% and 45% overall homology, respectively. If the analysis focused on the 26-base sequence centered around the crosslinking site (only 19 of the 26 are within the 50-mer), then regions of 50-60% homology were identified. It appears that regions of 40-60% homology are likely both to be found in randomnly chosen DNA and to be sufficient for interaction between the RecA-coated oligomers and a duplex substrate.

We have demonstrated that the RecA protein can be used to introduce psoralen-monoadducted oligomers into duplex targets, yielding three-stranded complexes that can be covalently fixed by photochemical conversion of the monoadducts into crosslinks. Our data indicate that ATPγS is required as the cofactor, and that the length of the oligomeric substrate and position of the psoralen within the substrate both affect reaction efficiency. RecA is able to utilize a 30-mer in our two-step process, but a 50-mer is needed for 50% efficiency, and incorporation is facilitated if the psoralen is located away from the 5'-end of the molecule. Uncrosslinked product is detected, particularly with the longer oligomers, but crosslink formation is necessary for an efficient yield of three-stranded complexes. Finally, we observed that 40-60% homology may be sufficient for complex formation.

Successful psoralen crosslinkage within RecA-DNA complexes indicates that the invading DNA strand is at least locally base paired to its complement, and that any psoralen-induced structural distortions are tolerated by the RecA. The ability of the RecA protein to mediate hybridization under nondenaturing conditions may prove useful for probe technology using crosslinkable probes. One application that has been envisioned for use in mapping is the coupling of RecA with a cleaving agent for DNA. RecA would be used to introduce probes, and the resulting multistranded regions would then serve to localize the action of a cleavage agent. Such agents could be attached to the probes themselves, either to the DNA directly or through a psoralen, or added separately. Our results with pBR322 demonstrate the need to develop stringency conditions for applications in probe technology.

ACKNOWLEDGEMENTS

We thank Dr. Harrison Echols and Chi Lu for their generous gift of RecA, and Dr. Bruce Alberts for the gift of Gp32. We also thank Tom Kodadek for early discussions on recombinatory enzymes, David Koh for his assistance in synthesizing the DNA oligonucleotides, and David N. Cook for his comments on the manuscript. SC was supported by an NSF Graduate Fellowship.

REFERENCES

Cimino GD, Gamper HB, Isaacs ST, and Hearst JE (1985). 'Psoralens as photoactive probes of nucleic acid structure and function: organic chemistry, photochemistry, and biochemistry' *Ann Rev Biochem* **54**:1151-1193.

Cox MM and Lehman IR (1987). 'Enzymes of general recombination' *Ann Rev Biochem* **56**:229-262.

Gamper HB, Piette J, and Hearst JE (1984). 'Efficient formation of a crosslinkable HMT monoadduct at the KpnI recognition site' *Photochem Photobiol* **40**:29-34.

Gonda DK and Radding CM (1983). 'By Searching Processively RecA Protein Pairs Molecules That Share a Limited Stretch of Homology' *Cell* **34**:647-654.

Honigberg SM, Gonda DK, Flory J, and Radding CM (1985). 'The Pairing Activity of Stable Nucleoprotein Filaments Made from recA Protein, Single-stranded DNA, and Adenosine 5'-(γ-Thio)triphosphate' *J Biol Chem* **260**:11845-11851.

Leahy MC and Radding CM (1986). 'Topography of the Interaction of recA Protein with Single-stranded Deoxyoligonucleotides' *J Biol Chem* **261**:6954-6960.

Muniyappa K, Shaner SL, Tsang SS, and Radding CM (1984). 'Mechanism of the concerted action of recA protein and helix-destabilizing proteins in homologous recombination' *Proc Natl Acad Sci USA* **81**:2757-2761.

Radding CM (1982). 'Homologous pairing and strand exchange in genetic recombination' *Ann Rev Genet* **16**:405-437.

Tsang SS, Muniyappa K, Azhderian E, Gonda DK, Radding CM, Flory J, and Chase JW (1985). 'Intermediates in Homologous Pairing Promoted by recA protein' *J Mol Biol* **185**:295-309.

THE USE OF PSORALEN-PHOTOCROSSLINKING FOR THE ANALYSIS OF THE
CHROMATIN STRUCTURE DURING TRANSCRIPTION.

J.M. Sogo, A.Conconi and R.M.Widmer
Institute for Cell Biology
Swiss Federal Institute of Technology
CH-8093 Zurich, Switzerland

ABSTRACT. The structure of transcribing chromatin was studied by
DNA-DNA and DNA-RNA photocrosslinking with psoralen. In the
hyperactive ribosomal genes of Dictyostelium discoideum and Physarum
polycephalum, no nucleosomes could be detected within the coding
region. Two distinct types of ribosomal chromatin co-exist in Friend
cells, one that contains nucleosomes which represent the inactive
gene copies, and the other one which is free of nucleosomes and
corresponds to the actively transcribed genes. In contrast,
transcribing SV40 minichromosomes appeared quite similar to the bulk
of minichromosomes. Nucleosomes or nucleosome-like structures
appeared close to the sites of RNA synthesis. We propose that the
association of histones with DNA does not prevent transcription
elongation by RNA polymerase II on SV40 minichromosomes.

INTRODUCTION

It is well established that the bulk of eukaryotic DNA is organized
in structures called nucleosomes which are formed by 145 bp of DNA
wrapped around the core histones (H2A, H2B, H3 and H4) separated by
linkers with a variable length. Histone H1 is probably localized in
the entry and exit site of the nucleosome DNA (for reviews see Butler
1983, Pederson et al. 1986, Thoma and Sogo 1988). However, there is
still no consensus with regard to the chromatin organization during
transcription and replication. Analysis of sites of action of
topoisomerase I in camptothecin treated nucleoli of Xenopus laevis
oocytes (Sollner-Webb et al. 1988) suggests that the transcribed
sequences of the ribosomal genes are organized in periodic,
eventually nucleosome-like structures. In contrast, Labhart and
Koller (1982) by using a modified Miller spreading technique,
observed in Xenopus oocytes not only the lack of nucleosomal beads in
ribosomal transcribing chromatin, but even there was no visible
difference between coprepared naked DNA and the filaments connecting
putative RNA polymerase. These findings are in agreement with those
found in rapidly growing Dictyostelium discoideum cells (Ness et al.
1983). Again, the filaments associated with putative RNA polymerases
examined by electron microscopy appeared indistinguishable from
coprepared free DNA. Attempts were made to isolate active chromatin.
Davis et al. (1983) after mild micrococcal nuclease digestion

179

P. E. Nielsen (ed.), Photochemical Probes in Biochemistry, 179–194.
© 1989 by Kluwer Academic Publishers.

obtained an insoluble fraction enriched in ribosomal transcribing
chromatin lacking the nucleosomal repeat.

The data reported for the chromatin structure of genes
transcribed by RNA polymerase II also do not reveal a unified
picture. In activated heat shock genes of Drosophila melanogaster and
hyperactive Balbiani Ring 2 genes in the salivary glands of
Chironomus tentans, the transcribed sequences appear to be devoid of
nucleosomes (Wu et al. 1979, Levy and Noll 1981, Widmer et al. 1984,
Karpov et al. 1984). However, Solomon et al. (1988) after treatment
of intact Drosophila cells with formaldehyde, found that histone H4
remains bound to hsp 70 DNA suggesting that although transcription
perturbs nucleosome structure, at least H4 remains bound to actively
transcribed DNA sequence "in vivo". Biochemical and electron
microscopical methods provide evidence that genes transcribed by RNA
polymerase II at low rates are packed in nucleosomes (McKnight et al.
1978, Scheer 1978, Gariglio et al. 1979, Wurtz and Fakan 1983,
Barsoum and Varshavsky 1985, Sargan and Butterworth 1985), containing
hyperacetylated core histones (Allegra et al. 1987, Stern et al.
1987, Chen et al. 1987). But all the observations described still
leave the question open how the DNA sequences involved in transcription or those in close vicinity of the transcription machinery, are
structurally organized in chromatin. We have analyzed the structure
of "in vivo" transcribing chromatin using psoralen-crosslinking. It
allows to analyze the extent of crosslinking either by direct
visualization of the DNA under denaturing conditions in the electron
microscope or by gel electrophoresis of defined DNA fragments. Here
we summarize our data on ribosomal chromatin and transcribing Simian
virus 40 minichromosomes obtained with this method.

MATERIALS AND METHODS

Preparation of nuclei, nucleoli and DNA

Growing of Dictyostelium discoideum cells, purification of nuclei and
nucleoli, and extraction of DNA were performed as described (Ness et
al. 1983). Nucleoli of Physarum polycephalum were isolated as
described by Lucchini et al. (1987). Growing of Friend
erythroleukemia cells and enrichment of mitotic cells was done as
described by Homberger and Koller (1988). Stationary cells were kept
for at least 24 h at the maximal concentration of about 3×10^6 cells
per ml. Nuclei were prepared as described by Marzluff and Huang
(1984). The same procedure was followed to isolate methaphase plates
from mitotic cells. TC-7 monkey cells were infected with SV40 as
described by De Bernardin et al. (1986).

Labelling of nascent preribosomal RNA in run-on experiments

The protocol for RNA synthesis in isolated nuclei described by
Marzluff and Huang (1984) was followed, except that 100 µM cold ATP
and 100 µCi of each γ^{32}P-labelled CTP, GTP and UTP (each 410
Ci/mMol, Amersham) without supplementary cold triphosphates was
added. The reaction was performed with 10^7 nuclei of exponentially
growing cells in 200 µl of reaction volume and in presence of 5 µg/ml
α-amanitin. After 5 min incubation at 4°C and 15 min. at 25°C the

reaction mixture was diluted with 2 vol of cold storage buffer (Marzluff and Huang, 1984) and immediately psoralen crosslinked (see below).

Psoralen-crosslinking

Psoralen-crosslinking of nuclei, nucleoli or purified DNA of D.discoideum and P.Polycephalum were performed as described by Sogo et al. (1984) and Lucchini et al. (1987). Photoreaction of nuclei of Friend cells was performed in storage buffer (Marzluff and Huang, 1984) as described by Widmer et al. (1988). Briefly, the organelle surpensions were placed in open Petri dishes (diameter 3.5 cm) on ice; 0.05 vol of 4,5',8 trimethylpsoralen (stock solution 200 µg/ml in ethanol) was added 3 or 4 times during total irradiation time of 4 to 6 h. The U.V. lamp (366 nm) was placed at a distance of 7 cm above the Petri dishes. After the photoreaction, the DNA was purified and analysed by gel electrophoresis or by electron microscopy.

Photocrosslinking of nuclei from SV40 infected TC-7 cells and of isolated SV40 minichromosomes was done with 8-methoxypsoralen as described by De Bernardin et al. (1986)

Gel electrophoresis, transfer and hybridization

Electrophoresis of DNA fragments was done in 0.8% agarose in TAE buffer and gels were stained with ethidium bromide. DNA was transferred to Zeta-Probe membranes (BioRad, Bulletin 1233). Before blotting the photocrosslinked DNA, the gels were irradiated with short wavelength ultraviolet light as described by Sogo et al. (1984) in order to reverse the psoralen-crosslinking. Prehybridization and hybridization of the membranes was done as described by Widmer et al. (1984), except that tRNA was used as carrier.

Exonuclease assay

For the exonuclease assay, the DNA resolved into two differently migrating bands (see Fig. 3) was eluted from gel slices. To 100 ng of the eluted DNA, 15 µg of crosslinked lambda-DNA was added as carrier to a final volume of 80 µl. The exonuclease assay was performed according to procedure 1 described by Widmer et al. (1988). Lambda-DNA at a concentration of 0.1 µg/ml was crosslinked for 2 h and two additions of psoralen as previously described were done.

Electron Microscopy

Psoralen-crosslinked DNA was prepared for electron microscopy under denaturing conditions as described by Sogo et al. (1984, 1986).

RESULTS

I. Psoralen crosslinking of DNA in transcriptionally active chromatin

i. Characterization by Electron microscopy. Isolated nucleoli of exponentially growing Dictyostelium discoideum cells were extensively

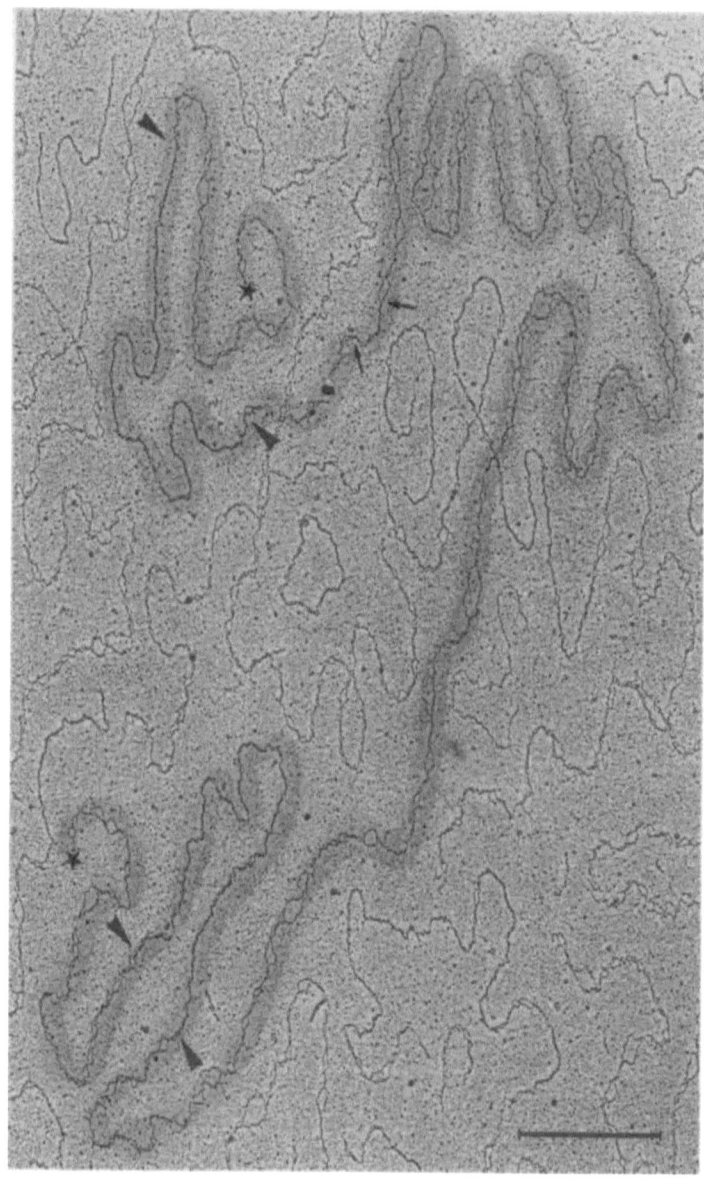

Figure 1. Micrograph of psoralen-crosslinked ribosomal DNA in isolated nucleoli of Dictyostelium. Asterisks indicate the ends of the rDNA
palindrom. Coding regions heavily crosslinked are visible as duplex
DNA (between arrowheads). The remaining DNA corresponding to the DNA
protected by nucleosomes (central spacer and terminal regions) are
organized in single-stranded bubbles. The bar represents 2 kb.

psoralen-crosslinked. The DNA was purified and analyzed in the electron microscope by spreading under denaturing conditions. Fig.1 shows an entire ribosomal DNA palindrome. The central region as well as the terminal stretches are organized into single-stranded bubbles as it is characteristic for DNA crosslinked with psoralens in bulk chromatin. Adjacent to the central segment two homogeneously crosslinked regions (indicated between arrowheads) identified as mostly double-stranded DNA are symmetrically located in the molecules. Statistical analysis of such crosslinked rDNAs reveals that the continuously crosslinked stretches correspond to the known position of the coding regions (Fig. 2b). In about 40% of the molecules, a second continuously crosslinked DNA region of approximately 0.8 kb was found further 5'upstream (Fig. 1 between arrows). The lengths of the DNA stretches visualized as rows of single-stranded bubbles are compatible with the values reported for the length of the central and the terminal non-transcribed regions. In order to confirm that the different degrees of crosslinking between the coding and the non-coding regions are mainly due to the presence of intact nucleosomes in the non-coding region, we treated isolated nucleoli with 2M NaCl (before and during photocrosslinking). This treatment is known to dissociate the histones from the DNA. After spreading under denaturing conditions the DNA appeared entirely crosslinked, and coding and non-coding segments could not be distinguished (data not shown).

The pattern of homogeneously crosslinked regions (coding sequences) alternating with regions of single-stranded bubbles (non-coding sequences) has been confirmed by using the same technique in isolated nucleoli from exponentially growing microplasmodia of Physarum polycephalum.

Lucchini et al. (1987) observed features in Physarum ribosomal palindromes not seen so far in other systems. Three well-defined DNA stretches on each half of the molecule exhibit non-crosslinked sequences even when the photoreaction was performed in the presence of 2M NaCl. These are located at the 5' and 3' ends of the coding sequences as well as in the close vicinity of the origins of replication. Since the DNA sequences of these regions (Ferris 1985) show a frequency of pyrimidines similar to that in random DNA, we suggest that somehow these sequences adopt an unusual conformation which is not compatible with psoralen crosslinking.

ii. Characterization by electrophoretic mobility of restriction fragments. Upon digestion with Eco RI and agarose gel electrophoresis nucleolar DNA of Dictyostelium shows seven bands (Fig.2a). The four largest bands (fragments I to IV) correspond to fragments of the non-transcribed region, fragments VII and IX (two) arise from the coding region and fragment V contains the 5' end of the ribosomal gene (Fig.2a, tracks 2 and 7 and Fig.2b). We observed that with increasing extent of photoreaction the mobility of fragments V, VII and IX continuously decreased, whereas the electrophoretic behaviour of fragments I, II, III and IV remained rather unaffected and therefore their migration was practically indistinguishable from that of uncrosslinked control DNA (Fig. 2a, tracks 3,4 and 5). DNA from nucleoli crosslinked in the presence of 2M NaCl was also digested with Eco RI. Under these histone-dissociating conditions all the Eco RI fragments exhibited a lower mobility than those obtained from not

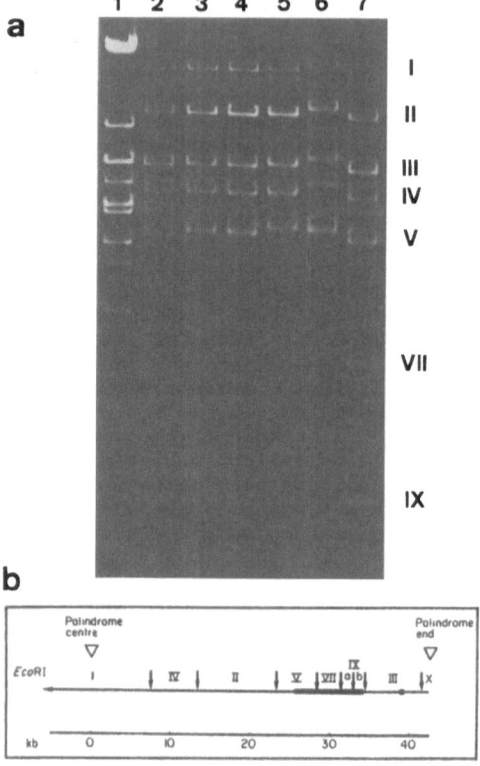

Figure 2. a) Photocrosslinking of the ribosomal DNA in isolated nucleoli of Dictyostelium discoideum and electrophoretic analysis of restriction fragments. Nucleoli were photoreacted with psoralen, DNA was extracted, digested with EcoRI and analysed by electrophoresis in 0.8% agarose gels. Track 1, a mixture of EcoRI-digested Hind III digested DNA. Tracks 2 and 7, non-crosslinked control DNA. Tracks 3,4 and 5 correspond to DNA from nuclei photoreacted at low salt for 10 min, 1 h and 4 h respectively. Track 6, corresponds to DNA from nucleoli photoreacted for 4 h in the presence of 2 M NaCl. b) EcoRI restriction map of rDNA of Dictyostelium discoideum. One half of the rDNA palindrom is shown. Filled box indicates the sequence of 17S and 26S rRNAs. Data taken from Ness et al. (1983).

crosslinked control DNA (compare tracks 6 and 7 in Fig. 2a). In order to confirm the difference in mobility induced by psoralen crosslinks, we mixed on the one hand uncrosslinked DNA with DNA crosslinked at or near saturation in nuclei and on the other hand DNA extensively crosslinked in nuclei either in the presence or in the absence of 2M NaCl. As expected, in the first experiment fragments V, VII and IX appeared as double bands, whereas in the second one fragments I, II, III and IV showed double bands (data not shown but see Sogo et al.

1984). The shift in electrophoretic mobility after psoralen-crosslinking unambiguously confirms the data obtained by electron microscopy described above. Fragments originating from the non-coding region incorporate psoralen to a low extent (due to the protection of the nucleosomal DNA, Hanson et al 1976, Conconi et al. 1984) and therefore can be partially denatured and their mobility in gel electrophoresis is not or insignificantly affected as compared to uncrosslinked control DNA. However, fragments arising from the coding region are heavily crosslinked and hence visualized as double-stranded (although spread under denaturing conditions) and migrating with a reduced mobility in gel electrophoresis.

The ribosomal chromatin of Physarum was also analysed by the gel retardation assay after DNA crosslinking in isolated nucleoli. The results obtained with a battery of restriction enzymes (Taq I, Sal I, Bgl II, EcoR I and Pst I which efficiently cleave the crosslinked DNA) and probes (from the central spacer, coding regions and terminal segments) fully confirmed the structure of active ribosomal chromatin as already described in the electron microscopy section (for more details see Lucchini, Ph.D. thesis).

The results presented so far (Dictyostelium and Physarum) are examples of the power of the psoralen technique for the study of chromatin structures at the DNA level. As models both systems have the advantage that the ribosomal genes are extrachromosomal and therefore can be isolated in mass. Since for higher eukaryotic organisms the isolation of distinct gene(s) at the chromatin level is not yet possible (except for minichromosomes like SV40), one aim of our laboratory during the last years has been to expand the application of the psoralen technique by combining it with hybridization techniques. Encouraging was the observation of double bands in the mixing experiments described above, which offers in principle the possibility to distinguish at the level of DNA restriction fragments between the active and the inactive state of a gene. From the data accumulated in our laboratory as well as from those available in literature we anticipated that from the about 100 copies of ribosomal genes in mouse cells only part of it is active. To test this idea, we decided to analyze the ribosomal chromatin from Friend erythroleukemia cells. Isolated nuclei from exponentially growing cells, stationary cells and isolated metaphase plates were extensively crosslinked with trimethylpsoralen. The purified DNA was digested with EcoRI and electrophoresed in 0.8% agarose gels. After blotting and hybridization with pUCmr100 a fragment of 6.7 kb which comprises about half of the transcribed sequence (Fig. 3b) can be analyzed. In uncrosslinked control DNA, as expected, one band of 6.7 kb is visualized (Fig. 3a, tracks 1 and 6). However, when the DNA was psoralen photoreacted in nuclei of exponentially growing cells two bands were resolved (Fig. 3a, track 2). The mobility of the lower band of the doublet is similar to that of uncrosslinked control DNA, whereas the mobility of the upper band is significantly retarded. When the crosslinking was done in the presence of heparin (which efficiently dissociates histones from DNA) only one clearly retarded band was visualized (Fig.3a, track 5). We interpreted the doublet (the ratio of the two bands is about one to one) of track 2 in Fig. 3a to originate from two distinctly different classes of chromatin structures (active and inactive gene copies).

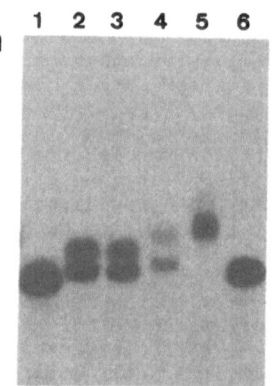

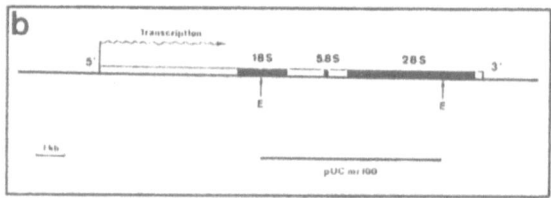

Figure 3. Photocrosslinking of the ribosomal chromatin in purified
nuclei of Friend cells. a) Nuclei of exponentially growing,
stationary and metaphase arrested cells were photoreacted with
psoralen for 6 h, DNA was extracted, digested with EcoRI, separated
by gel electrophoresis (0.8% agarose gels), blotted and hybridized
with the probe pUCmr100 (see b). Tracks 1 and 6, non-crosslinked
control DNA. Track 2, DNA from exponentially growing cells. Track 3,
DNA from stationary cells. Track 4, DNA from metaphase chromosomes.
Track 5, DNA from exponentially growing nuclei photoreacted for 6 h
in the presence of heparin (500 µg/ml). b) Map of the transcribed
region of mouse ribosomal DNA. Filled boxes indicate the sequence of
18S, 5.8S and 28S rRNAs. The wavy arrow indicates the start and
direction of the transcription for the primary transcript (pre-rRNA).
E, EcoRI cutting sites.

In order to test for a possible disappearance of one of the two
bands or a change in the ratio of them, additional experiments were
performed with isolated nuclei from stationary cells as well as with
isolated metaphase plates. From run-on experiments we know that in
stationary and metaphase cells the RNA polymerase I loading is about
six times lower than in exponentially growing cells (Conconi,
unpublished). Surprisingly, the two bands were again resolved in both
stationary and in metaphase arrested cells (Fig.3a, tracks 3 and 4).

iii. Characterization by exonuclease assay. When bulk chromatin is
photoreacted with psoralen, crosslinking preferentially occurs in the
linker DNA between nucleosomes. The pattern of these crosslinks can be

analyzed by exonuclease digestion, since exonucleases stop at sites of psoralen crosslinks. The digestion products are resolved as DNA fragments of nucleosomal and polynucleosumal sizes. When, however, the DNA is more or less homogeneously crosslinked, the remaining DNA digestion fragments are heterogeneous in length and therefore visible as a smear (Widmer et al. 1988). When DNA from each band of the doublet described above was isolated from a preparative agarose gel and subjected to the exonuclease assay, the DNA arising from the lower band (fast migrating) produced fragments of nucleosomal and polynucleosomal size whereas the DNA from the upper band produced a homogeneous smear (data not shown).

II. DNA-RNA hybrids on transcribing genes

i. Identification by band shift. We have shown (De Bernardin et al. 1986) that in in vitro transcription experiments about 10 % of the nascent RNA molecules were crosslinked to the template DNA with trimethylpsoralen. To answer the question which band(s) of the doublet described above respresents the transcribing ribosomal chromatin, we crosslinked RNA labelled in run-on experiments to the template. For this purpose, isolated nuclei from exponentially growing cells were, before psoralen crosslinking, incubated in the presence of ^{32}P labelled ribonucleoside triphosphates and α-amanitin. After crosslinking the total nucleic acids were isolated and digested with EcoRI. The autoradiogram obtained from direct exposure of a dried agarose gel shows that only the upper band of the doublet and bands corresponding to partial digests could be detected (Fig. 4, track 2). As a control an aliquot of the sample was run in

Figure 4. Psoralen-crosslinking of nascent preribosomal RNA to DNA in ribosomal chromatin of Friend cells. Isolated nuclei of exponentially growing cells were incubated with ^{32}P labelled precursors and then photoreacted for 6 h under transcriptional run-on conditions. Total nucleic acids were extracted, digested with EcoRI and electrophoresed in 0.8% agarose. One half of the gel was blotted and hybridized with pUCmr100, the other half was dried and exposed. Track 1, autoradiograph after blotting and hybridization; the EcoRI fragments are indicated by horizontal bars. Track 2, autoradiograph of the exposed dried gel; note that only the upper band (and its partial digests) of the doublet is visible.

parallel on the same gel, cut out, blotted and hybridized, and as
expected, the double band was visualized (Fig. 4 track 1). Although
in the experiment presented no treatment with exogenous RNase was
performed, the activity of contaminating RNase was sufficient to
reduce the length of the crosslinked RNA-molecules to such an extent
that the mobility of the EcoRI fragment was not influenced. This
experiment shows that only the fragments forming the retarded band
arise from the transcribing genes.

ii. Visualization by electron microscopy. In order to see whether the
chromatin structure described for ribosomal genes is specific for
genes transcribed by RNA polymerase I or whether it is a general
feature of active chromatin independent of the acting polymerase, we
decided to analyse the structure of "in vivo" transcribing SV40
minichromosomes using electron microscopy. We crosslinked the nascent
RNA to the template DNA with trimethyl and 8-metoxy psoralens and
searched for individual molecules undergoing transcription. The
hybrid molecules resulting from "in vivo" crosslinked SV40
minichromosomes were purified in sucrose gradients and those from the
fast migrating peak were spread under denaturing conditions. The
panel of Fig. 5 shows three representative DNA-RNA complexes. In a)
the nascent RNA is crosslinked to a covalently closed SV40 DNA
template. In order to visualize the chromatin structure of such
crosslinked minichromosomes (i.e. single-stranded bubbles), it is
necessary to introduce at least one nick per circle. Otherwise the
crosslinked covalently closed duplex DNA appears as double-stranded
circles although it is prepared for electron microscopy under
denaturing conditions (for further explanations see Sogo et al.
1986). Transcribing nicked molecules containing common features are
shown in b) and c). They are organized in regularly spaced single-

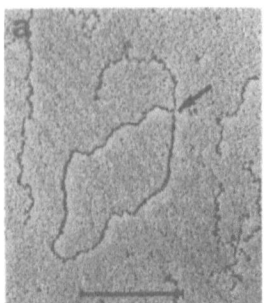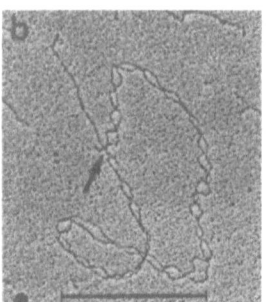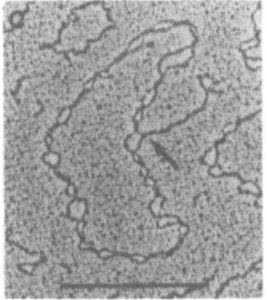

Figure 5. Crosslinking of nascent RNA in SV40 minichromosomes.
Isolated SV40 minichromosomes or SV40 infected nuclei were
photoreacted with 8-methoxy-psoralen, deproteinized, DNA-RNA
complexes partially purified and spread under partially denaturing conditions
for electron microscopy. Arrows point to the attachment sites of
nascent RNA to DNA. In a) SV40 DNA is supercoiled and appears
covalently closed. In b) and c) nicked circular SV40 molecules
organized in single-stranded bubbles corresponding to nucleosomal DNA
are shown. In b) the RNA has been crosslinked to the linker DNA
between two adjacent bubbles. In c) the RNA has been crosslinked to
nucleosomal DNA (single-stranded bubble).The bars represent 1 kb.

stranded bubbles and the size distribution of the bubbles is similar
to that of non-transcribed bulk minichromosomes. In about half of the
transcribing molecules the RNA is attached to double-stranded
stretches between two adjacent bubbles (linkers, Fig. 5b) whereas in
the other half the nascent RNA strand is attached to the template DNA
within a single-stranded bubble (Fig. 5c). The size of the single-
stranded bubbles with crosslinked RNA is not significantly different
from that determined for the bubbles of SV40 bulk chromatin (De
Bernardin et al., 1986).

DISCUSSION

I.Chromatin structure of ribosomal genes

Psoralen-crosslinking of nucleoli of the slime molds Dictyostelium
discoideum and Physarum polycephalum shows single-stranded bubbles
of nucleosomal size in the electron microscope in the DNA of the non-
transcribed regions (central and terminal non-transcribed spacers)
and therefore suggests that these DNA regions are packed in nucleosomes (see also Ness et al. 1983). However, in the coding sequences
of exponentially growing cells, the chromatin DNA is crosslinked to
such an extent that it appears as duplex because of fairly uniform
crosslinking. Especially in Physarum we observed irregularly distributed small or very small bubbles originating probably from non
nucleosomal components involved in the transcription process and/or
from disrupted or partially depleted nucleosomes (Johnson et al.
1987, Lucchini et al. 1987, Solomon et al. 1988). These observations
made with the psoralen technique are fully compatible with the
nuclease digestion studies (Ness et al. 1983, Widmer et al. 1984,
Lucchini et al. 1987). In Dictyostelium the continuously crosslinked
stretches extend over 11 kb into the coding sequences, and the 0.8 kb
crosslinked stretch in the central non-transcribed spacer correlates
with sites of activity of topoisomerase I (Ness et al. 1988 and Lin
et al. 1988). In Physarum the low crosslinking efficiency in the
5'and 3' flanking regions which was observed after crosslinking of
nucleoli as well as of the corresponding deproteinized DNA, suggests
that these sequences (tandemly repeated sequences, region B, Ferris
1985) probably adopt a peculiar conformation. Similar observations
were made at the putative origins of replication (Lucchini et al.
1987) containing the same repeated sequence as in the 5'flanking
region (region B and H, Ferris, 1985). These analyses in the electron
microscope are in agreement with mobility shifts observed by gel
electrophoresis of restriction fragments. Fragments arising from the
non-coding region had an electrophoretic mobility close to that of
the non-crosslinked control, but those containing transcribed
sequences had a significantly reduced mobility. The more psoralen is
incorporated in a restriction fragment the slower it migrates in
agarose gel electrophoresis. This may be due to the increased
molecular weight and stiffness (Carlson et al. 1982), to the
increased length (Weisehan and Hearst 1978), and/or to changes in the
helicity (Sindedn and Haggerman 1984, Widmer et al. 1988) rather than
to kinking of DNA (Kim et al. 1982).
    Earlier psoralen-crosslinking experiments in ribosomal genes
performed under non-saturating conditions with Tetrahymena

thermophila (Cech and Karrer 1980) and Physarum (Judelson and Vogt 1982) did not reveal the detailed picture emerging from our data. The proposed model by Cech and Karrer (1980) in which the chromatin structure of transcriptionally active rDNA consists of a series of "half nucleosomes" as deduced from the size distribution of single-stranded bubbles localized in the coding region, is most likely due to the low extent of crosslinking performed in these experiments.

With regard to the ribosomal genes in Friend cells we demonstrated that there are two distinct classes of chromatin structures, because an EcoRI fragment containing a large part of the coding region of the precursor rRNA appears as two bands after psoralen crosslinking of cells or isolated nuclei. As deduced from the exonuclease assay, the DNA of the faster moving band must be packed into nucleosomes. The observed smear with the DNA of the slower moving band suggests uniform crosslinking similar to the observations made in the transcriptionally active coding region of Dictyostelium and Physarum. This was confirmed by labelling the nascent RNA in run-on experiments. Only the upper band (and its partial digests) is labelled after DNA-RNA crosslinking. It is interesting to note that in stationary as well as in metaphase arrested cells the proportion of both bands remains fairly constant, although their in vitro run-on transcription rate is 6 times lower than in exponentially growing cells. This shows that the active ribosomal chromatin structure is stably propagated through the cell cycle and cells growing under adverse conditions, and suggests that it is independent of the transcriptional activity, opposite to observations in the ribosomal chromatin of Dictyostelium (Ness et al. 1983) or in the Balbiani ring 2 genes of Chironomus larvae (Björkroth et al. 1988).

Finally we emphasize that the psoralen does not allow us to distinguish between perturbed nucleosomes (Solomon et al. 1988) and the absence of histones (Labhart and Koller 1982) in coding region. However, the absence of stable nucleosomes in active ribosomal chromatin is clearly demonstrated.

II.Structure of genes transcribed by RNA polymerase II

Psoralen-crosslinking of nascent RNA strands to their template DNA has been used "in vitro" (Shen and Hearst 1978) and in nuclear extracts (Sargan and Butterworth 1985). We have combined this method with electron microscopy of transcribing SV40 minichromosomes (De Bernardin et al. 1986). The associated single-stranded tails were interpreted to represent nascent RNA strands, because after treatment with RNAse the tails disappeared, and because the length of the putative template DNA (after cutting it with restriction enzymes) and the length of the nascent RNAs were compatible. The nascent RNA was either crosslinked in the linkers between nucleosomes or within single-stranded bubbles. Because of their size and because they disappeared upon treatment of the samples with 2M NaCl before and during crosslinking, it is concluded that the bubbles are due to nucleosome or nucleosome-like particles. That the transcription machinery itself protects DNA from being crosslinked in a similar way as nucleosomes is unlike since about 50% of the nascent RNA strands are attached to the crosslinked linkers between nucleosomal bubbles. We suggest that transcription elongation in SV40 minichromosomes may proceed in the presence of nucleosomes or nucleosome-like particles (De Bernardin et al. 1986). Similar conclusions were reached by Petryaniak and Lutter (1987) using a different biochemical approach.

ACKNOWLEDGEMENTS. We thank Th. Koller and F. Thoma for useful discussions and critical reading of the manuscript and H. Mayer-Rosa for excellent technical assistance. This work was supported by Schweizerischer Nationalfonds zur Förerung der wissenschaftlichen Forschung to J.M.S.

REFERENCES

Allegra, P., Steiner, R., Clayton, D.F. and Allfrey, V.G.: 'Affinity chromatographic purification of nucleosomes containing transcriptionally active DNA sequences'. *J.Mol.Biol.*, **196**, 379-388 (1987).

Barsoum, J. and Varshavsky, A.: 'Preferential localization of variant nucleosomes near the 5`end of the mouse dihydroflate reductase gene'. *J.Biol.Chem.* **260s**, 7688-7697.

Björkroth, B., Ericsson, C., Lamb, M.M. and Daneholt, B.: 'Structure of the chromatin axis during transcription'. *Chromosoma (Berl.)* **96**, 333-340 (1988).

Butler, P.J.B.: 'The folding of chromatin'. *CRC Critical Reviews in Biochemistry* **15**, 57-91 (1983).

Carlson, J.O., Pfenninger, O., Sinden, R.R., Lehman, J.M. and Pettijohn, D.E.: 'New procedure using a psoralen derivate for analysis of nucleosome associated DNA sequences in chromatin of living cells'. *Nucl.Acids Res.* **10**, 2043-2063 (1982).

Cech, T.R. and Karrer, K.M.: 'Chromatin structure of the ribosomal RNA genes of Tetrahymena thermophila as analyzed by trimethylpsoralen crosslinking "in vivo"'. *J.Mol.Biol.* **136**, 395-416 (1980).

Chen, T.A., and Allfrey, V.G.: 'Rapid and reversible changes in nucleosome structure accompany the activation, repression and superinduction of murine fib roblast protonco genes c-fos and c-myc'. *Proc.Natl.Acad.Sci.USA* **84**, 5252-5256 (1987).

Conconi, A., Losa, R., Koller, Th. and Sogo, J.M.: 'Psoralen-crosslinking of soluble and H1-depleted soluble rat liver chromatin'. *J.Mol.Biol.* **178**, 920-928 (1984).

Culotta, V. and Sollner-Webb, B.: 'Sites of topoisomerase I action on X. laevis ribosomal chromatin: Transcriptionally active rDNA has an 200 bp repeating structure'. *Cell* **52**, 585-597 (1988).

Davis, A.M., Rendelhuber, T.L. and Garrard, W.T.: 'Variegated chromatin structures of mouse ribosomal RNA genes'. *J.Mol.Biol.* **167**, 133-155 (1983).

De Bernardin, W., Koller, Th. and Sogo, J.M.: 'Structure of "in vivo" transcribing chromatin as studied in SV40 minichromosomes'. *J.Mol.Biol.* **191**, 469-482 (1986).

Ferris, P.J.: 'Nucleotide sequence of the central non-transcribed spacer region of Physarum polycephalum rDNA'. *Gene* **39**, 203-211.

Gariglio, P., Llopis, R., Oudet, P. and Chambon, P.: 'The template of the isolated native simian virus 40 transcriptional complexes in a minichromosome'. *J.Mol.Biol.* **131**, 75-105 (1979).

Homberger, H. and Koller, Th.: 'The integrity of the histone-DNA complex in chromatin fibres is not necessary for the maintenance of the shape of mitotic chromosomes'. *Chromosoma (Berl.)* **96**, 197-204 (1988).

Hanson, C.V., Shen, C.J. and Hearst, J.E.: 'Cross-linking of DNA "in situ" as a probe for chromatin structure'. *Science* **193**, 62-64 (1976).

Johnson, E.M., Sterner, R. and Allfrey, V.G.: 'Altered nucleosomes of active nucleolar chromatin contain accessible histone H3 in its hyperacetylated forms'. *J.Biol.Chem.* **262**, 6943-6946 (1987).

Judelson, H.S. and Vogt, V.M. : 'Accessibility of ribosomal genes to trimethyl psoralen in nuclei of Physarum polycephalum'. *Mol.Cell.Biol.* **2**, 211-220 (1982).

Kim, S.-H., Peckler, S., Graves, B., Kanne, D., Rapoport, H. and Hearst, J.E.: 'Sharp kink of DNA at psoralen-crosslink site deduced from crystal structure of psoralen-thymine monoadduct'. *Cold Spring Harbor Symp.Quant.Biol.* **47**, 361-365 (1982).

Labhart, P. and Koller, Th.: 'Structure of the active nucleolar chromatin of Xenopus laevis oocytes'. *Cell* **28**, 279-292 (1982).

Levi, A. and Noll, M.: 'Chromatin fine structure of active and repressed genes'. *Nature* **289**, 198-203 (1981).

Lucchini, R., Pauli, U., Braun, R. and Koller, Th. : 'Structure of the extrachromosomal ribosomal RNA chromatin of Physarum polycephalum'. *J.Mol.Biol.* **196**, 829-843 (1987).

Llopis, R., Perrin, F., Bellard, F. and Gariglio, P.: 'Quantitation of transcribing native SV40 minichromsomes extracted from CV1 cells late in infection' *J.Virol.* **38**, 82-90 (19821).

Marzluff, W. and Huang, R.C.C.: 'Transcription of RNA in isolated nuclei. Transcription and translation. A practical approach'. Ed. by B.D. Hames and S.J. Higgins. IRL Press. 89-129 (1984).

McKnight, S.L., Bustin, M. and Miller, O.L., Jr.: 'Electron microscopic analysis of chromosome metabolism in the Drosophila melanogaster embryo'. *Cold Spring Harbor Symp.Quant.Biol.* **42**, 741-754 (1978).

Ness, P.J., Labhart, P., Banz, E., Koller, Th. and Parish, R.W.:
'Chromatin structure along the ribosomal DNA in Dictyostelium'.
J.Mol.Biol. **166**, 361-381 (1983).

Ness, P.J., Koller, Th. and Thoma, F.: 'Topoisomerase I cleavage
sites identified and mapped in the chromatin of Dictyostelium
ribosomal RNA genes'. *J.Mol.Biol.* **200**, 127-139 (1988).

Ohlenbusch, H.H., Olivera, B.M., Tuan, D. and Davidson, N.:
'Selective dissociation of histones from calf thymus nucleoprotein'.
J.Mol.Biol. **25**, 299-315 (1967).

Pederson, D.S., Thoma, F. and Simpson, R.: 'Core particle, fiber and
transcriptionally active chromatin structure'. *Ann.Rev. of Cell Biol.*
2, 117-147 (1986).

Petryaniak, B. and Lutter, L.C.: 'Topological characterization of the
simian virus 40 transcription complex'. *J.Mol.Biol.* **48**, 289-295
(1987).

Sargan, D.R. and Butterworth, P.H.W.: 'Eukaryotic ternary
transcription complexes: Transcription complexes of RNA polymerase II
are associated with histone-containing, nucleosome-like particles "in
vivo"'. *Nucl.Acids Res.* **13**, 3805-3822 (1985).

Scheer, U.: 'Changes of nucleosome frequency in nucleolar and non-
nucleolar chromatin as a function of transcription: An electron
microscope study'. *Cell* **13**, 535-549 (1978).

Shen, C.-K.J. and Hearst, J.E.: 'Photochemical crosslinking of
transcription complexes with psoralen. I. Covalent attachment of in
vitro SV40 nascent RNA to its double-stranded DNA template'.
Nucl.Acids Res. **5**, 1429-1441 (1978).

Sinden, R.R. and Hagemann, P.J.: 'Interstrand psoralen crosslinks do
not introduce appreciable bends in DNA'. *Biochem.* **23**, 6299-6303
(1984).

Sogo, J.M., Ness, P.J., Widmer, R.M., Parish, R.W. and Koller, Th.:
'Psoralen crosslinking of DNA as a probe for the structure of active
nucleolar chromatin'. *J.Mol.Biol.* **178**, 897-928 (1984).

Sogo, J.M., Stahl, H., Koller, Th. and Knippers, R.: 'Structure of
replicating simian virus 40 minichromosomes. The replication fork,
core histone segregation and terminal structures'. *J.Mol.Biol.* **189**,
189-204 (1986).

Solomon, M.J., Larsen, P.L. and Varshavsky, A.: 'Mapping protein-DNA
interactions "in vivo" with formaldehyde: Evidence that histone H4 is
retained on a highly transcribed gene'. *Cell* **53**, 937-947 (1988).

Sterner, R., Boffa, L.C., Chen, T.A. and Allfrey, V.G.: 'Cell cycle-
dependent changes in conformation of nucleosomes containing human
histone genes sequences'. *Nucl.Acids Res.* **15**, 4375-4391 (1987).

Thoma, F. and Sogo, J.M.: 'Structures of bulk and transcriptionally active chromatin revealed by electron microscopy'. *Chromosomes and Chromatin* Vol I, ed. by K.W. Adolph. CRL Press, 85-107 (1988).

Wasilik, B., Thevenin, G., Oudet, P. and Chambon, P.: 'Transcription of "in vitro" assembled chromatin by Escherichi coli RIVA polymerase'. *J.Mol.BViol.* **128**, 409-438 (1979).

Weisehahn, G.P., Hyde, J.E. and Hearst, J.E.: 'The photoaddiction of trimethylpsoralen to Drosophila melanogaster nuclei: A probe for chromatin structure'. *Biochem.* **16**, 925-932 (1977).

Widmer, R.M., Lucchini, R., Lezzi, M., Meyer, B., Sogo, J.M., Edström, J.-E. and Koller, Th.: 'Chromatin structure of a hyperactive secretory protein gene (in Balbiani ring 2) of Chironomus'. *EMBO J.* **3**, 1635-1641 (1984).

Widmer, R.M., Koller, Th. and Sogo, J.M.: 'Analysis of the psoralen-crosslinking pattern in chromatin DNA by exonuclease digestion'. *Nucl.Acids Res.* **16**, 7013-7024 (1988).

Wu, C., Wong, Y.-C. and Elgin, S.C.R.: 'The chromatin structure of specific genes. II. Disruption of chromatin during gene activity'. *Cell* **16**, 807-814 (1979).

Wurtz, T. and Fakan, S.: 'Isolation and characterization of a transcribing polynucleosomal chromatin fraction'. *Biol.Cell.* **48**, 109-120 (1983).

Zhang, H., Wang, J.C. and Liu, L.F.: 'Involvement of DNA topoisomerase I in transcription of human ribosomal RNA genes'. *Proc.Nat. Acad.Sci.USA* **85**, 1060-1064 (1988).

Metal Complexes as Photochemical Probes of DNA Structure

Jacqueline K. Barton, Department of Chemistry,
Columbia University, New York New York 10027 U.S.A.

Transition metal chemistry provides a versatile means to develop reagents which with photoactivation cleave DNA either in a sequence-specific or sequence-uniform fashion. In my laboratory we have exploited the photochemistry of tris(phenanthroline) metal complexes and their derivatives in developing such photocleaving molecules.[1] The tris(phenanthroline) complexes are illustrated schematically below.

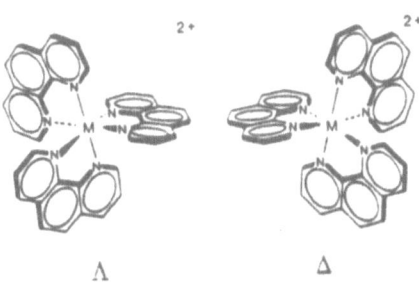

Λ   Δ

The complexes all are rigid in structure, with the metal center defining not only the photochemical properties but also the coordination geometry of the complexes, all of which bind to DNA in a non-covalent fashion.

Chiral tris(phenanthroline)ruthenium complexes and their derivatives have been useful in designing non-covalently bound spectroscopic probes for different DNA conformations.[2] Based upon the symmetry of the complex versus that of the DNA helix, complexes have been developed which recognize preferentially the left-handed Z-conformation.[3] Based upon shape-selective considerations, a complex has furthermore been obtained which binds preferentially to the A-like double-stranded conformation.[4] The simple complex tris(phenanthroline)ruthenium(II), $Ru(phen)_3^{2+}$, binds to DNA non-specifically, through intercalation with the preferential binding of the Δ-isomer to right-handed B-form DNA, and through a surface-bound interaction in the minor groove for which the Λ-isomer is preferred.[5]

But these DNA-binding molecules may be converted to DNA-cleaving molecules through photochemical activation. Upon photoactivation, the analogous complexes of cobalt(III) and rhodium(III) cleave specifically at their bound sites in an oxygen independent reaction.[6] These reactions appear to target the DNA sugar, and hence the photocleavage itself is sequence-neutral. The cleavage then reflects either the sequence

P. E. Nielsen (ed.), Photochemical Probes in Biochemistry, 195–197.
© *1989 by Kluwer Academic Publishers.*

uniformity or sequence-specificity of the binding interaction. In contrast, ruthenium(II) polypyridyl complexes efficiently sensitize the formation of singlet oxygen, and thus, upon irradiation, ruthenium complexes bound to DNA yield cleavage of the strands in a reaction with DNA bases which is mediated by the diffusible singlet oxygen species. Singlet oxygen reacts preferentially with guanine residues. Hence photocleavage reactions obtained with ruthenium complexes reflect not only the DNA-binding characteristics of the molecules but the inherently sequence-dependent reactions of singlet oxygen at sites along the strand.

By mixing different recognition characteristics of metal complexes to different kinds of photochemistry, a variety of photochemical DNA cleavage probes are obtained. $Rh(phen)_3^{3+}$, for example, with photoactivation cleaves efficiently and in a sequence-uniform manner along double-stranded DNA.[1] Λ-$Co(DIP)_3^{3+}$ has been shown to cleave at altered conformations such as those in the Z-conformation.[7] $Ru(TMP)_3^{2+}$, where TMP = tetramethylphenanthroline, cleaves DNA upon photoactivation in a singlet oxygen mediated reaction at sites in an A-like conformation.[8] Most recently we have found that tris(diphenylphenanthroline)rhodium(III), upon photolysis, cleaves DNA in a double-stranded fashion adjacent to cruciform structures which differ in their sequence.[9] The complex therefore provides a remarkably specific probe for this unique structure.

In summary, then, by matching shapes and symmetries of coordinatively saturated transition metal complexes to those of local DNA conformations, simple metal complexes may be designed which target sites along the DNA strand, and by coupling the photochemistry of these transition metal complexes to these binding interactions, a range of sequence-specific and sequence -uniform DNA cleaving molecules are obtained. This family of transition metal complexes are now being developed to probe DNA structure both in vitro and in vivo.[10]

References
1.	M.B. Fleisher, H.Y. Mei and J.K. Barton, *Nucleic Acids and Molecular Biology,* **2,** 65 (1988).

2.	C.V. Kumar, J.K. Barton, and N.J. Turro, *Journal of the American Chemical Society,* **107,** 5518 (1985).

3.	J.K. Barton, L.A. Basile, A. Danishefsky and A. Alexandrescu, *Proceedings of the National Academy of Sciences USA,* **81,** 1961 (1984).

4.	H.Y. Mei and J.K. Barton, *Journal of the American Chemical Society,* **108,** 7414 (1986).

5.	J.K. Barton, J.M. Goldberg, C.V. Kumar, and N.J. Turro, *Journal of the American Chemical Society,* **108,** 2081 (1986).

6.	M.B. Fleisher, K.C. Waterman, N.J. Turro, and J.K. Barton, *Inorganic Chemistry,* **25,** 3549 (1986).

7.	J.K. Barton and A.L. Raphael, *Journal of the American Chemical Society,* **106,** 2466 (1984); J.K. Barton and A.L. Raphael, *Proceedings of the National Academy of Sciences USA,* **82,** 6460 (1985).

8.   H.Y. Mei and J. K. Barton, *Proceedings of the National Academy of Sciences USA*, **85**, 1339 (1988).

9.   M.R. Kirshenbaum, R. Tribolet, and J.K. Barton, *Nucleic Acids Research* **16**, 7943 (1988).

10.   L.B. Chapnick, L.A. Chasin, A.L. Raphael and J.K. Barton, *Mutation Research*, **201**, 17 (1988).

LASER-INDUCED STRAND BREAK FORMATION
OF POLYURIDYLIC ACID
IN THE PRESENCE AND ABSENCE OF
TRIS(2,2'-BIPYRIDYL)RUTHENIUM CHLORIDE AND $K_2S_2O_8$

D. SCHULTE-FROHLINDE, A. B. TOSSI, AND H. GöRNER
Max-Planck-Institut für Strahlenchemie
Stiftstr. 34-36, D-4330 Mülheim a.d. Ruhr, FRG

ABSTRACT. The effect of light (steady state and laser pulse) on the system poly(U), $Ru(bpy)_3^{2+}$ (Ru(II)) and $S_2O_8^{2-}$ has been studied and the following mechanism for the formation of single-strand breaks (ssb) is proposed: Electronically excited $Ru(bpy)_3^{2+}$ ions, bound to poly(U), react with $S_2O_8^{2-}$ by electron transfer resulting in $Ru(bpy)_3^{3+}$ (Ru(III)) and SO_4^{-}. Both the ruthenium(III) complex and SO_4^{-} react with the bases of poly(U) leading to base radicals which in turn leads to ssb formation by H abstraction from position 2' of the sugar moiety. The reaction scheme is complicated by reactions of the base radicals with Ru(III) or with $S_2O_8^{2-}$, reactions which compete with ssb formation. The rate and yield of ssb formation was measured by transient conductivity.

The rate of the $Ru(bpy)_3^{2+}/S_2O_8^{2-}$ photosensitized ssb formation is not pH dependent (range 4-7), in contrast to that under OH radical-induced conditions. The explanation is that the ruthenium complex is so strongly bound to poly(U), due to its double positive charge, that protons do not compete in binding even at pH4. Therefore the rate is not accelerated by the presence of protons, as is observed in the case of the OH-induced ssb formation where only K^+ ions serve as counterions. The relative yield and the rate of ssb formation depend on the nucleotide/sensitizer ratio (N/S). The rate constant for ssb formation is 4.5 s^{-1} at N/S = 3 and 14 s^{-1} at N/S = 10. The reason for the influence of the N/S ratio is not clear as yet. The formation and decay of Ru(III) has been measured spectroscopically. The decay rate constant at N/S = 10 is very similar to that for ssb formation in agreement with the postulated reaction scheme.

P. E. Nielsen (ed.), Photochemical Probes in Biochemistry, 199–208.
© 1989 by Kluwer Academic Publishers.

Introduction

The reaction of metal cations with polynucleotides and DNA has been subject of several studies and strand break formation has often been observed.[1-3] However, details of the chemical mechanism leading to single-strand breaks (ssb) and to other kinds of damage are in most cases unknown. The elucidation of such mechanisms appears difficult due to two problems. The first is that cations are electrostatically attached to nucleic acids, as these are highly charged polyanions. According to Manning's theory[4] the cations are "condensed" on the surface of the polyanion under certain conditions and neutralize a high percentage of the negative charges. Therefore, homogeneous kinetics may not apply. The second problem is that free radical damage of a nucleic acid produces several different classes of free radicals at the strand each with different redox properties. This leads one to expect that secondary reactions of metal cations in their various oxidation states with the different DNA radicals should result in a rather complex reaction scheme.

We simplified the first problem by choosing conditions of total binding of the heavy metal complex to the strand.[5] Equilibration between bound and unbound complexes therefore does not complicate the interpretation of the experimental data. The second problem is the subject of the present report. We studied the properties of the system polyuridylic acid (poly(U)), $Ru(bpy)_3^{2+}$ (Ru(II)), and $S_2O_8^{2-}$ under the influence of UV and visible light by steady state and ns laser flash irradiations. Conditions were chosen such that the ruthenium complex was completely electrostatically bound to the strand.[5,6] Poly(U) was studied since it is single-stranded at room temperature, it does not show base stacking and much is already known concerning the reaction mechanisms for OH radical-induced ssb formation[7,8] and the consequences of laser excitation.[9,10] Poly(U) appears therefore to be a good model compound. With this system we were able to study the chemical steps leading to ssb formation and, to a certain extent, secondary reactions of bound ruthenium complexes with poly(U) free radicals.

Results and Discussion

The experiments and the results are described in detail in reference 11. Here we summarize the main features.

1. Flash photolysis, conductivity studies

The electrical conductivity of a solution of $Ru(bpy)_3^{2+}$ in the presence of $S_2O_8^{2-}$ following a laser pulse (λ_{exc} = 353 nm, pulse length 20 ns) is shown in Fig. 1. Such curves have been interpreted as being due to a fast reaction of the electronically excited $Ru(bpy)_3^{2+}$ ($^*Ru(II)$) with $S_2O_8^{2-}$ (eq. 2) followed by reaction (3).[12] In contrast to Bolletta et al.[13] we found a side reaction in which protons are produced. Our interpretation of the proton formation is shown by reaction (4) in which the SO_4^- radical anion reacts with the ligand (bipyridyl) leading, after electron transfer and hydrolysis, to the ligand-OH-adduct radical and a proton. The contribution of this reaction is 40% of the total reaction of SO_4^- with the Ru(II) complex (in the absence of organic additives).[6] The ligand radicals will not further react with $S_2O_8^{2-}$ at a concentration as low as 1 mM, because the bimolecular disappearence of the ligand radicals is relatively fast due to the small molecular weight of the ruthenium complex. Bound to poly(U) the lifetime of the ligand radical is expected to be much longer.

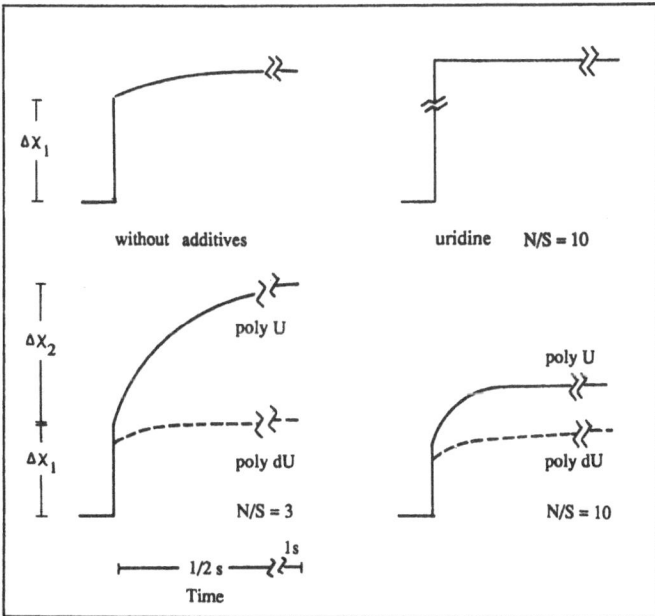

Figure 1. Conductivity signals after the laser pulse in argon-saturated aqueous solution at pH5 and 25 °C using 50 μM $Ru(bpy)_3^{2+}$ and 1.0 mM $K_2S_2O_8$ in the absence of additives and in the presence of uridine (0.5 mM), poly(U), and poly(dU), both at N/S = 3 and 10; $\Delta\kappa_1$ and $\Delta\kappa_2$ are the relative yields for conductivity increase at 10 μs and 1 s, respectively.

In the presence of an organic compound (uridine) the conductivity after the pulse shows only a fast increase and no further change (Fig. 1). We interpret this result by assuming that reaction (1) is followed by reactions (2) - (5). Most of the SO_4^- radical anions are scavenged analogously to reaction (5) since, among other arguments, the formation of protons is larger in the presence of uridine than in its absence.

Addition of poly(U) as the organic additive leads to the expected fast conductivity increase (relative yield, $\Delta\kappa_1$) but this is followed by a large slow increase ($\Delta\kappa_2$) in the range of 1/2 s. The magnitude of the increase in $\Delta\kappa_2$ is much smaller in the case of polydeoxyribouridylic acid (poly(dU)) and is negligible for uridine (Fig. 1) and uridine-5'-monophosphate (5'-UMP). This slow conductivity increase is mainly ascribed to counterion liberation on ssb formation. This interpretation is based on a comparison with the results of OH radical-induced ssb formation (which is in the same order of magnitude) and on the comparison with uridine, 5'-UMP and poly(dU). For poly(dU) it is now known that the yield of ssb formation by OH radicals is much lower than for poly(U).[14] Since, however, in poly(U) and in poly(dU) the same base is present, the uracil moiety, we expect that SO_4^- produces primarily the same base radicals in both polynucleotides (SO_4^- radical anions react predominantly with the base and not with the sugar). The large difference in the slow conductivity increase of the two polynucleotides cannot stem from the base radicals. We interpreted the difference to be the result of a much larger yield of ssb formation in poly(U) in contrast to the smaller yield in poly(dU), analogously to the OH radical-induced case. In fact the 2'-pathway, which has been proven[15] to be the pathway to ssb formation in poly(U), does not function in poly(dU) due to the absence of the 2'-OH group in the deoxyribose of poly(dU).

The rate constant of the slow conductivity increase depends on the ratio of the number of bases divided by the number of bound ruthenium complexes (N/S) (Table 1). The influence of the N/S ratio on the results is not fully understood as yet. The smaller yield and the higher rate of ssb formation at N/S = 10 may be explained by assuming a more efficient reaction of Ru(III) with the base radicals (see section 4.). At N/S = 3 all binding sites available for Ru(II) and/or Ru(III) are occupied and the mobility of bound ions is low.[5] At N/S = 10 the majority of sites is not populated, the mobility is higher, and the scavenging of base radicals by Ru(III) (eq. 9) is more effective.

A pH change has no influence on the rate constant of ssb formation (range 4-7) in contrast to ssb formation induced by OH radicals (Fig. 2). This is due to the strong binding of the ruthenium complex to the strand. The ruthenium complex is not released from the strand even at pH4 so that protons cannot increase their concentration at the surface of the strand and cannot catalyse the ssb formation as in the case with OH radicals when only potassium ions serve as counterions,[7] which readily exchange with protons.

Table 1.
Relative yield ($\Delta \kappa_2$) and rate constant for the slow conductivity increase reflecting ssb formation for poly(U) following a laser pulse ($\lambda_{exc} = 353$ nm).[a]

N/S	3	10
relative yield ($\Delta \kappa_2$)	0.09 $(0.03)^b$	0.04, 0.09^c $(0.03)^{b,c}$
k (s^{-1})	4.5 $(6.6)^b$	14 $(10)^{b,c}$

[a]Concentrations, $Ru(bpy)_3^{2+}$ 50 μM and $S_2O_8^{2-}$ 0.4 mM.
[b]Value in parenthesis refer to poly(dU). [c]Concentration of $S_2O_8^{2-}$ 1.0 mM.

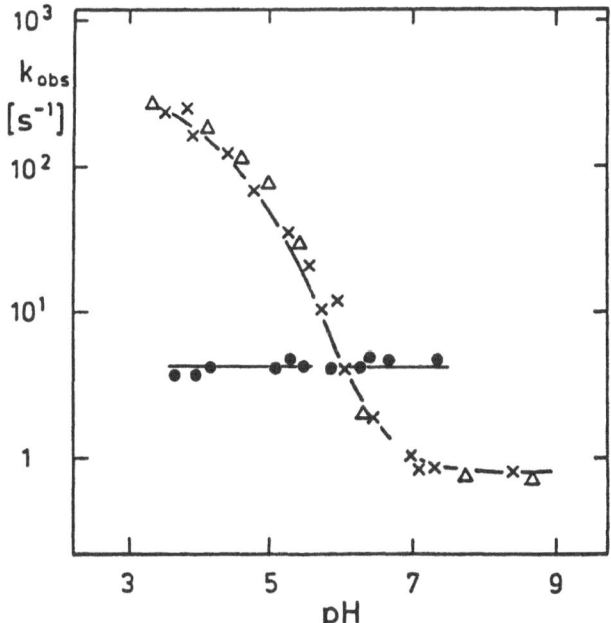

Figure 2. Comparison of the rate constants for ssb formation (conductivity increase) of poly(U) in argon-saturated aqueous solution versus pH under the conditions: OH radical-induced by pulse radiolysis (Δ),[7] laser-photoionization (x),[9] and photosensitized by $Ru(bpy)_3^{2+}/S_2O_8^{2-}$ (o).[11]

2. Flash photolysis studies, absorption measurements

At λ_{max} = 452 nm the extinction coefficient of Ru(bpy)$_3^{2+}$ is much higher than that of Ru(bpy)$_3^{3+}$. This allows us to measure the conversion of Ru(II) into Ru(III) and vice versa. Following laser pulse excitation of the system containing Ru(bpy)$_3^{2+}$ bound to poly(U) and $S_2O_8^{2-}$, the disappearance and reappearance of Ru(II) can be measured (Fig. 3). The conversion of Ru(II) into Ru(III) occurs in all cases in the μs range. In the absence of organic additives Ru(III) is stable within minutes.[6] In the presence of poly(U), but not of uridine. Ru(II) slowly reappears in the range of 1/2 s (Fig. 3). The rate constants for this process is k = 6.7 s^{-1} with poly(U) at N/S = 3 and k = 14 s^{-1} at N/S = 10. The reappearance is complete at N/S = 10 but incomplete at N/S = 3 (Fig. 3). The reactions which Ru(III) can undergo are shown in the eqs. (7) and (9). These reactions include that of the bound Ru(III) with base radicals of poly(U) (eq. 9). The detailed reason why the decay of Ru(III) is incomplete at N/S = 3 is unknown as yet.

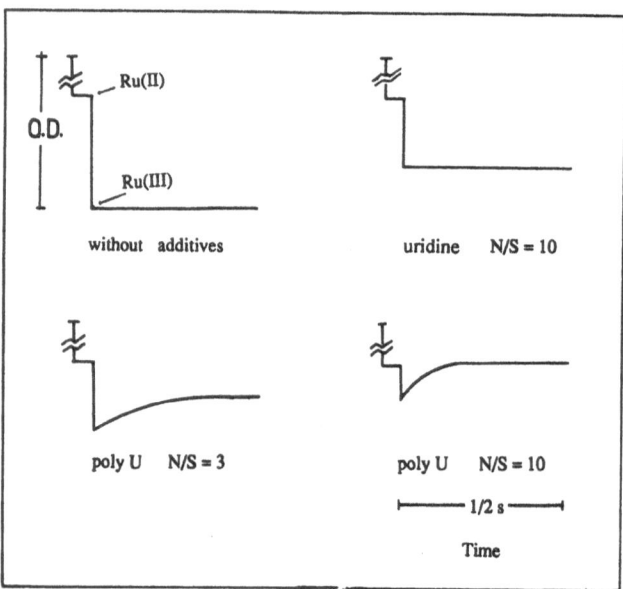

Figure 3. Transient absorption at 452 nm after the laser pulse in argon-saturated aqueous solution at pH5 and 25 °C using 50 μM Ru(bpy)$_3^{2+}$ and 1.0 mM K$_2$S$_2$O$_8$ in the absence of additives and in the presence of uridine (0.5 mM) and poly(U) at N/S = 3 and 10.

3. Mixing studies

Irradiation of a solution containing $Ru(bpy)_3^{2+}$ (50 μM) and $S_2O_8^{2-}$ (1 mM) with visible light leads to the formation of $Ru(bpy)_3^{3+}$ (eqs. 1 - 3) and ruthenium complexes carrying OH ligand adduct radicals (eq. 4). If this solution is mixed in the dark after a few seconds, when no free radicals are present any more, with solutions containing various concentrations of poly(U), an increase of the Ru(II) concentration at the expense of the Ru(III) concentration is observed spectroscopically at 452 nm. Why does a reaction take place at all? The redox potential of $Ru(bpy)_3^{3+}$ is only 1.26 V[16], whereas 1.73 V[17] is needed to oxidize thymine. The redox potential for uracil may be even higher. Since we have no evidence up to now that another oxidizing agent besides Ru(III) is present in the solution we have to assume that Ru(III) is able to oxidize the base in poly(U). The reason for this reaction to occur may be that the ruthenium complex is constantly bound to poly(U) which may change the activity of both agents. Furthermore, the redox potential of the base, which refers to the free base, may be considerably modified in the polymer. Both effects may lead to a shift in the redox potentials which may allow this reaction to proceed.

Another peculiarity is that on mixing at N/S < 10 the reaction does not go to completion. A portion of Ru(III) is not reconverted to Ru(II) depending on N/S. At the moment we cannot explain this result. It may be that either the mobility of Ru(III) along the strand or the redox properties of the system depend on the N/S ratio.

4. Reaction mechanism

As discussed above the results are interpreted by the following reaction scheme (eqs. 1 - 10), where polyU -OH and polyU(OH)$_2$ denote a base radical of poly(U) and a modified base (a glycol), respectively. The abbreviation (bpy$\cdot$-OH) denotes a ligand radical.

Reaction Scheme

[1] $Ru(II)/polyU \xrightarrow{h\nu} {}^*Ru(II)/polyU$

[2] ${}^*Ru(II)/polyU + S_2O_8^{2-} \longrightarrow Ru(III)/polyU + SO_4^{2-} + SO_4^{\cdot-}$

[3] $Ru(II)/polyU + SO_4^{\cdot-} \longrightarrow Ru(III)/polyU + SO_4^{2-}$

[4] $Ru(II)/polyU + SO_4^{\cdot-} \longrightarrow Ru(bpy)_2(bpy\dot{-}OH)/polyU + SO_4^{2-} + H^+$

[5] $Ru(II,III)/polyU + SO_4^{\cdot-} \longrightarrow Ru(II,III)/polyU^{\cdot+} + SO_4^{2-}$

[6] $\qquad\qquad\qquad\qquad\qquad \downarrow H_2O$

$\qquad\qquad\qquad Ru(II,III)/polyU\dot{-}OH + H^+$

[7] $Ru(III)/polyU + H_2O \longrightarrow Ru(II)/polyU\dot{-}OH + H^+$

[8] $Ru(II,III)/polyU\dot{-}OH + S_2O_8^{2-} \longrightarrow Ru(II,III)/polyU-(OH)_2 + SO_4^{\cdot-} + SO_4^{2-} + H^+$

[9] $Ru(III)/polyU\dot{-}OH + H_2O \longrightarrow Ru(II)/polyU-(OH)_2 + H^+$

[10] $Ru(II)/polyU\dot{-}OH \longrightarrow$ **single strand break and release of sensitizer**

The base radical is produced by reaction with $SO_4^{\cdot-}$ (eqs. 5,6) or Ru(III) (eq. 7). Model studies with 1,3-dimethyluracil and $SO_4^{\cdot-}$ have shown that 6-yl radicals are mainly produced in this reaction (eqs. 11 and 12).[18] With uracil derivatives a contribution of a 5-yl radical, which has oxidizing properties, has also to be considered.

[11]

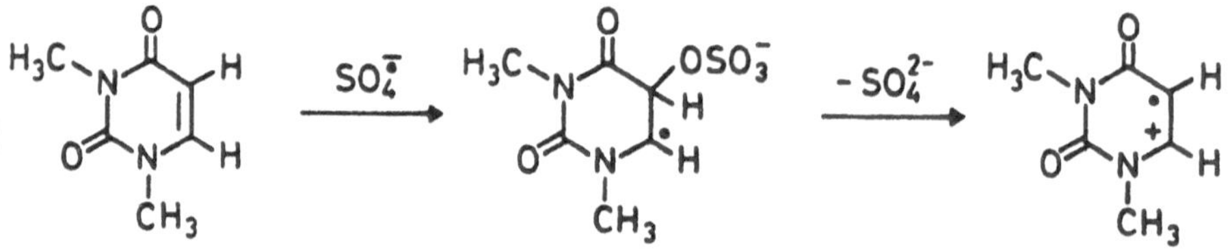

[12]

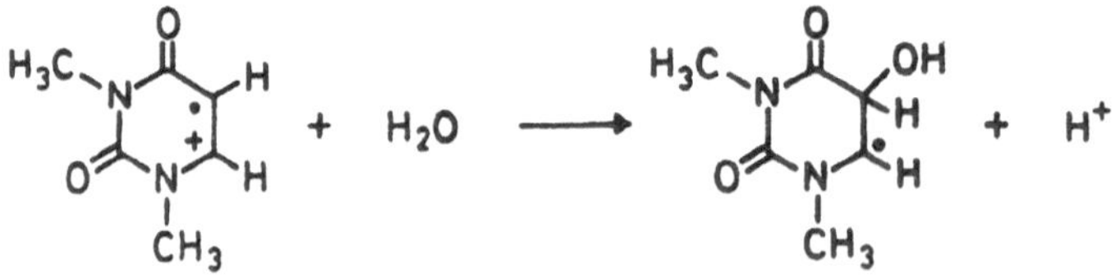

With Ru(III) similar base radicals as with $SO_4^{\cdot}$ are expected to be produced since Ru(III) is also oxidizing. The 6-yl radical has reducing properties and will easily react with Ru(III) by electron transfer, which should finally lead to Ru(II), H^+, and a uracil glycol. We have not as yet made an attempt to identify the glycol. However, the glycol has been observed in the above mentioned model system with $SO_4^{\cdot}$ as oxidizing species.[18]

Since the rate constant for ssb formation in the system $Ru(bpy)_3^{2+}$/ poly(U)/ $S_2O_8^{2-}$ at pH7 is higher than that induced by OH radicals, we assume that the reason for this difference is the reaction of Ru(III) with the base radical (mainly the 6-yl radical) by electron transfer (eq. 9), thereby preventing ssb formation and accelerating the rate. An indication for this reaction is the result that the decay rate constant of Ru(III) at N/S = 10 and the rate constant for ssb formation are very similar.

Strand break formation in this system is assumed to occur by a similar pathway (eq. 10) as induced by OH radicals. The base radical, in the photosensitized system produced by reactions (6) and (7), in the pulse radiolysis case by addition of OH radicals to the base, abstracts an H atom from the sugar at position 2' followed by a heterolytic splitting of the 3'-phosphoric acid ester bond.[15]

The complexity of the reaction scheme arises from the fact that the reducing base radical, which is an intermediate in ssb formation, can also react with $S_2O_8^{2-}$ and Ru(III). The reaction with $S_2O_8^{2-}$ should lead to a chain reaction (8). The reactions of $S_2O_8^{2-}$ and Ru(III) with the 6-yl radical prevent strand break formation, whereas the reaction of $SO_4^{\cdot}$ and of Ru(III) with the bases initiate strand break formation. The exact contribution of the various types of reactions are not known at present. The work shows, however, the involvement of secondary reactions with the metal ion complexes in processes leading to ssb formation and the complexity of systems containing heavy metals.

208

References

1. J.K. Barton, J.M. Goldberg, C.V. Kumar and N.J. Turro, J. Am. Chem. Soc. 1986, 108, 2081.

2. M.B. Fleisher, K.C. Waterman, N.J. Turro and J.K. Barton, Inorg. Chem. 1986, 25, 3549.

3. J.M. Kelly, A.B. Tossi, D.J. McConnell and C. OhUigin, Nucleic Acids Res. 1985, 13, 6017.

4. G.S. Manning, Quart. Rev. Biophys. 1978, 11, 179.

5. C. Stradowski, H. Görner, L.J. Currell and D. Schulte-Frohlinde, Biopolymers, 1987, 26, 189.

6. H. Görner, C. Stradowski and D. Schulte-Frohlinde, Photochem. Photobiol. 1987, 47, 15.

7. E. Bothe and D. Schulte-Frohlinde, Z. Naturforsch. 1982, 37c, 1191.

8. E. Bothe, G.A. Qureshi and D. Schulte-Frohlinde, Z. Naturforsch. 1983, 38c, 1030.

9. D. Schulte-Frohlinde, J. Opitz, H. Görner and E. Bothe, Int. J. Radiat. Biol. 1985, 48, 397.

10. D.N. Nikogosyan, A.A. Oraevsky, V.S. Letokhov, Z. Kh. Arbieva and E.N. Dobrov, Chem. Phys. 1985, 97, 31. J. Opitz and D. Schulte-Frohlinde, J. Photochem. 1987, 39, 145.

11. A.B. Tossi, H. Görner and D. Schulte-Frohlinde, to be published.

12. R. Humphry-Baker, J. Lilie and M. Grätzel, J. Am. Chem. Soc. 1982, 104, 422.

13. F. Bolletta, A. Juris, M. Maestri and D. Sandrini, Inorg. Chim. Acta, 1980, 44, L175.

14. M. Adinaryana, E. Bothe and D. Schulte-Frohlinde, to be published.

15. H. Hildenbrand and D. Schulte-Frohlinde, to be published.

16. A. Juris, V. Balzani, F. Barigelletti, S. Campagna, P. Belser and A. v. Zelewsky, Coord. Chem. Rev. 1988, 84, 85.

17. L. Kittler, G. Löber, F.A. Gollmick and H. Berg, Bioelectrochem. Bioenerg. 1980, 7, 503.

18. H.-P. Schuchmann, D.J. Deeble, G. Olbrich and C. v. Sonntag, Int. J. Radiat. Biol. 1987, 51, 441.

PHOTONUCLEASES

Ole Buchardt, Gunnar Karup, Michael Egholm, Troels Koch, Ulla Henriksen, Morten Meldal, Claus Jeppesen (a) and Peter E. Nielsen (a) Research Center for Medical Biotechnology, The H. C. Ørsted Institute, University of Copenhagen, Universitetsparken 5, DK-2100 Copenhagen Ø, Denmark, (a) Department of Biochemistry B, The Panum Institute, Blegdamsvej 3, DK-2200 Copenhagen

ABSTRACT. Synthetic reagents which cleave DNA in light-induced reactions ("photonucleases") are useful as tools in biochemistry and of potential interest in medicine. We have designed and tested a wide variety of new as well as known reagents as "photonucleases", and in this paper we review some of the results. Photochemical modification of DNA with azidoacridines, azidobenzamides, nitrobenzamides or 2-diazocyclopentadienylcarbonyloxy derivatives results in DNA cleavage, which is several fold increased upon subsequent piperidine treatment and which in some cases is A + G or G-specific.

1. INTRODUCTION

"Photonucleases" have a wide potential in both biological chemistry and medicine. A priori, they should be superior to analogous thermal reagents [1-4] because they will only be operating when light of the correct wavelengths is employed.

The following applications, in decreasing order of practicality, are possible at present for photonucleases:

1.   Photofootprinting of ligand binding sites on DNA.
2.   Attachment to substances whose specific interaction with DNA can subsequently be examined.
3.   DNA-sequencing.
4.   Attachment to ligands which are DNA-site specific for use as:
4a. "Restriction enzymes",
4b. Photodrugs.

P. E. Nielsen (ed.), Photochemical Probes in Biochemistry, 209–218.
© 1989 by Kluwer Academic Publishers.

We have so far tried the first three areas with some success.

Most of our experiments have been based on a design, where a DNA-intercalator (or more than one), in our hands always a 9-acridinylamino group, is attached to the DNA-cleaving ligand via a polymethylene linker, *e.g.*, 1.

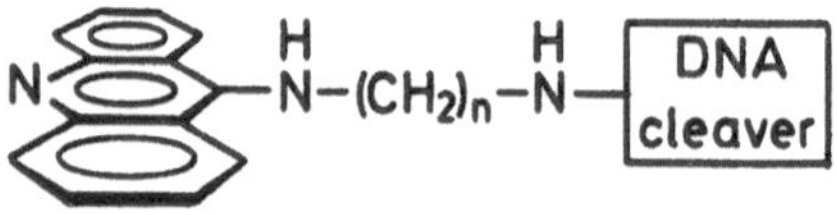

1

2. RESULTS & DISCUSSION

Somewhat to our initial surprise, the 4-azidobenzamido derivative 1a turned out to be DNA-photocleaving [5] as well as photobinding [6]. The mechanism presumably involves initial DNA photoalkylation, almost exclusively on A + G, which increases their lability. On treatment with base, depurination takes place and subsequent β-elimination results in purine specific strand breakage with formation of 5'-phosphates. The alkylation is believed to take place via singlet excitation, ring expansion and electrophilic attachment on N or conceivably O of the bases.

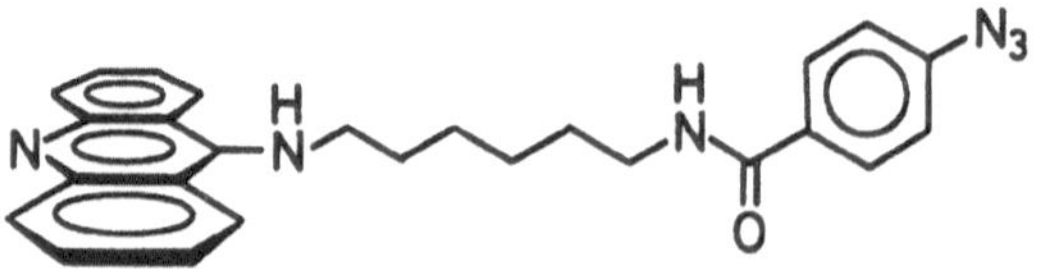

1a

Related results were obtained by C. Helene et al. [7,8] (cf. chapter 17) at about the same time with 4-azidophenacyl and azidoproflavine derivatives.

Subsequently, one of the ligands which was previously used in reagents for protein-DNA crosslinking [9], namely 6-azido-2-methoxy-9-aminoacridinyl was examined in the form of reagents 2 and 3 [10]. It is almost certain that the azidoacridine ligand intercalates (UV/VIS-DNA-titration), and we can assume that it is intercalated with the 9-position in the minor groove. The mechanism of photoreaction is not known, but most likely it involves ring

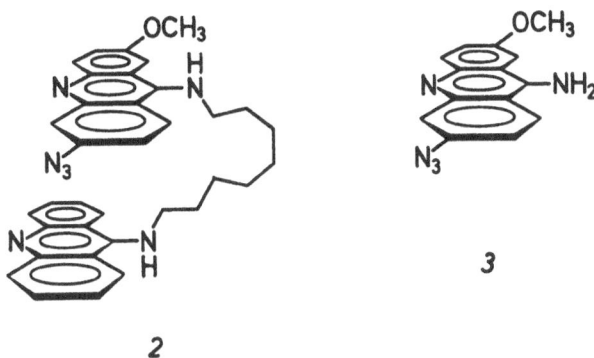

2

3

expansion like above. Some spontaneous photonicking is observed, and subsequent piperidine treatment results in a ten fold increase of the DNA cleavage, which is strongly sequence dependent and somewhat base specific (Table 1). These reagents are useful as photofootprining reagents.

TABLE 1. Activity of DNA "Photonucleases".

Reagent	Photo-nicking activity	Estimated quantum yield[a] Φ_{nick}	Base in-duced photo-cleavage yield[b]	Base specifi-city
1a	0.02	0.02	0.2	A + G
1b	0.03	0.03	0.09	none[d]
1c	>0.01[e]	10^{-5}	0.04	G>T>C>>A[c]
2	0.02	0.02	0.2	T·G>C>>A[c]
3	0.02	0.02	0.2	T·G>C>>A[c]
5	0.01	-	0.1	G
UO_2Ac_2	>0.01[e]	10^{-4f}	-	none[d]

Single strand nicks per reagent molecule assayed in a plasmid relaxation assay.

[a] Assuming that Φ_{dec} = 1 (decomposition) (cf. Geiger et al. [11]) for reagents 1a, b, 2 and 3.

[b] Estimated from the magnitude of the cleavage-enhancement by piperidine treatment.

[c] Highly sequence dependent.

[d] Somewhat sequence dependent.

[e] The photoreaction was still linear at 60 min of ir-radiation.

[f] At 300 nm.

212

By using the 2-diazocyclopentadienylcarbonyl ligand in the form of reagent **1b**, it was expected to obtain a non

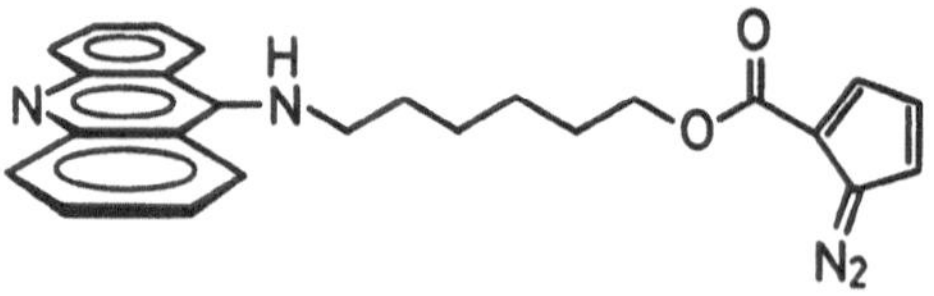

1b

selective reactivity, *i.e.*, sequence and base independent DNA photoreactivity. Indeed, no base specificity was observed, and little site specificity. The mechanism is not understood, but most probably non-discriminate alkylation takes place either at backbone or bases of the DNA, or at both. By adding di- or tri-9-acridinylamino containing reagents which are bis or tris intercalating, an interesting change in the specificity occurred leading to efficient selective G-specific cleavage. This phenomenon is not understood.

None of these reagents appear to execute their action through free radicals just as all evidence is against singlet oxygen. Thus their strength lies in only reacting where they are situated. However, they all have the inherent weakness that their photochemical reaction takes place irrespective of the availability of a biological target, *i.e.*, DNA. Thus the efficiency, *i.e.*, target modification per reagent molecule can of necessity be expected to be quite low until the ideal reagent is produced, where all or most photochemical transformations result in the warranted reaction. By combining nitrobenzamido [12-14] or nitrosulfamido ligands with DNA-affinity probes such reagents may have been found since these ligands appear to be photochemically inactive in the absence of DNA.

In principle, the reagent design was analogous to that previously discussed, with a 9-acridinylamino ligand to secure DNA-binding, and with a linker protruding into the minor groove, to which the nitroaryl ligand was attached either *via* a carboxamido link or a sulfonamido one, *e.g.* **1c**.

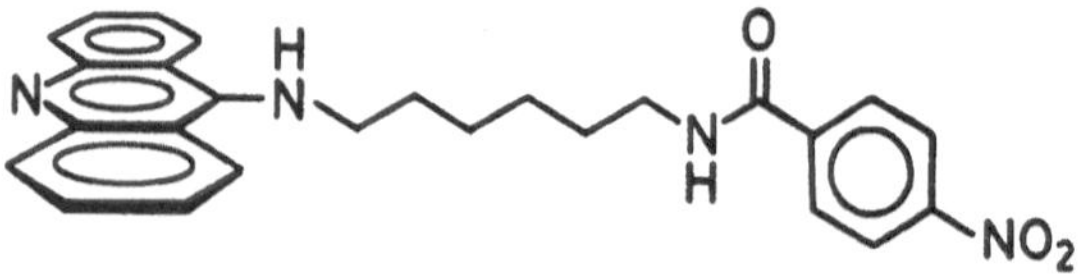

1c

The results with these nitro compounds can be summarized as follows [12]:

2- and 4-nitrobenzamido derivatives and 4-nitrosulf-amido derivatives appear to function analogously, whereas the 3-nitrobenzamido derivatives behave in a more compli-cated way. The presence of a second nitro group does not enhance the efficiency. Removal of the acridine group reduces the efficiency by a factor 10, as does moving the nitro group from position 4 to 2. Moving to position 3 causes a reduction in efficiency by a factor 30.

The efficiency of 2- and 4-, but not 3-nitro deri-vatives is directly related to their absorption spectrum (only $\lambda > 300$ nm tried). Φ for the best compound, *i.e.*, 1c at best wavelength (300 nm) = 10^{-5} nick/photon. The 3-nitrobenzamido derivatives are relatively efficient also in the acridinyl absorption region (λ 420 nm). Some linker length dependency was observed, but the dependency is complex. By subsequent piperidine treatment, the photo-cleavage is four fold increased. 3-Nitrobenzamido deriva-tives behave almost analogously to the 2- and 4-deriva-tives at 300 nm.

Photocleavage of DNA with the 3-nitrobenzamido de-rivatives at 420 nm gives DNA fragments which are nick translation substrates for DNA polymerase (up to at least 50%) (Fig. 1). The DNA products from photocleavage by 2-

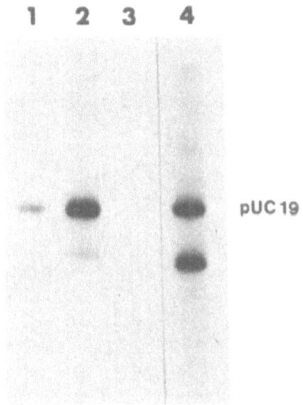

Fig. 1. Nick translation of photocleaved plasmid DNA. Supercoiled pUC 19 DNA was nicked (>90%) with either 4-nitrobenzamido-aminoacridine 1c (lane 3) the corresponding 3-nitro-5-methoxy reagent (lane 2) or DNase I (lane 4). Lane 1 is an untreated control. The DNA was subsequently incubated with E. coli DNA polymerase and α-^{32}P-dATP and analyzed by gel electrophoresis in 1% agarose followed by autoradiography. pUC 19 denotes the position of released circular pUC 19 DNA. The faster migrating labeled DNA in lane 4 corresponds to linearized pUC 19.

or 4-nitrobenzamido derivatives do not function as sub-strates for DNA polymerase (Fig. 1). Furthermore, the reaction is pH- (more efficient at acidic pH) and concen-tration dependent, the latter in a not straightforward manner. Electron donating substituents (-OMe, -OH) in the 5-position of the 3-nitrobenzamides increase the cleavage yield at 420 nm.

These observations have led us to suggest the follow-ing mechanism for 2- and 4-nitrobenzamido DNA-cleavage and for 3-nitro when excited directly, e.g., at 300 nm (Scheme 1), where the triplet $-NO_2$ abstracts a hydrogen atom from the 1'-, 3'- or 4'-position of the deoxyribose of the DNA backbone followed by O-transfer, and strand scission.

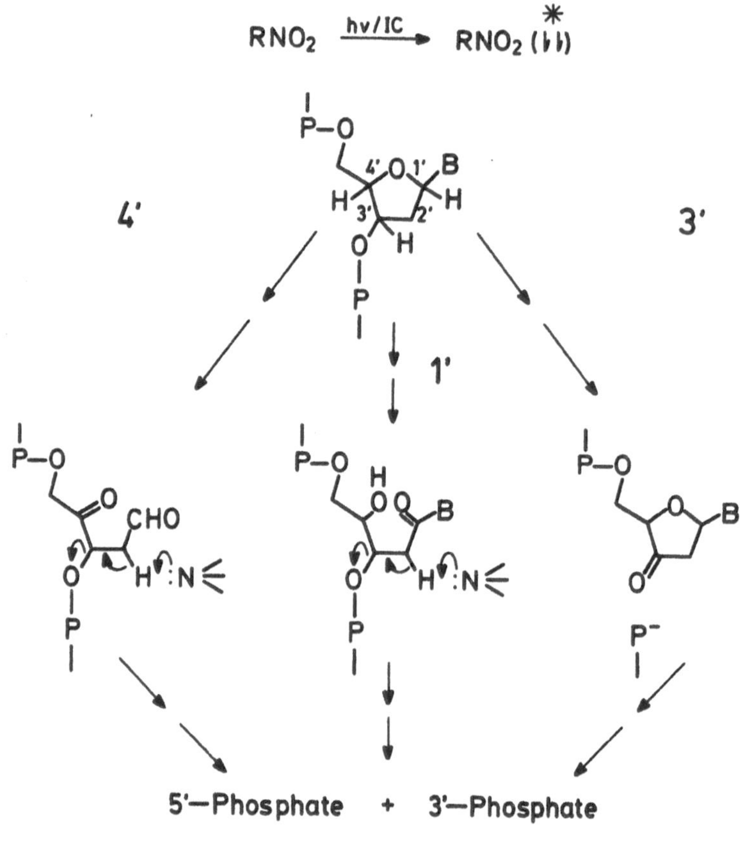

Scheme 1. Proposed mechanism for photocleavage of DNA by 4-NO_2-benzamides.

At 420 nm, a different mechanism takes over for the 3-nitro derivatives, where cleavage beyond the background

value only takes place when the reagent concentration
exceeds a threshold value (saturation of all intercalation
sites). We suggest a cleavage mechanism, where intermolecular electron transfer to the 3-nitrobenzamido ligand
from a non intercalated acridine generates the 3-nitro-
benzamido anion radical, which interacts with the phosphate linkage, thereby rendering it more susceptible to
acid hydrolysis (Scheme 2).

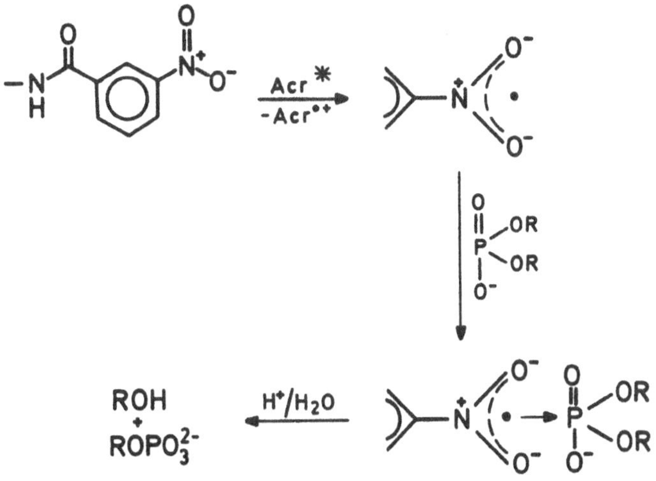

Scheme 2. Proposed mechanism for DNA photocleavage by
NO_2-benzamido-9-acridinylamines.

In order to exploit such DNA-photocleaving ligands
for the construction and characterization of DNA sequence
specific peptides we have prepared a nitrosulfamido derivative, 4, which can be used in automated peptide syn-

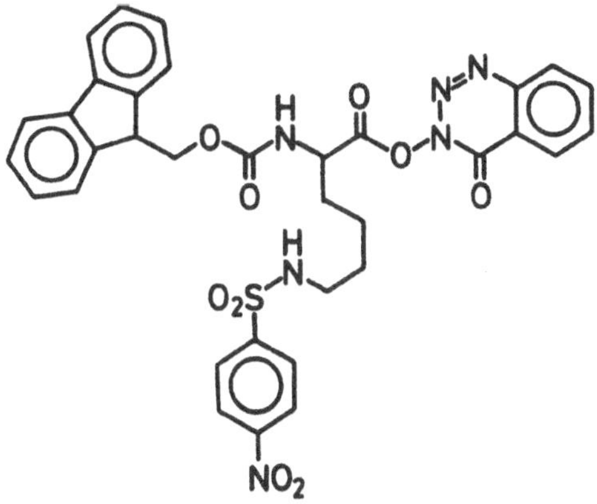

4

thesis, and we are now able to synthesize any peptide sequence containing its own DNA-cleaver [15].

All the reagents discussed so far have been designed to contain either a DNA-intercalator or another DNA-affinity probe. However, there is also a demand for "photonucleases" which perturb the DNA as little as possible. Several such reagents have now been found and are described below:

Uranyl acetate ($UO_2(AcO)_2$) and uranyl nitrate ($UO_2(NO_3)_2$) serve as photonucleases at long wavelengths ($300 < \lambda < 450$ nm) [16]. The cleaving efficiency is not due to singlet oxygen, but it may be due to the photochemical uranyl mediated formation of HO-radicals at the DNA backbone to which UO_2^{2+} is attached by electrostatic forces very much behaving like Mg^{2+} (H. Eggert, personal communication). The DNA-cleavage is practically sequence and base independent, and the reagent(s) appear to have considerable potential as photofootprinting reagents, etc. Furthermore, the ubiquitous presence of uranyl in seawater raises the question of its importance in "natural" mutagenesis.

Finally, it was most recently shown that S-arylmethylthiouronium salts photoinduce DNA cleavage and guanine specific alkaline sensitive sites _via_ a carbonium ion mechanism (Fig. 2, Scheme 3) [17].

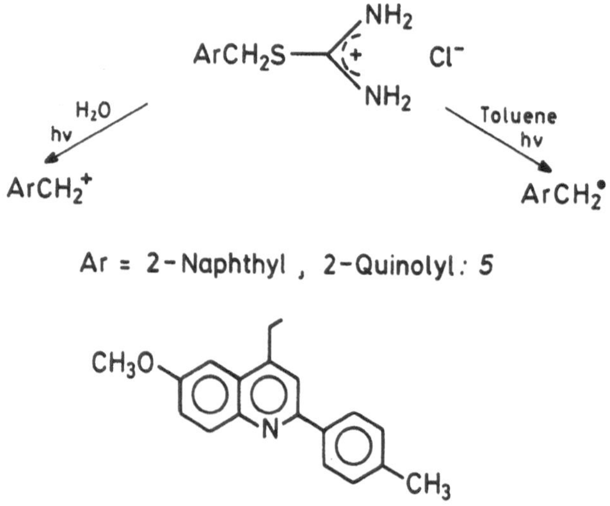

Ar = 2-Naphthyl , 2-Quinolyl: 5

Scheme 3. Structure of S-arylmethylthiouronium salts and photoreaction mechanism.

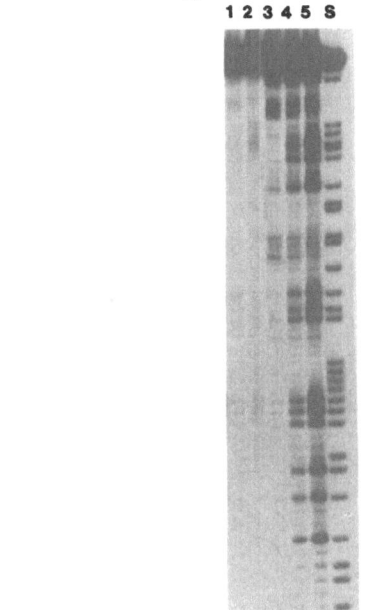

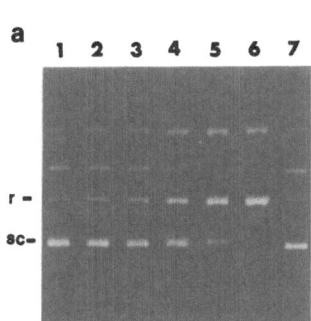

Fig. 2a. Photonicking of pUC 19 DNA by reagent 5. Each sample contained 0.3 µg DNA in 10 µl 10 mM Tris-HCl 1 mM EDTA pH 7.5 (TE). All samples except that of lane 6 were irradiated for 30 min at 300 nm (Philips TL 20W/12 fluorescent light tube) and the following concentrations of reagent 5 were used 0, 03, 1, 3, 10 and 30 µg/ml (lanes 1-6, respectively). The samples were analyzed by gel electrophoresis in 1% agarose and the DNA visualized by staining with ethidium bromide. Sc: supercoiled DNA, r: relaxed (nicked) DNA.

Fig. 2b. Sequence dependency of the photomodification of DNA by reagent 5. All samples contained 0.1-0.2 pmol ^{32}P-(EcoRI) endlabeled pUC 19 fragment (Pun II, 90 base pairs and 0.25 µg calf thymus DNA in 100 µl TE and all samples except that of lane 1 were irradiated for 30 min at 300 nm. Samples of lanes 1, 3, 4 and 5 were treated with 1 M piperidine (20 min, 90°C). The concentrations of 5 used were 2, 2, 0, 2 or 10 µg/ml (lanes 1-5, respectively). S is an A+G sequence reaction. The strong bands in lane 5 correspond to cleavage at G-residues.

ACKNOWLEDGEMENTS. This work was supported by the NOVO Foundation (a Hallas Møller fellowship to P. E. Nielsen), the Danish Biotechnology Programme and the Danish Natural Science Research Council (a candidate fellowship to CJ), which is gratefully acknowledged.

REFERENCES
1. Dervan, P. B. *Science* **232**, 464-471 (1986).
2. Sigman, D. S. *Acc. Chem. Res.* **19**, 180-186 (1986).
3. Moser, M. E. and Dervan, P. B. *Science* **238**, 645-650 (1987).
4. Sluka, J. P., Morvath, S. J., Bruist, M. F., Simon, M. I. and Dervan, P. B. *Science* **238**, 1129-1132 (1987).
5. P. E. Nielsen, C. Jeppesen, M. Egholm and O. Buchardt, *Nucl. Acids Res.* **16**, 3877-3888 (1988).
6. P. E. Nielsen, J. B. Hansen, T. Thomsen and O. Buchardt, *Experientia*, **39**, 1063-1072 (1983).
7. Praseuth, D., Chassignol, M., Takasugi, M., Le Doan, T., Thuong, N. T. and Helene, C., *J. Mol. Biol.* **196**, 939-942 (1987).
8. Le Doan, T., Perrouault, L., Praseuth, D., Mabhoub, N., Decout, J.-L., Thuong, N. T., Lhomme, J. and Helene, C. *Nucleic Acids Res.* **15**, 7749-7760 (1987).
9. P. E. Nielsen, J. B. Hansen, and O. Buchardt, *Biochem. J.* **223**, 519-526 (1984).
10. P. E. Nielsen, C. Jeppesen, U. Henriksen, and O. Buchardt, *Biochem. Pharmacol.* **37**, 1831-1832 (1988); C. Jeppesen, O. Buchardt, U. Henriksen, and P. E. Nielsen, *Nucl. Acids Res.* **16**, 5755-5770 (1988).
11. Geiger, M. W., Elliot, M. M., Karocostas, V. D., Moricone, T. J., Salmon, J. B., Sideli, V. L. and St. Onge, M. A. *Photochem. Photobiol.* **40**, 545-548 (1984).
12. O. Buchardt, M. Egholm, G. Karup and P. E. Nielsen, *J. Chem. Soc., Chem. Commun.* **1987**, 1696-1697.
13. P. E. Nielsen, C. Jeppesen, M. Egholm and O. Buchardt, Biochemistry, **27**, 6338-6343 (1988).
14. O. Buchardt, M. Egholm, G. Karup, and P. E. Nielsen, (in preparation).
15. G. Karup, M. Meldal, P. E. Nielsen, and O. Buchardt, *Int. J. Pep. Prot. Res.*, in press.
16. P. E. Nielsen, C. Jeppesen, and O. Buchardt, *FEBS Lett.* **235**, 122-124 (1988).
17. O. Buchardt, G. Karup, U. Henriksen, C. Jeppesen, and P. E. Nielsen (in preparation).

SEQUENCE-TARGETED PHOTOCHEMICAL REACTIONS IN SINGLE-STRANDED AND DOUBLE-STRANDED NUCLEIC ACIDS BY OLIGONUCLEOTIDE-PHOTOSENSITIZER CONJUGATES

C. Hélène[*], T. Le Doan[*] and N.T. Thuong[+]
[*]Laboratoire de Biophysique
Muséum National d'Histoire Naturelle
INSERM U.201 - CNRS UA.481
43 rue Cuvier - 75005 PARIS (FRANCE)

[+]Centre de Biophysique Moléculaire
CNRS - 45071 ORLEANS CEDEX 05 (FRANCE)

ABSTRACT. Several photoactive groups have been covalently attached to the 3'- or 5'-ends of oligodeoxynucleotides. These include proflavine, azidoderivatives (azidophenacyl, azidoproflavine), porphyrins, 9-aminoacridine, ellipticine and diazapyrene derivatives. These derivatized oligonucleotides bind to their complementary sequences on single-stranded nucleic acids. Upon visible or near-UV irradiation of these complexes the oligonucleotide becomes covalently crosslinked to the target nucleic acid. Upon treatment by 1 M piperidine at 90°C the photocrosslinked species are converted to cleaved products. The location of the cleavage sites identifies the nucleotides that were involved in the photocrosslinking reactions. Only ellipticine and diazapyrene derivatives led to cleavage reactions at neutral pH.
Homopyrimidine oligodeoxynucleotides carrying a photoactive group can bind to the major groove of duplex DNA at homopurine.homopyrimidine sequences. A local triple helix is formed. Upon visible or near-UV irradiation crosslinking of the oligonucleotide is observed on both strands of the double helix. Cleavage reactions occur at the photocrosslinking sites upon alkali treatment. The homopyrimidine oligonucleotide is bound (hydrogen-bonded) in a parallel orientation with respect to the purine-containing strand of double-helical DNA.
Oligodeoxynucleotides can be synthesized with the $[\alpha]$-anomers of nucleoside units instead of the natural $[\beta]$-anomers. Oligo-$[\alpha]$-deoxynucleotides are resistant to nucleases and form hybrids with complementary natural ($[\beta]$) sequences. Photocrosslinking and cleavage reactions were used to demonstrate that the two strands of an $[\alpha]$(DNA)-$[\beta]$(DNA) double helix adopt a parallel orientation. An $[\alpha]$-octathymidylate is able to bind to the major groove of duplex DNA containing a $d(A_8).d(T_8)$ sequence. Its orientation is parallel to the purine-containing strand as is the corresponding $[\beta]$ analogue.

P. E. Nielsen (ed.), Photochemical Probes in Biochemistry, 219–229.
© 1989 by Kluwer Academic Publishers.

Oligonucleotide-photosensitizer conjugates open new possibilities to photoinactivate specific messenger RNAs or specific genes and therefore to control gene expression in a highly selective way. They can also be used to induce site-directed mutations in predetermined regions of a cellular genome.

1. INTRODUCTION

Artificial control of gene expression can be achieved by oligodeoxynucleotides (oligo dN) complementary to specific messenger RNAs (see Hélène, 1987 ; Toulmé and Hélène, 1988, for reviews). When the oligonucleotide binds upstream of or covers the AUG initiation codon it may prevent ribosome assembly. Ribosomes possess an unwinding activity that will release an oligonucleotide bound to the coding region of the mRNA. However oligodeoxynucleotide-mRNA duplexes are substrates for a ribonuclease, RNase H, which selectively cleaves the mRNA and therefore induces an irreversible inhibition of mRNA translation (Cazenave et al., 1987). Such an RNase H activity is present, e.g., in the cytoplasm of **Xenopus laevis** oocytes, which explains the high efficiency of oligodeoxynucleotides at inhibiting translation of microinjected mRNAs.

In order to increase the stability of the complexes formed by an oligodeoxynucleotide with a complementary sequence we have covalently attached intercalating agents to one (or both) end(s) of the oligonucleotide (Asseline **et al.**, 1984). An additional binding energy is provided by the intercalating agent which inserts its aromatic ring between the terminal base pairs of the duplex structure. Such oligonucleotide-intercalator conjugates have been shown to inhibit the cytopathic effect of influenza virus on cells in culture (Zérial **et al.**, 1987) or to kill trypanosomes in culture (Verspieren **et al.**, 1987). These biological effects are very specific to the oligonucleotide sequence which must be complementary either to the 3'-end of influenza virus RNAs or to the common 39 nucleotides at the 5'-end of all trypanosomal mRNAs.

In order to increase the efficacy of oligonucleotides it is necessary to increase their lifetime in living cells, in particular to make them resistant to endogenous nucleases. Several strategies have been developed : i) a phosphonate or a phosphorothioate can be substituted to the phosphate group of phosphodiester linkages (Blake **et al.**, 1985 ; Marcus-Sekura **et al.**, 1987) ; ii) the natural $[\beta]$ anomer of the nucleoside units can be replaced by its synthetic $[\alpha]$ anomer (Morvan **et al.**, 1987 ; Thuong **et al.**, 1987). In the first case two diastereoisomers are present for each linkage. Therefore an oligonucleotide with n phosphonates or phosphorothioate linkages contains 2^n isomers which bind with different affinities to the complementary sequence. Such an heterogeneity is not encountered with oligo-$[\alpha]$-deoxynucleotides.

The efficiency of oligonucleotides at inhibiting biological processes should be increased if the oligonucleotide could induce irreversible reactions in its target sequence. This can be achieved by covalent attachment of a reactive group at one (or both) end(s) of the oligonucleotide. This group can be chemically activated to induce either covalent attachment of the oligonucleotide to its target sequence or cleavage of the target nucleic acid. Alternatively, the substituent can be activated by light irradiation to induce irreversible reactions. The results obtained with this photochemical approach are surveyed in the present review.

2. SEQUENCE-SPECIFIC PHOTOCHEMICAL REACTIONS IN SINGLE-STRANDED NUCLEIC ACIDS

Some of the photoactive groups covalently attached to oligonucleotides are shown in figure 1. The linkers used to tether the photosensitizers to the oligonucleotides can be found in the original publications. We have developed a general method which consists in attaching a thiophosphate group to the 5'- or 3'-end of the oligonucleotide followed by reaction of the thiophosphate group with a bromoalkyl derivative of the photosensitizer (Asseline and Thuong, 1988).

2.1. Photocrosslinking reactions

Two derivatives carrying an azido group (1 and 2, figure 1) were shown to induce crosslinking reactions with a nucleic acid containing the complementary sequence of the oligonucleotide upon irradiation at wavelengths longer than 300 nm (Praseuth et al., 1987 ; 1988a ; Le Doan et al., 1987). The chemistry of the crosslinking reaction has not been investigated yet. The main reaction occurred at the thymine immediately adjacent to the complementary sequence on the target nucleic acid. Upon treatment of the crosslinked samples with 1 M piperidine at 90°C cleavage reactions were observed at the crosslinked bases. This cleavage allowed us to characterize the orientation of the two chains in the double helix. When the oligodeoxynucleotide was synthesized with the natural β-anomers of nucleotide units the two chains were found to run in an antiparallel orientation, as expected on the basis of the Watson-Crick model for a double helix. When the oligodeoxynucleotide was synthesized with the α-anomers, the double helix formed with a natural DNA single-stranded sequence was characterized by a parallel orientation of the two chains (Praseuth et al., 1987 ; 1988a).

Oligonucleotides carrying psoralen derivatives (7, 8) have been synthesized (Lee et al., 1988 and unpublished results from our laboratory). Photocrosslinking occurred to a thymine base adjacent to the complementary sequence. A psoralen derivative can also be photochemically attached to an internal thymine at a TpA sequence inside an oligonucleotide (Gamper et al., 1987). Upon further irradiation of the complex formed by the derivatized oligonucleotide with a complementary sequence, crosslinking of the two chains can be induced. This is due to the property of psoralen derivatives to form two covalent adducts with thymines involving both the pyrone and the furane double bonds. In both cases a cyclobutane ring is formed.

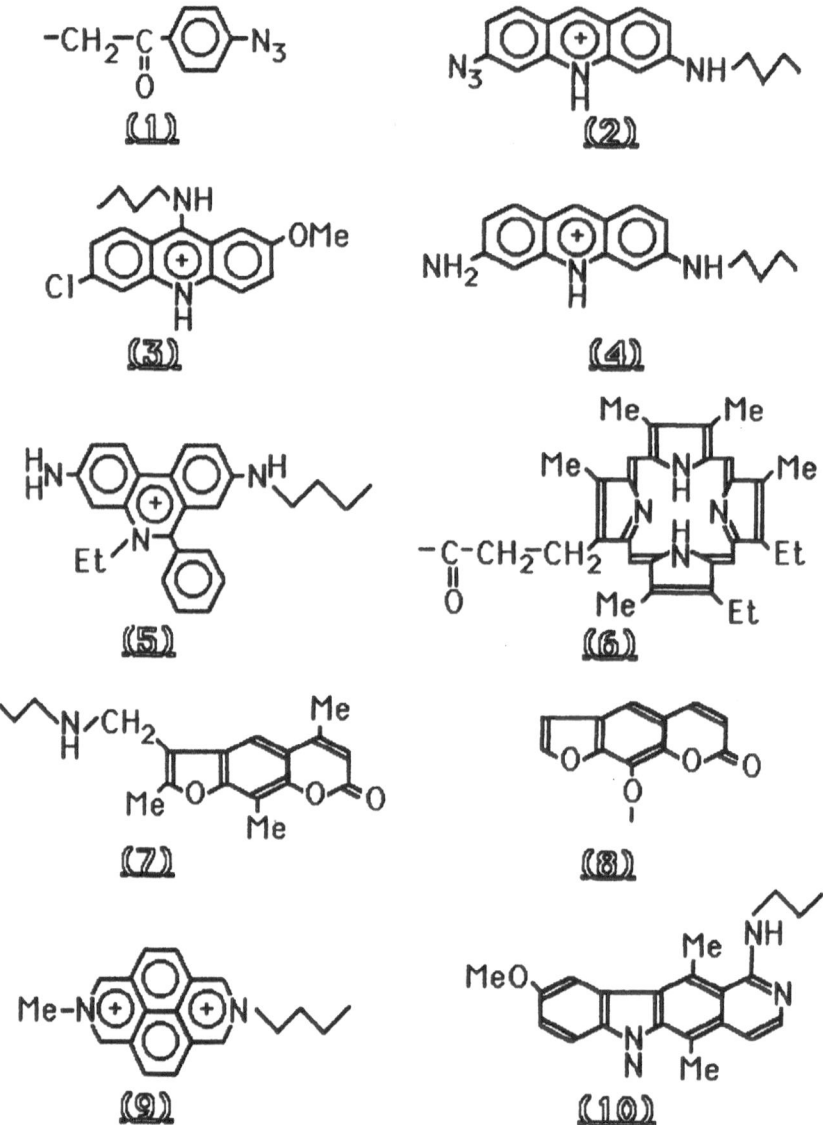

Figure 1. Chemical structure of photoactive groups which have been covalently attached to oligodeoxynucleotides. The site of attachment to the terminal phosphate (5' or 3') of the oligonucleotide is indicated. Different linkers have been used to tether the photoactive groups (see text for references).

2.2. Photocrosslinking and photooxidation reactions

When compounds 3, 4, 6, 9 and 10 were covalently linked to oligonucleotides visible light irradiation induced crosslinking of the oligonucleotide to the target nucleic acid. In addition photooxidation
reactions were induced in the target, when the oligonucleotide carried
proflavine (4) (Praseuth et al., 1988b), porphyrin (6) (Le Doan et
al., unpublished results) or diazapyrene (9) (unpublished results).
The photooxidation reactions involve both singlet oxygen and direct
electron transfer from guanines to the dye. The relative contribution
of these two mechanisms depends on the photosensitizer. Singlet oxygen
is produced by energy transfer from excited proflavine (4) and porphyrin (6) to ground-state molecular oxygen. The diazapyrene derivative (9) is a strong electron acceptor which leads to base oxidation
products after electron transfer reactions.

The photooxidized and crosslinked samples could be separated
by gel electrophoresis and separately treated with 1 M piperidine at
90°C. Both gave rise to cleavage reactions at crosslinked and photooxidized sites. The location of the cleavage sites indicated whether
singlet oxygen was involved in the photooxidation reactions. Singlet
oxygen is a diffusing species with a long lifetime (4 μs in H_2O at
20°C) and can therefore react far away from the site where it is generated. In contrast, electron transfer reactions require a close
contact between the two partners.

2.3. Photoinduced cleavage

Of all photosensitizers tested until now, only the diazapyrene derivative (9) and the ellipticine derivative (10) induced site-directed
cleavage reactions at neutral pH when covalently linked to an oligonucleotide. Direct cleavage was accompanied by the crosslinking and
oxidation reactions described above, which in turn led to cleavage
after alkaline treatment. It should be noted that high intensity
laser irradiation of an oligonucleotide-dye (5) conjugate also led to
direct cleavage of a complementary nucleic acid at neutral pH
(Benimetskaya et al., 1988). This reaction involves biphotonic processes in contrast to the oligonucleotide-diazapyrene (9) or ellipticine (10) conjugates which cleaves via a monophotonic process (unpublished results from our laboratory). The latter compounds provide
the first examples of a sequence-specific photoendonuclease.

2.4. Triple-strand formation

In order to test the relative efficiencies of different photoactive
groups at inducing sequence-specific reactions these groups were
tethered to an oligo(dT)$_8$ sequence. The target was a 27-mer single-
stranded fragment containing an oligo(dA)$_8$ sequence. The location of
the photocrosslinking reactions described above (after cleavage under
alkaline conditions) revealed the position of the photoactive group
with respect to the bases of the target nucleic acid. At low ionic

concentration, photocrosslinking occurred only on one side of the oligo(dA)$_8$ sequence as expected from an antiparallel orientation of the two strands. In contrast, at high ionic concentration, photocrosslinking was observed on both sides of the target sequence (Praseuth et al., 1988a). This observation indicated that two molecules of oligo(dT)$_8$ carrying the photocrosslinker were able to form a triple helix with the oligo(dA)$_8$ sequence. The two oligo(dT)$_8$ had opposite orientations with respect to the oligo(dA)$_8$ sequence as revealed by the location of the photocrosslinking sites. One oligo(dT)$_8$ formed Watson-Crick hydrogen bonds and ran antiparallel to the complementary sequence while the second oligo(dT)$_8$ formed Hoogsteen hydrogen bonds with the Watson-Crick A-T base pairs and ran parallel to the oligo(dA)$_8$ strand.

3. SEQUENCE-SPECIFIC PHOTOCHEMICAL REACTIONS IN DOUBLE-STRANDED NUCLEIC ACIDS

Watson-Crick A.T and G.C base pairs can be recognized in the major groove of DNA by thymine and cytosine, respectively. Thymine forms two Hoogsten-type hydrogen bonds with adenine in an A.T base pair. Cytosine must be protonated to form two hydrogen bonds with a G.C base pair. The double helix is a polyelectrolyte characterized by a condensation of cations, including H$^+$. The local pH around the double helix is therefore lower than in the bulk solution. The difference depends on the total ionic concentration. Consequently, cytosine can be protonated at bulk pH values much higher than its pK (4.1).

In order for an oligonucleotide to bind to the major groove of DNA, oligonucleotides containing pyrimidines only were synthesized. The recognized sequence contains all purines on one strand. The homopyrimidine oligonucleotide binds in a parallel orientation with respect to the purine-containing strand of the double helix. More recently it has been shown that a homopurine oligonucleotide could also bind to a homopurine.homopyrimidine sequence of DNA in a parallel orientation with respect to the purine-containing strand (Cooney et al., 1988).

A homopyrimidine oligonucleotide carrying a photoactive group could bind to the major groove of DNA at homopurine.homopyrimidine sequences and induce photosensitized reactions on either strand of the double helix. We demonstrated that an oligo(dT)$_8$ covalently linked to an azidophenacyl (1) (Praseuth et al., 1988a) or an azidoproflavine (2) (Le Doan et al., 1987) derivative could be photocrosslinked to both strands of a 27-mer duplex DNA fragment containing a (dA)$_8$.(dT)$_8$ sequence. The crosslinked bases could be identified after treatment of the DNA fragment with hot piperidine which led to cleavage reactions. The location of the crosslinked bases indicated that the oligo(dT)$_8$ was bound in a parallel orientation with respect to the oligo(dA)$_8$ sequence.

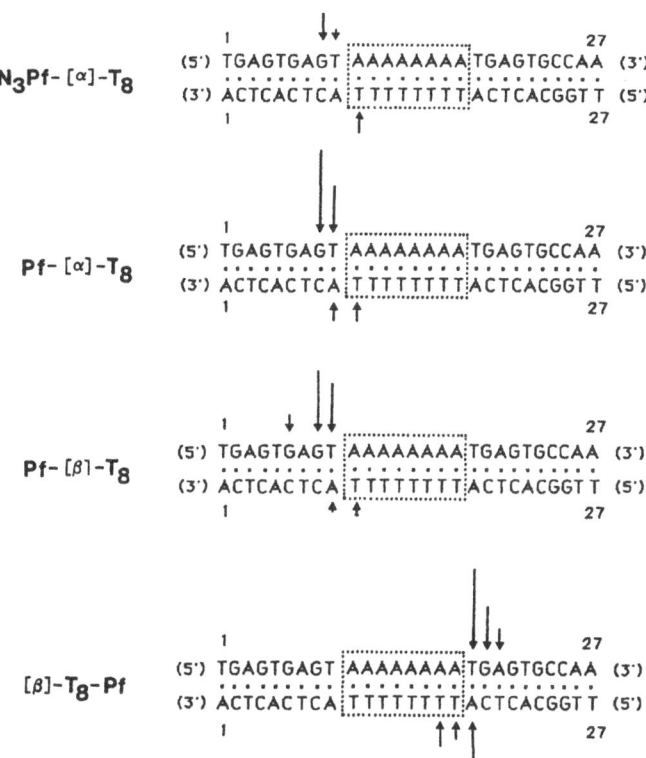

Figure 2. Sites of cleavage observed on double-stranded DNA after irradiation and treatment by 1 M piperidine at 90°C. The target duplex sequence is boxed. The octathymidylate carrying either proflavine (Pf, 4 on figure 1) or azidoproflavine (N$_3$Pf, 2 on figure 1) was synthesized with either the natural β-anomers or the synthetic α-anomers of thymidine. The site of attachment is the 5'-phosphate for the top three compounds and the 3'-phosphate for the last one.

An oligo-[α]-deoxynucleotide ([α]-(dT)$_8$) was covalently linked to azido derivatives 1 and 2. Upon visible light irradiation crosslinking of the oligo-[α]-(dT)$_8$ was observed on both strands of the DNA fragment. Surprisingly this oligo-[α]-nucleotide was also bound in a parallel orientation with respect to the oligo(dA)$_8$ sequence as was the oligo(dT)$_8$ synthesized with the natural [β] anomers (it should be remembered that when the target was a single-stranded DNA fragment the two oligonucleotides (α and β) adopted a reverse orientation (see above)). The same conclusion was reached independently of the photoactive group used to photocrosslink the oligonucleotide. As shown in figure 2, proflavine and azidoproflavine could be used as photocrosslinkers. In all cases the oligothymidylate was oriented <u>parallel</u> to the adenine-containing strand of the double helix, independently of the anomeric configuration of the nucleotide units. However this conclusion seems to be a peculiar property of oligothymidylate sequences : an oligo-[α]-nucleotide of sequence d(TCTCCTCCTTT) was recently shown in our laboratory to bind to the major groove of a DNA double helix in an <u>antiparallel</u> orientation with respect to the purine-containing strand whereas an oligo-[β]-deoxynucleotide was bound to the same sequence only when it was synthesized with a <u>parallel</u> orientation. Oligo-[α]-thymidylates also behaved differently from other sequences when the target was a single-stranded <u>RNA</u>. An oligo-[α]-(dT)$_8$ was bound in an antiparallel orientation as was the natural oligo-[β]-(dT)$_8$ (Sun et al., 1988). In contrast a 17-mer oligo-[α]-deoxynucleotide complementary to rabbit globin messenger RNA was bound only when it was synthesized in a parallel orientation with respect to the complementary sequence (Cazenave et al., unpublished results). These results based upon photocrosslinking and cleavage reactions are summarized in Table 1.

TABLE 1. Relative orientation of the strands in double or triple helices involving oligodeoxynucleotides synthesized with the natural [β] or synthetic [α] anomers of deoxynucleotides.

	Oligonucleotide	Target[1]	Orientation of oligonucleotide
double helix	β-(dT)$_8$	ssDNA	antiparallel
	α-(dT)$_8$	ssDNA	parallel
	β-(dT)$_8$	ssRNA	antiparallel
	α-(dT)$_8$	ssRNA	antiparallel
	α-d(ACACCTTCTT CAACCAC)	ssRNA[2]	parallel
triple helix	β-(dT)$_8$	dsDNA	parallel to (dA)$_8$
	α-(dT)$_8$	dsDNA	parallel to (dA)$_8$
	β-d(TTTCCTCCTCT)	dsDNA	parallel to (Pu)$_n$[3]
	α-d(TCTCCTCCTTT)	dsDNA	antiparallel to (Pu)$_n$[3]

(1) ss and ds refer to single-stranded and double-stranded nucleic acids.
(2) rabbit β-globin mRNA.
(3) The dsDNA contained the sequence (5') AAAGGAGGAGA (3') refered to as (Pu)$_n$.

4. CONCLUSION

Covalent attachment of a photoactive group to an oligonucleotide allows one to target photochemical reactions to specific sequences on single-stranded DNA or RNA. Any accessible sequence can thus be modified. Homopyrimidine oligonucleotides can be used to target photochemical reactions to specific sequences on double-stranded DNA. This is presently limited to homopurine.homopyrimidine sequences but it is likely that oligonucleotide modifications will provide access to more diverse sequences on duplex DNA.

The sequence-specific photochemical reactions that have been described involve photocrosslinking, photooxidation and cleavage reactions. They can be used to control gene expression. For example, irreversible photochemical reactions on messenger RNAs should block the translation process. When occurring on duplex DNA these reactions might inhibit gene transcription. The same photochemical reactions could be used to induce site-directed mutagenesis. Further studies should indicate whether they can be achieved inside living cells. The development of sequence-specific photoendonucleases should also open new possibilities for the control of gene expression and provide new tools for molecular and cellular biology.

5. REFERENCES

U. ASSELINE, M. DELARUE, G. LANCELOT, F. TOULME, N.T. THUONG, T. MONTENAY-GARESTIER & C. HELENE
'Nucleic acid-binding molecules with high affinity and base sequence specificity : intercalating agents covalently linked to oligodeoxynucleotides'
Proc. Natl. Acad. Sci. USA, 81 (1984) 3297-3301.

U. ASSELINE & N.T. THUONG
'Solid-phase synthesis of modified oligodeoxyribonucleotides with an acridine derivative or a thiophosphate group at their 3' end'
Tetrahedron letters (in press).

L.Z. BENIMETSKAYA, N.V. BULYCHEV, A.L. KOZIONOV, A.A. KOSHKIN, A.V. LEBEDEV, S. YU. NOVOZILOV & M.I. STOCKMANN
'Highly effective complementary addressed laser modification (scission of oligodeoxyribonucleotides)'
Bioorg. Khim., 14 (1988) 48-57.

K.R. BLAKE, A. MURAKAMI, S.A. SPITZ, S.A. GLAVE, M.P. REDDY, P.O.P. TS'O & P.S. MILLER
'Hybridization arrest of globin synthesis in rabbit reticulocyte lysates and cells by oligodeoxyribonucleoside methylphosphonates'
Biochemistry, 24 (1985) 6139-6145.

C. CAZENAVE, N. LOREAU, N.T. THUONG, J.J. TOULME & C. HELENE
'Enzymatic amplification of translation inhibition of rabbit β-globin mRNA mediated by anti-messenger oligodeoxynucleotides covalently linked to intercalating agents'
Nucleic Acids Res., 15 (1987) 4717-4736.

M. COONEY, G. CZERNUSZEWICZ, E.H. POSTEL, S.J. FLINT & M.E. HOGAN
'Site-specific oligonucleotide binding represses transcription of the human c-myc gene in vitro'
Science, 241 (1988) 456-459.

H.B. GAMPER, G.D. CIMINO & J.E. HEARST
'Solution hybridization of crosslinkable DNA oligonucleotides to bacteriophage M13 DNA. Effect of secondary structure on hybridization kinetics and equilibria'
J. Mol. Biol., 197 (1987) 349-362.

C. HELENE
'Specific gene regulation by oligodeoxynucleotides covalently linked to intercalating agents'
DNA-ligand interactions, Edited by W. Guschlbauer and W. Saenger (Plenum Publishing Corporation) (1987) pp. 127-140.

T. LE DOAN, L. PERROUAULT, D. PRASEUTH, N. HABHOUB, J.L. DECOUT, N.T. THUONG J. LHOMME & C. HELENE
'Sequence-specific recognition, photocrosslinking and cleavage of the DNA double helix by an oligo-[α]-thymidylate covalently linked to an azidoproflavine derivative'
Nucleic Acids Res., 15 (1987) 7749-7760.

B.L. LEE, A. MURAKAMI, K.R. BLAKE, S.B. LIN & P.S. MILLER
'Interaction of psoralen-derivatized oligodeoxyribonucleoside methyl-phosphonates with single-stranded DNA'
Biochemistry, 27 (1988) 3197-3203.

C.J. MARCUS-SEKURA, A.M. WOERNER, K. SHINOZUKA, G. ZON & G.V. QUINNAN Jr.
'Comparative inhibition of chloramphenicol acetyltransferase gene expression by antisense oligonucleotide analogues having alkyl phos-photriester, methylphosphonate and phosphorothioate linkages'
Nucleic Acids Res., 15 (1987) 5749-5763.

F. MORVAN, B. RAYNER, J.L. IMBACH, S. THENET, J.R. BERTRAND, J. PAOLETTI, C. MALVY & C. PAOLETTI
'α-DNA II. Synthesis of unnatural α-anomeric oligodeoxyribonucleotides containing the four usual bases and study of their substrate activi-ties for nucleases'
Nucleic Acids Res., 15 (1987) 3421-3437.

D. PRASEUTH, M. CHASSIGNOL, M. TAKASUGI, T. LE DOAN, N.T. THUONG & C. HELENE
'Double helices with parallel strands are formed by nuclease-resistant oligo-[α]-deoxynucleotides and oligo-[α]-deoxynucleotides covalently linked to an intercalating agent with complementary oligo-[β]-deoxynucleotides'
J. Mol. Biol., 196 (1987) 939-942.

D. PRASEUTH, L. PERROUAULT, T. LE DOAN, M. CHASSIGNOL, N. THUONG & C. HELENE
'Sequence-specific binding and photocrosslinking of α and β oligodeoxynucleotides to the major groove of DNA via triple-helix formation.
Proc. Natl. Acad. Sci. USA, 85 (1988) 1349-1353.

D. PRASEUTH, T. LE DOAN, M. CHASSIGNOL, J.L. DECOUT, N. HABHOUB, J. LHOMME, N.T. THUONG & C. HELENE
'Sequence-targeted photosensitized reactions in nucleic acids by oligo-α-deoxynucleotides and oligo-β-deoxynucleotides covalently linked to proflavin'
Biochemistry, 27 (1988) 3031-3038.

J.S. SUN, J.C. FRANCOIS, R. LAVERY, T. SAISON-BEHMOARAS, T. MONTENAY-GARESTIER, N.T. THUONG & C. HELENE
'Sequence-targeted cleavage of nucleic acids by oligo-[β]-thymidylate-phenanthroline conjugates : parallel and antiparallel double helices are formed with DNA and RNA, respectively.
Biochemistry, 27 (1988) 6039-6045.

N.T. THUONG, U. ASSELINE, V. ROIG, M. TAKASUGI & C. HELENE
'Oligo(α-deoxynucleotide)s covalently linked to intercalating agents : differential binding to ribo- and deoxyribopolynucleotides and stability towards nuclease digestion'
Proc. Natl. Acad. Sci. USA, 84 (1987) 5129-5133.

J.J. TOULME & C. HELENE
'Antimessenger oligodeoxyribonucleotides : an alternative to antisense RNA for artificial regulation of gene expression. A review.
Gene, 72 (1988) 51-58.

P. VERSPIEREN, A.W.C.A. CORNELISSEN, N.T. THUONG, C. HELENE & J.J. TOULME
'An acridine-linked oligodeoxynucleotide targeted to the common 5' end of trypanosome mRNAs kills cultured parasites'
Gene, 61 (1987) 307-315.

A. ZERIAL, N.T. THUONG & C. HELENE
'Selective inhibition of the cytopathic effect of type A influenza viruses by oligodeoxynucleotides covalently linked to an intercalating agent'
Nucleic Acids Res. 15 (1987) 9909-9919.

PHOTOFOOTPRINTING ANALYSIS OF PROTEIN-DNA INTERACTIONS

Peter E. Nielsen & Claus Jeppesen

    Research Center for Medical Biotechnology,
    Department of Biochemistry B,
    The Panum Institute,
    University of Copenhagen,
    Blegdamsvej 3c,
    DK-2200 Copenhagen N,
    Denmark

ABSTRACT. A series of photochemical probes that can be used to study ligand-DNA interactions has been developed and tested. The probes include cleavable protein-DNA photocrosslinking reagents, protein photolabeling reagents with DNA affinity (for photolabeling of DNA-associated proteins), and DNA photofootprinting reagents. The latter reagents such as psoralens, azidoacridines and diazo-derivatives of 9-aminoacridine can be used to study the sequence specific binding of proteins (e.g. λ-repressor or RNA polymerase) or low molecular weight drugs (e.g. dis-tamycin or echinomycin) to DNA. Finally, we have found that protein contacts with the DNA backbone can be probed by photofootprinting using uranyl salts. This technique has been used to study the E.coli deoP1 promoter RNA polymerase open complex.

1. INTRODUCTION

Protein-DNA interactions play a central role for both struc-ture the function of the genetic material. The packing of DNA in the nucleus in eukaryotic cells (to a packing ratio exceeding 10^4) is predominantly accomplished through histone-DNA interactions forming the nucleosome fiber, which is further coiled into solenoids, loops and finally into chromosomes (van Holde, 1988). Histone-like proteins have likewise been found in prokaryotic cells (Schmid, 1988). Transcriptional regulation of genes is to a large extent due to proteins, gene-repressors or transcription factors. These proteins bind to regulatory DNA sequences such as promoters, enhancers a.o. thereby switching genes off or on by inhibit-ing or facilitating the access of RNA polymerase to the promoter sequence.
    Therefore much effort is being devoted to the detailed study of protein-DNA interactions. Questions may be asked at

P. E. Nielsen (ed.), Photochemical Probes in Biochemistry, 231–240.
© *1989 by Kluwer Academic Publishers.*

several levels. One can ask which sequences in the DNA bind proteins and which proteins. One may also study the specific molecular interactions such as hydrogen-bonding, electrostatic or hydrofobic interactions that are responsible for the protein-DNA recognition, and thus gain information of the structure of protein-DNA complexes. These structural questions can be asked both in vitro with purified materials, or in vivo under functional conditions in order to study the biological role of the protein-DNA interactions. Finally, it is possible to perform time resolution experiments to study the dynamics of such interactions.

Specific protein DNA interactions are profitably studied by footprinting techniques (Fig.1) by which various types of protein contacts with the DNA can be assessed, using either biological DNA cleavers such as DNaseI (Galas & Schmitz, 1978) or chemical DNA cleavers such as transition metal complexes (EDTA/FeII) (Dervan, 1986, Tullius et al., 1987,) or phenanthroline/CuI (Sigman, 1986). While methidium propyl-EDTA/FeII (MPE) (Dervan, 1986) only reports the extent of a protein binding on the DNA, DNaseI (Galas & Schmitz, 1978) and phenanthroline/CuI in addition reports changes in the DNA conformation (Sigman, 1986) and EDTA/FeII reports protein DNA-backbone contacts as well as structural variations in the DNA (Tullius et al., 1987).

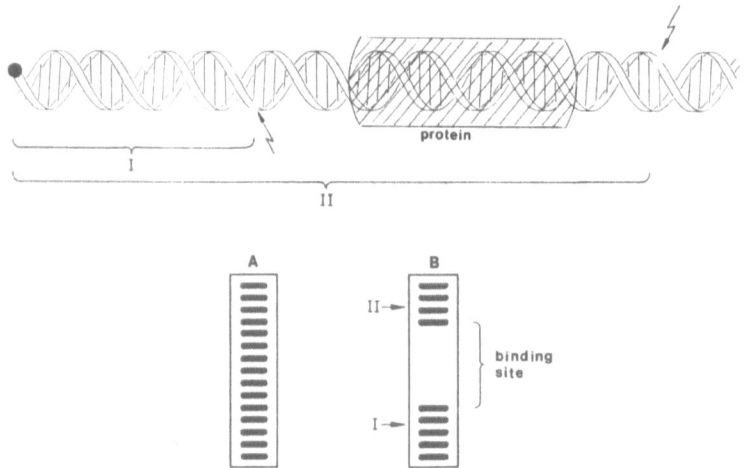

Figure 1. Principle of a footprinting experiment. An end-labeled DNA molecule is treated with a DNA cleaving agent giving less than one cleavage (arrow) per DNA molecule. Due to the end labeling, each cleavage site corresponds to a labeled DNA fragment when analyzed by gel electrophoresis (A) and DNA regions protected by bound ligands will thus turn up as "holes" in the gel pattern (B).

Photofootprinting is ideally suited for time resolution experiments due to the activation by an external source, light, that can be administered in high doses (even at low temperature) and within a short time ($<10^{-6}$ sec by flash irradiation). Photofootprinting may also prove advantageous for in vivo photofootprinting studies.

We have developed a range of photochemical reagents that can be used to study protein DNA interactions. These reagents may be divided into three types. Photocrosslinking, photoaffinity labeling and photofootprinting reagents. Most of the reagents contain a ligand that binds to DNA by inter-calation between the base pairs. A typical reagent is thus composed of a photoprobe, a linker and a DNA intercalator. The principle is shown schematically in Fig.2. In the case where both the photoprobe (ph) and the intercalator are photoactive, photocrosslinking of proteins to DNA results. When the reagent contains an intercalator and a photoprobe, photoaffinity labeling of DNA associated proteins results if the photoreation is with protein, and photofootprinting of the protein DNA binding site if the photoreaction is with DNA. It is, however, a prerequisite for photofootprinting that the photoreaction results in strand cleavage, or that the positions of the adducts can be determined.

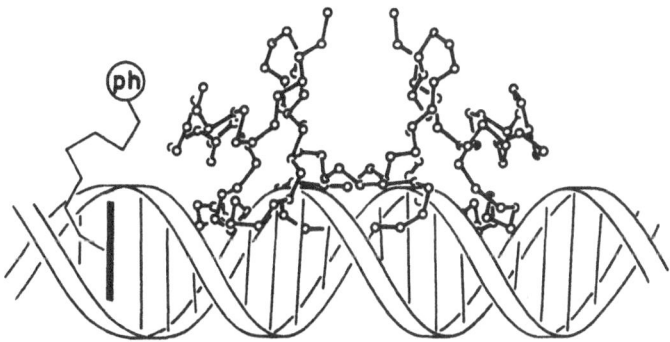

Figure 2. Principle of photoreagent. The figure shows a piece of DNA to which a protein (repressor) and a reagent molecule (intercalator, linker and photoprobe (ph)) are bound.

2. MATERIALS AND METHODS

2.1 Photofootprinting. For details see Nielsen et al. (1988a,b) and Jeppesen et al. (1988) (and Jeppesen & Nielsen (manuscript in preparation)).

3. RESULTS & DISCUSSION

3.1 Photocrosslinking

Our protein-DNA photocrosslinking reagents consist of an azidobenzoyl ligand for photoreaction with protein, a cleavable methylene-disulfide linker and either a psoralen (1) or an azidoacridine (2) moiety for binding to and photoreaction with DNA. We have shown that both reagents 1 (Elsner et al., 1985) and 2 (Nielsen et al., 1984) can be used to induce cleavable photocrosslinks between histones and DNA in chromatin at a yield of 0.01-0.03 crosslinks per reagent molecule.

3.2 Photoaffinity Labeling

DNA intercalating reagents (e.g. 3,) that photoreact with proteins have also been prepared (Nielsen et al., 1983, Nielsen, 1982, 1985). These reagents can be used to photolabel DNA associated proteins in a process probably most correctly termed affinity mediated photolabeling since the photolabeling of the proteins is mediated through the DNA affinity of the reagents. Using reagent 3 we have shown that this compound can be used as a probe for changes in the structure of protein-DNA complexes exemplified by the salt or urea induced changes in histone-DNA interactions in chromatin (Nielsen, 1982, 1985, 1988, Nielsen et al., 1985). It is furthermore feasible to conduct such studies on intact nuclei or whole cells (Nielsen, 1985).

3.3 Photofootprinting

Psoralens were the first compounds to be used extensively
for photofootprinting of nucleosome positions in chromatin.
This method which is still yielding valuable new information
(e.g., chapter 12 in this volume) is based on the detection
of psoralen-DNA interstrand crosslinks by electron microscopy. We (Zhen et al., 1986) and others (Sage & Moustacchi,
1987) have recently developed nuclease assays for the detection of psoralen-DNA interstrand crosslinks (and
monoadducts) at the nucleotide level, and we have shown that
4,5',8-trimethyl psoralen (TMP) can also be used as
photofootprinting reagents to study sequence specific
protein-DNA interactions exemplified by the binding of λ-
repressor to operator DNA and E.coli RNA polymerase to
promoter DNA (Zhen et al., 1988). Analogously, the binding
of lac-repressor to the lac-operator can be assayed in vivo
(unpublished).

Recently, we have described a number of reagents which
photochemically induce single strand nicks as well as base
labile adducts and/or sites in DNA. These reagents include
azido-9-aminoacridines (e.g. 4 & 5) (Jeppesen et al.,
1988a), nitrobenzamido- (Nielsen et al., 1988c),
azidobenzamido- or diazocyclopentadienyl ligands (Nielsen et
al., 1988a) linked to 9-aminoacridine via polymethylene
linkers (e.g. reagent 6), as well as arylthioronium salts
(Buchardt et al., in preparation) and uranyl salts (Nielsen
et al., 1988c) (see also chapter 16). Several of these are
useful as photofootprinting reagents. For instance, we have
shown that DNA binding sites of small ligands such as dis-
tamycin and echinomycin are reported by reagents 5 and 6

(Jeppesen & Nielsen, 1988) (Fig.3). Analogously reagent <u>4</u> reports an extension footprint of <u>E.coli</u> RNA polymerase bound to the <u>deo</u>P1 promoter (Jeppesen <u>et al</u>., 1988a) (Fig-.5), while uranyl salts report protein DNA backbone contacts (Fig.4) in much the same way as found for EDTA/FeII (Tullius & Dombroski, 1986)

```
                20                40                60
5'AATTCGAGCTCGGTACCCGGGGATCCTCTAGAGTCGACCTGCAGGCATGCAAGCTTGGCGTAA
 3'AGCTCGAGCCATGGGCCCCTAGGAGATCTCAGCTGGACGTCCGTACGTTCGAACCGCATT

                80                100               120
   TCATGGTCATAGCTGTTTCCTGTGTGAAATTGTTATCCGCTCACAATTCCACACAACATA
   AGTACCAGTATCGACAAAGGACACACTTTAACAATAGGCGAGTGTTAAGGTGTGTTGTAT

                140               160               180
   CGAGCCGGAAGCATAAAGTGTAAAGCCTGGGGTGCCTAATGAGTGAGCTAACTCACATTA
   GCTCGGCCTTCGTATTTCACATTTCGGACCCCACGGATTACTCACTCGATTGAGTGTAAT
```

Figure 3. Photofootprinting of the binding sites of dis-tamycin on the <u>Eco</u>RI/<u>Pvu</u>II 232 base pair pUC19 DNA fragment. The boxes indicate regions of the DNA which are protected by distamycin from photoreacting with reagents <u>5</u> or <u>6</u>.

In this connection it should be mentioned that photochemical reactions of the bases in DNA can also be used to probe protein-DNA interactions (Becker & Wang, 1984, Selleck & Majors, 1987, 1988, Jeppesen <u>et al</u>., 1988b). The photoreactions include pyrimidine-pyrimidone 6-4 adducts, thymine-lysine adducts and other less well characterized reactions, and these can either be enhanced or quenched upon binding of protein to the DNA.

3.4 RNA polymerase-promoter complex

Prior to initiation of RNA transcription in Escherichia coli, RNA polymerase binds to the promoter DNA sequence upstream from the gene to be transcribed. Following binding of the RNA polymerase to the promoter a conformational change takes place in which the DNA is unwound ~ 1.5 turns and 12 base pairs in the ~ -9 to +3 region become suscep-tible to chemical modification, maybe due to formation of a single stranded DNA loop. This RNA polymerase-promoter com-plex which is ready to commence RNA synthesis when ribonucleotide triphosphates are supplied is referred to as the "open complex"(for reviews on RNA polymerase promoter interaction, see Reznikoff <u>et al</u>., 1985 & Travers, 1987).

Using some of above mentioned photofootprinting reagents we have studied the complex between <u>E.coli</u> RNA polymerase and the <u>deo</u>P1 promoter DNA.

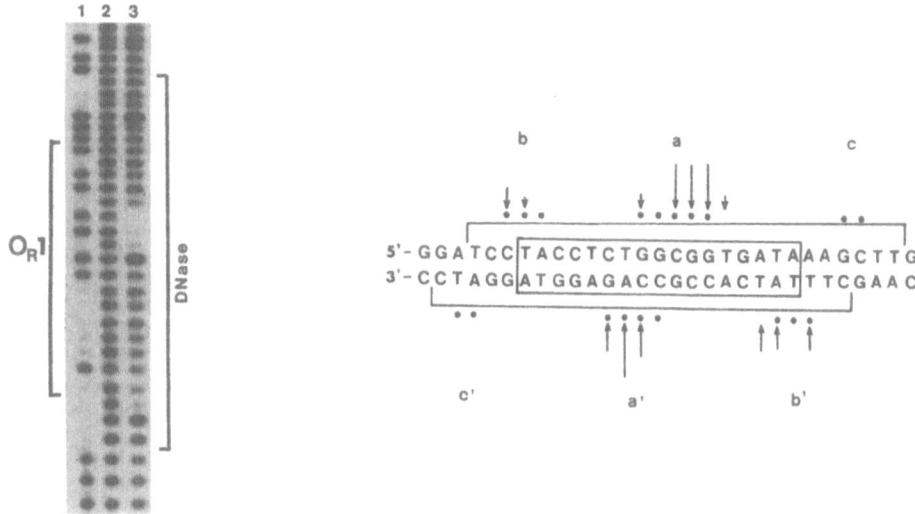

Figure 4. Photofootprinting of protein-DNA backbone contacts of the λ-repressor operator DNA complex. Lane 1: A+G sequence reaction. Lanes 2 & 3: operator containing DNA fragment cleaved by uranyl in the absence (2) or presence of λ-repressor (3). The operator sequence and the DNaseI footprint are indicated by brackets. b. Schematic representation of these results. The arrows indicate positions which are protected from photocleavage by $UO_2(NO_3)_2$ in the presence of λ-repressor. The dots are analogous results obtained with EDTA(FeII) (Tullius & Dombroski, 1986). The operator sequence is boxed in and the DNaseI footprint is shown by brackets.

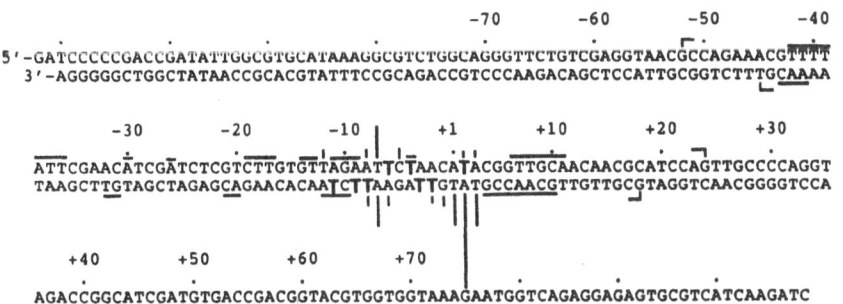

Figure 5. Footprinting of the binding of <u>E.coli</u> RNA polymerase to <u>deoP1</u> promoter DNA. Underlined sequences are protected against uranyl photocleavage. The bars denote uranyl hypersensitive sites and DNAseI footprint is shown by brackets. Transcription start is indicated by +1. Thymines reacting with permanganate are in bold.

Reagent 4, probably by probing access for intercalation in the DNA, reports an extension footprint very much like that obtained with DNaseI (Fig.5) except for the absence of hyperreactive sites. Reagent 5, on the other hand, only produces a weak footprint whereas two hyperreactive sites are found in the -10 to +1 region (A_{-7} & T_{+1}). (Jeppesen et al., 1988a).

We have shown that uranyl salts report DNA-backbone contacts of the λ-repressor bound to operator DNA (Nielsen et al., 1988b), and we believe that uranyl in contrast to EDTA/FeII is reacting with the DNA from the major groove. Uranyl photofootprinting of the E.coli RNA polymerase deoP1 promoter open complex yielded the results shown in Fig.5. (Jeppesen & Nielsen, in preparation). Several regions of protection, most notably from +4 to +11, as well as sites of enhanced reactivity (-9 to +3) are observed. The hypersensitive sites are all situated within the unwound "DNA-loop" of the open complex, which is also susceptible to C-methylation (Siebenlist, 1979, Kirkegaard et al., 1983) and to thymidine oxidation (Fig.5).

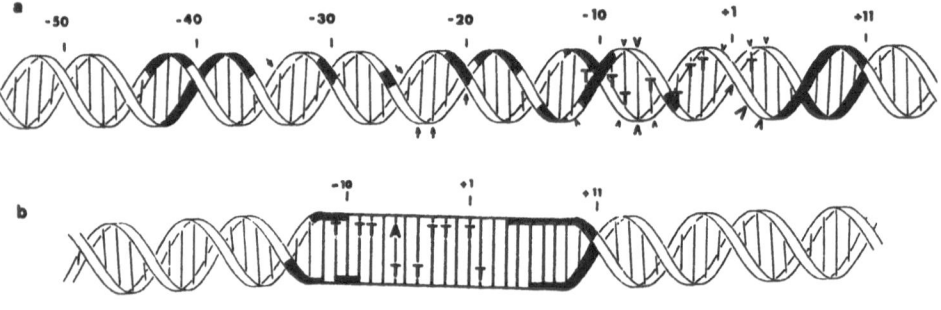

Figure 6. a) Model of the deoP1 DNA helix showing RNA polymerase DNA backbone contacts (heavy shade) (uranyl photofootprint), uranyl (arrow heads), DNaseI (arrows) and $KMnO_4$ (written thymidines (T)) hypersensitive sites. b) Same as a) showing unwound -10 to +5 region.

These results are consistent with the view that RNA polymerase is bound predominantly to one side of the promoter DNA helix (Fig.6). In particular the contacts in the -35 and the -15 (TATA box) regions are also observed by base-methylation and phosphate ethylation interference studies (Siebenlist et al., 1980). The strong backbone contacts in the +4 to +11 region are noteworthy. Upon binding of RNA polymerase to promoter DNA the DNA is unwound ~ 1.5 turns (Gamper & Hearst, 1982, Amouyal & Buc, 1987,), and this unwinding is believed to occur predominantly in the -9 to + 3 region, i.e the region in which the bases are susceptible to chemical modification. The uranyl photofootprinting results give a clue as to the mechanism of this DNA unwinding. It is observed that the RNA polymerase has a firm "grip" on both sides of the unwound loop and is thus able to "lock" the DNA into this energetically unfavourable conformation. Mechanistically this can take place via an active involvement of the RNA polymerase, in which the −10 and +10 contacts are first established and the DNA is subsequently unwound through a conformational change of the RNA polymerase. Alternatively, the RNA polymerase remains in a fixed conformation that is able to "catch" and "freeze" the unwound DNA-conformation. The latter mechanism requires that a fraction of the promoter DNA molecules actually adopt the unfavourable open loop DNA conformation in solution.

Indeed, recent results have indicated that the DNA secondary structure in the TATA box region is relatively unstable (Lefevre et al., 1988). The slow process of open complex formation (Spassky et al., 1985) could likewise reflect the "catching" of DNA molecules from a scarcely populated conformational state. Possibly the RNA polymerase first binds the promoter DNA using the upstream +1 to −40 recognition sequences (closed complex) and that this complex then subsequently "waits" for the DNA to adopt (by dynamic fluctuation) the unwound conformation and then locks it there using the +1 to +10 backbone contacts.

3.5 CONCLUDING REMARKS

Our present applications of photochemical probes for studying protein DNA interactions have only been concerned with static situations where in principle thermal probes, when available, could just as well have been employed. It is our hope and expectation that photochemical probes will prove their superiority for analysis of dynamic systems in time resolution studies. Such experiments are now in progress.

ACKNOWLEDGEMENTS

The financial support of the NOVO Foundation (PEN is a Hallas-Møller fellow) and the Danish Natural and Medical Science Research Councils (a fellowship to CJ) is gratefully acknowledged.

REFERENCES

Amouyal,M. & Buc,H. (1987) J. Mol. Biol. 195, 795-808.
Becker,M.M. & Wang,J.C. (1984) Nature 309, 682-687.

240

Dervan,P.B. (1986) Science, 232, 464-471.
Elsner,H., Buchardt,O., Møller,J. & Nielsen,P.E. (1985) Anal. Biochem. 149, 575-581.
Galas,D.J. & Schmitz,A. (1978) Nucl. Acids Res. 5, 3157-3170.
Gamper,H.B. & Hearst,J.E. (1982) Cell 29, 81-90.
Holde,H.van, ed. (1988) Chromatin, Springer.
Jeppesen,C. & Nielsen,P.E. (1988) (submitted).
Jeppesen,C., Buchardt,O., Henriksen,U. & Nielsen,P.E. (1988a) Nucleic Acids Res. 16, 5755-5770.
Jeppesen,C., Jensen,K.F. & Nielsen,P.E. (1988b) Nucleic Acids Res. 16, 9545-9555.
Kirkegaard,K., Buc,H., Spassky,A. & Wang,J.C. (1983) Proc. Natl. Acad. Sci. USA 80, 2544-2548.
Lefevre,J.-F., Lane,A.N. & Jardetzky,O. (1988) Biochemistry 27, 1086-1094.
Nielsen,P.E. (1982) Eur. J. Biochem. 122, 283-289.
Nielsen,P.E., Hansen,J.B., Thomsen,T. & Buchardt,O. (1983) Experientia 39, 1063-1072.
Nielsen,P.E., Hansen,J.B. & Buchardt,O. (1984) Biochem. J. 223, 519-526.
Nielsen,P.E. (1985) Biochemistry 24, 2298-2303.
Nielsen,P.E., Matsuoka,Y. & Norden, B. (1985) Eur. J. Biochem. 147, 65-68.
Nielsen,P.E. (1988) (in preparation).
Nielsen,P.E. Jeppesen,C., Egholm, M. & Buchardt,O. (1988a) Nucleic Acids Res. 16, 3877-3888.
Nielsen,P.E., Jeppesen,C. & Buchardt,O. (1988b) FEBS Lett. 235, 122-124.
Nielsen,P.E., Jeppesen,C., Egholm,M. & Buchardt,O. (1988c) Biochemistry 27,
Reznikoff,W.S., Siegele,D.A., Cowing,D.W. & Gross,C.A. (1985) Ann. Rev. Genet. 19, 355-387.
Sage,E. & Moustacchi,E. (1987) Biochemistry 26, 3307-3314.
Selleck,S.B. & Majors,J. (1987) Nature 325, 173-177.
Selleck,S.B. & Majors,J. (1988) Proc. Natl. Acad. Sci. USA 85, 5399-5403.
Siebenlist,U. (1979) Nature 279, 651-652.
Siebenlist,U., Simpson,R.B. & Gilbert,W. (1980) Cell 20, 269-281.
Sigman,D.S. (1986) Acc. Chem. Res. 19, 180-186.
Spassky,A., Kirkegaard,K. & Buc,H. (1985) Biochemistry 24, 2723-2731.
Tullius,T.D., Dombroski,B.A., Churchill,M.E.A. & Kam,L. (1987) Meth. Enzymol. 155, 537-558.
Tullius,T.D. & Dombroski,B.A. (1986) Proc. Natl. Acid. Sci. USA 83, 5469-5473.
Travers,A.A. (1987) CRC. Crit. Rev. Biochem. 22, 181-219.
Zhen,W.-P., Buchardt,O., Nielsen,H. & Nielsen,P.E. (1986) Biochemistry 25, 6598-6603.
Zhen, W.-P., Jeppesen,C. & Nielsen,P.E. (1988) FEBS Letters 229, 73-76.

STUDIES OF THE BINDING AND BIOLOGICAL ACTIONS OF ETHIDIUM

K. Lemone Yielding
The University of Texas Medical Branch
Galveston
Texas

ABSTRACT Photosensitive derivatives have been prepared corresponding to the biologically active analogs of the phenanthridine
and acridine classes of DNA binding drugs. These were used as
photoaffinity probes to study DNA binding, trypanocidal action,
frameshift mutagenesis, mitochondrial mutagenesis, and DNA repair.
In each instance there was good evidence that photoaffinity
labeling provided covalent adducts corresponding to the non-
covalent binding which led to the biological actions of the parent
compounds, and the resulting covalent adduct formation resulted in
significant enhancement of biological activity as confirmation.
These studies illustrated the power of photoaffinity labeling for
studying biological receptors of these important agents.

INTRODUCTION

*These studies were designed to identify by photoaffinity
labeling the putative action sites for phenanthridine and
acridine drugs. Each of these classes of drugs displays a
spectrum of biological properties, and each is known widely for
its classic mode of interaction with nucleic acids. Although
each is presumed to be biologically active because of its nucleic
acid binding properties, the biologic action mechanisms have not
been defined, and even the precise details of nucleic acid
binding in vivo, especially at low drug concentrations, have not
been defined.*

*Photoaffinity Labeling and Drug Action - The identification of
the actions of drugs selected empirically poses great
difficulties often due to the binding to multiple targets which
occurs at the drug concentrations which must be used in order to
permit convenient identification of reversible complexes.
Furthermore, the targeting to a particular biological process in
the intact organism requires confirmation, by independent
identification, of the unique essential nature of that process
for the target cell. Theoretically, some of these difficulties
may be addressed by photoaffinity labeling since the drug complex
may be converted to an irreversible adduct at relevant drug
concentrations within intact cells and retrieved for
identification, while correlating such adducts with biological
activity.*

P. E. Nielsen (ed.), Photochemical Probes in Biochemistry, 241–259.
© 1989 by Kluwer Academic Publishers.

Photoaffinity labeling is a uniquely important technique to be used in the elucidation of drug action. This approach provides an opportunity for immobilizing a drug in its biologically effective target site within an intact biological system on "command", allowing for the subsequent isolation and characterization of the drug-active site complex. Success in this approach requires first the design or identification of an appropriate active probe. A photoactivatable group is then substituted at a position on the probe which does not interfere with its "normal" activity, but results in generation by light of a highly reactive species which forms a covalent bond between the probe and its biological receptor. Since the spectral properties of the drug molecule itself determine the energy required for activation, substitution onto fairly large chromophores is desirable. The important concept here is that photoactivation merely converts a reversible complex into an irreversible one which is completely representative of the original drug complex. The steps in the design of a photoaffinity probe for biological studies involve: 1) clarification of the structure-function relationships for the parent compound to determine which substituents are essential to biological action; 2) preparation of light sensitive derivatives which preserve these structural and functional properties; 3) biological testing to verify that the light sensitive analog without light activation has the same biological activity as the active parent drug; 4) verification that the new probe shares the same initial non-covalent binding properties as the parent drug if these binding properties are known; 5) analysis of the biological consequences in an appropriate system following photoactivation to produce covalent adducts between drug and biological target; and 6) retrieval of the drug adducts with subsequent characterization of the biological receptors.

Study of the biological consequences of photoactivated binding of a probe to the drug receptor is important for two major reasons. First, a change in activity provoked by covalent binding may confirm that the probe is attached to the receptor. Second, the nature of the response modification may provide some clues to the nature of the drug action. The simplest model for drug action is a drug-receptor complex which has altered biological action due to equilibrium binding of the drug. The production of an irreversible complex may enhance the biological activity of the drug provided covalent attachment does not cause additional distortion of the receptor. Such a complex could be part of a catalytic molecule, a channel or gating mechanism, an informational molecule such as DNA, or a transport molecule; and the basic drug effect could be enhancement or inhibition. Alternatively, a drug may act in a more stoichiometric manner with binding of each drug molecule coupled to a specific evoked response; or there may be required cycling of each drug molecule between the bound and unbound state. In these latter cases, covalent drug attachment would prevent subsequent drug action. The inhibition of biological activity by a photoaffinity probe may be difficult to interpret because it may also represent

binding outside the actual receptor site proper with secondary
inhibition of function. The same may be true for "competitive"
blocking by the probe of drug binding to a conformationally
metastable receptor. The desirable features for a photoaffinity
probe, therefore, may be summarized as follows. First, the probe
must mimic fully the binding and biological properties of the
parent drug. Second, the photosensitive derivative must be
coupled with a 'good' chromophore with strong absorption outside
the UV range so that the biological system targeted will be
undisturbed by the light irradiation. Third, photoactivation
should generate, with high efficiency, a highly reactive,
non-selective, short-lived species so that covalent attachment
occurs before dissociation of the complex and the site of
reaction is determined only by the nature and position of the
non-covalent complex. Uncomplexed activated molecules should
simply react with bulk solvent. Finally, the adduct must be
readily identifiable so that it may be retrieved for study.

For identification of drug targets and mechanism of action
the complexities of probe design for photoaffinity labeling can
be avoided in some instances by recognizing that the parent drug
itself may be photoactivated to a reactive species. Such
opportunities, often heralded by recognition that a drug is
photosensitive (or photosensitizing) are often limited by the
fact that prolonged and/or intense irradiation may be required
(for example, see 1). Where feasible, however, the use of the
parent compound for photoaffinity labeling provides a direct and
unambiguous approach to target identification.

Bioactive Properties of Ethidium and Acridine Compounds -
Ethidium (3,8-diamino-5-ethyl-6-phenyl-phenanthridine) was
identified as a trypanocide and used in livestock more than 30
years ago to treat African trypanosomiasis (2,3); but its action
mechanism(s) have still not been explained. Meanwhile, it has
found widespread attention as an agent which interacts
characteristically with DNA in defined systems and is used as a
probe for nucleic acid structure and functions. Furthermore,
ethidium exhibits other biological properties of considerable
interest. First, it is a potent agent for production of
respiratory deficient mutants (RD mutants) in yeast, with
conversion of essentially an entire population of resting
Saccharomyces cells at a drug concentration of 1×10^{-6} M (4,5).
This process appears to involve some metabolic processing of the
agent by mitochondria (6,7), particularly since glucose
repression prevents mutagenesis (8), and mutation resistant
strains have been isolated (9). Interestingly in our own
studies, the ability of ethidium and acridine compounds to cause
RD mutants is correlated well with trypanocidal activity.
Second, ethidium, although a classic 'intercalator', can produce
frameshift mutations in Salmonella only upon metabolic activation
(10). These various biologic characteristics provide ample
opportunity to explore the authentic biological properties of any
photoaffinity probe selected for ethidium action. The present
studies, if successful, would provide a means of defining

biological actions of the drug and permit modeling of potential
new agents; could lead to better understanding of parasite
biology; could add great strength to the use of the agent for
probing nucleic acid structure and function; and could serve as a
model for the application of photoaffinity labeling for study of
drug action.

Acridines were discovered over a hundred years ago, and have
evoked widespread interest as antibacterial, antiparasitic, and
anticancer agents. They have also been used widely as biological
stains, and as reagents for studying nucleic acids. DNA
interactions have been of special interest in recent years, both
for defining types of drug interactions, and for elucidating
mechanisms of mutagenesis. Mutations in viruses were first
described in 1952 (11), and it was recognized subsequently that
acridines did not produce chemical changes in bases of the target
DNA, but resulted in modification of progeny DNA during the
replication processes. These changes were best rationized in
terms of frameshifts (12) through addition of extra bases or loss
of bases in newly synthesized DNA segments.   The proposal of
intercalation as a binding mode for acridines provided an
explanation for the frameshift mutations (13), and has been
extended as a concept subsequently for the binding of numerous
other drugs. Interestingly, it was this frameshift concept that
formed an important part of the argument for defining the triplet
nature of the genetic code (14). It was also recognized that
acriflavine produced respiratory deficient (petite) mutants by
modifying yeast mitochondrial DNA (15). In contrast to ethidium,
which produces petites in resting cells, acridines produce
petites only in growing cells. It was also recognized that
acriflavine, as a potent acridine, was toxic to trypanosomes and
led to the generation of dyskinetoplastic organisms (16,17). In
contrast to ethidium, for which all biological activities appear
to follow the same structure requirements, the acridines exhibit
rich substitution possibilities; and, in fact, the different
biological activities can be assigned to different structural
requirements. It is also ironic that, despite the pioneering
importance of the intercalation model for drug binding, it has
not been possible to identify the precise nature of the
acridine-DNA complex under the limiting conditions which
approximate the in vivo conditions of biological action. Thus,
photoaffinity labeling offers a valuable approach to the study of
this important series of compounds. It is particularly
interesting that ethidium provides only 2 convenient sites for
straightforward derivatization and limited structure-function
dictates for biological action; but acridine has 10 readily
derivatizable sites, and multiple combinations thereof and
displays a wealth of structure-function possibilities in relation
to its several biological effects.

PHOTOAFFINITY LABELING STUDIES WITH ETHIDIUM ANALOGS

Design of Probe - As outlined above, a photoaffinity probe
should exhibit identical dark binding properties as those of the
parent compound, and should be able to react covalently upon
light activation without any re-arrangement within the receptor
site. These requirements are quite stringent and limit the type
of derivatization which can lead to acceptable probes. Ideally,
the reactive species should be non-selective in its reactivity so
that all specificity is directed by the non-covalent (dark)
binding process. Alternatively, the reactivity of the receptor
site must be appropriate to the reactive species. In either
event, the lifetime of the active state must be short to prevent
'wandering reactions' resulting from dissociation of the drug
complex before attachment has occurred. The properties of the
chromphore are quite demanding for use with biological systems
since irradiation must be at a wavelength which does not lead to
destructive changes unrelated to drug binding. The primary aryl
amino groups of ethidium (figure 1) provided an opportunity for
derivatization and the absorptive properties of ethidium (max 470
nM) made it an ideal candidate for use in biological systems.
Thus, azido analogs of ethidium were synthesized corresponding to
3 and/or the 8 positions, ostensibly to provide for photolytic
formation of the nitrene at these positions for non-specific
insertion into the receptor site (18,19). (Other discussions at
this symposium have emphasized that such aryl azides are, in
fact, more selective photoaffinity probes for nucleophilic
sites.) Preliminary experiments, in fact, showed that
equilibration of DNA with either of the mono-azido analogs
followed by irradiation with visible light resulted in formation
of covalent DNA adducts (18,20). These findings prompted a
variety of studies with the azido analogs to explore their
authenticity as photoaffinity probes for ethidium action.

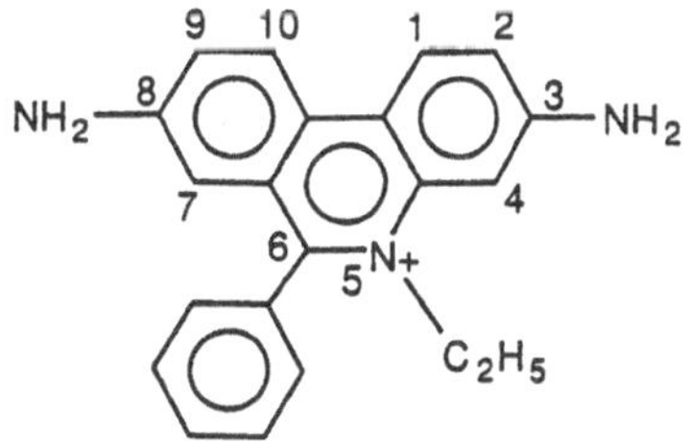

Figure 1: Structure of ethidium

TABLE 1. *Comparison of Ethidium and its Photosensitive analogs for non-covalent binding to DNA.* (25) (spectrophotometric titration and Scatchard analysis (53) in Tris-NaCl 0.015 M pH 7.5)

	n	Ka
3,8-diamino-5-ethyl-6-phenyl-phenanthridinium bromide (Ethidium bromide)	0.18	8×10^5
8-azido-3-amino-5-ethyl-phenyl-phenanthridinium chloride	0.18	5.4×10^5
3-azido-8-amino-5-ethyl-6-phenanthridinium chloride	0.21	4.7×10^5
3,8-diazido-5-ethyl-6-phenyl-phenanthridinium chloride	0.20	1.3×10^4
8-azido-5-ethyl-6-phenyl-phenanthridinium chloride	0.20	1×10^5
3-azido-5-ethyl-6-phenyl-phenanthridinium chloride	0.22	7.9×10^4

Comparison of ethidium and its azido analog for non-covalent binding to DNA - Ethidium binding to DNA is studied most often by using either the shift in absorption or fluorescence emisssion spectra resulting from complex formation as the basis for Scatchard analysis. Using Calf Thymus (ct) DNA which had been deproteinated thoroughly, it was determined that the Ka and the n values obtained for binding of the mono-azido analogs (3- or 8-) were essentially the same as for the parent compound, whether measured by absorption (21) or fluorescence spectroscopy (22). The diazido-compound, however, showed less favorable binding. Experiments performed with freshly prepared chromatin showed the same lack of deviation of non-covalent binding properties from those of the parent compound (23). Furthermore, estimation of binding parameters based on DNA repair synthesis provoked by photoaffinity labeling of human lymphocytes were also in good agreement with the values for ethidium (24). (Table 1.) These experiments were consistent with the conclusion that the conversion of ethidium to the mono-azido analog of ethidium did not change appreciably its non-covalent binding properties. It was established further that the extent of unwinding of helical DNA resulting from non-covalent binding of ethidium or the 8-or

3-azido analog was essentially the same (25). Thus, the non-covalent binding properties of the photoaffinity probe were indistinguishable from the parent compound. The contention that the photoaffinity probe was binding to the same site as the parent compound was tested additionally by studying whether ethidium could compete for the binding of the probe. Using DNA (26) or freshly prepared chromatin (23), binding was clearly competitive. Furthermore, ethidium could compete effectively for the in vivo binding in nuclei and mitochondria (27). Based on such identity of the non-covalent binding, it was concluded tentatively that covalent adduct formation from photoaffinity labeling was representative of the non-covalent binding site for ethidium.

Biological consequences of photoaffinity labeling with ethidium monoazide - A rigorous test of the authenticity of a drug photoaffinity probe is the comparison of its biologic effects with those of the parent compound. Without photoactivation, based on comparable binding properties, it would be expected that the photoprobe should mimic the parent compound. Upon photoactivation one should expect either enhancement of biologic effect or specific blockade of subsequent drug action depending on the end-point being assessed. The enchancement follows from the fact that reversible, equilibrium receptor binding is converted to an irreversible drug-receptor complex, resulting in greater extent of receptor saturation with the same level of drug exposure.

TABLE 2. Azido analogs of acridine (43)

Compound

1. *9-amino-2-azidoacridine*
2. *9-amino-2--azido-10-methyl-acridinium chloride*
3. *9-amino-3-azidoacridine*
4. *9-amino-3-azido-10-methyl-acridinium chloride*
5. *9-amino-3-azido-7-ethoxyacridine*
6. *9-azidoacridine*
7. *9-azido-10-methylacridinium chloride*
8. *9-azido-1-nitroacridine*
9. *9-azido-2-nitroacridine*
10. *9-azido-3-nitroacridine*
11. *1-azidoacridine*
12. *1-azido-10-methylacridinium chloride*
13. *2-azidoacridine*
14. *2-azido-10-methylacridinium chloride*
15. *3-azidoacridine*
16. *3-azido-10-methylacridinium chloride*
17. *3-azido-6-aminoacridine*
18. *3-azido-6-amino-10-methyl-acridinium chloride*
19. *3,6-diazidoacridine*
20. *3,6-diazido-10-methyl-acridinium chloride*

Production of Respiratory Deficient (RD) mutants (petites) in Saccharomyces by Ethidiumor by photo-labeling with Ethidium monoazide - Experiments were conducted comparing the rates and extents of RD mutant formation caused by incubating resting yeast cells with ethidium or with the 8-azido analog in the dark, i.e., without photoactivation. These results showed that the 8-azido derivative was only slightly less efficient than the parent compound in producing mutant cells. In contrast, when a period of dark equilibration was followed by photoactivation of the cells, the rate of RD mutant formation was considerably enhanced using equimolar concentrations of drug (18,28), with each attaining virtually 100% RD conversion with prolonged incubation and irradiation. Comparison of RD induction by ethidium and its photosensitive analog provided additional evidence for a requirement for metabolic activation of ethidium. First, glucose repression, which prevents ethidium mutagenesis (8), failed to modify RD induction by photoaffinity labeling with ethidium monoazide (29). Second, a mutant cell line resistant to ethidium induction of RD mutants (9) still showed active induction of RD mutants from photoaffinity labeling with ethidium monoazide (30) comparable to results with the parent strain. The ability to block RD induction in wild type cell lines by glucose repression without changing the effects of photoaffinity labeling also provided the basis for examining the competition between ethidium and the photoaffinity probe for the initial binding step. Thus, yeast cells were grown in 10% glucose to repress ethidium induction so that cells could be photoaffinity labeled in the presence of excess ethidium without the complication of ethidium induced mutants. These experiments showed clearly that an excess of ethidium provided straightforward competition for RD induction produced by photoaffinity labeling (28), presumably as a result of competition for the initial binding step.

Photoaffinity labeling with ethidium monoazide and induction of frameshift mutations - Metabolic activation of ethidium produces a specific class of frameshift mutations in *Salmonella* (10) (Reversion of TA 1538). In a completely analagous manner, photoactivation of ethidium monoazide after equilibration in the dark resulted in the same class of frameshift mutations, but without metabolic activation (31,32). Thus, the ethidium moiety, although considered a classic intercalator, apparently must be bound covalently to DNA in order to produce frameshift mutations. Such photoaffinity probes may prove useful in mechanistic studies of frameshift mutagenesis.

Trypanocidal Effects of ethidium and its photoaffinity analogs - Trypanocidal activity was tested using strain EATRO 110 of *Trypanosoma brucei brucei* obtained from Dr. Mary Rifkin at Rockefeller University, and EATRO 164 and a dyskinetoplastic derivative kindly provided by Dr. Kenneth Stuart of the Seattle Biomedical Research Institute, Seattle, Washington. In *vitro* testing was performed by exposing a measured suspension of parasites under appropriate conditions to selected drug concentrations followed by injection of 10^6 drug-treated

parasites into test animals. In the absence of drug treatment
animals died characteristically in 5-6 days. Total parasite kill
was required for survival beyond 30 days (33,34). For in vivo
testing, infected animals were given test drugs
intraperitoneally. In the initial studies using EATRO 110 under
in vitro conditions concentrations of drug in excess of 1×10^{-4}
M were required to achieve significant parasite kill using either
ethidium or the ethidium photoaffinity probe in the absence of
photoactivation. By contrast, complete parasite kill could be
achieved by photoaffinity labeling the parasites with the
ethidium photoprobe (8 azido or 3 azido) at a concentration of
1×10^{-6} M (35) and significant parasite kill could be observed
at probe concentrations of 1×10^{-7} M. Thus, photoaffinity
labeling enhanced the trypanocidal effect by more than 100 fold.

The fluorescent properties of ethidium compounds provided a
simple means to localize bound drug by fluorescence microscopy.
At 10^{-4} M ethidium, both nuclei and kinetoplasts were
fluorescently labeled. At 10^{-6} M however, fluorescent staining
was seen with parasites which had been photoaffinity labeled, and
fluorescence was distributed both in nuclei and kinetoplasts. At
concentrations for photoaffinity labeling between 10^{-6} and 10^{-7}
M, where limited parasite kill was obtained, drug localization
could not be achieved with simple fluorescence microscopy (36).

The next step in identification of bound drug was to develop
a more sensitive detection method. To this end, polyclonal
antibodies were produced in rabbits to the covalent ethidium -DNA
adduct resulting from photoaffinity labeling. Parasites were
then photoaffinity labeled with trypanocidal concentrations of
ethidium monoazide, washed, and exposed to the antiserum. Bound
drug was detectable both by direct and indirect fluorescence
staining. At 1×10^{-6} M, both kinetoplasts and nuclei were
stained; but at 1×10^{-7} M and 1×10^{-8} M drug only
kinetoplasts were antibody stained for the bound drug (37).

These experiments, therefore, were consistent with high
affinity ethidium sites in kinetoplasts which correlate with the
drug's trypanocidal action. Based on early observations of
kinetoplast fluorescence with high concentrations of ethidium,
kinetoplasts were considered possible targets for drug action.
It was argued, however, that the blood stream form of the
parasite does not require kinetoplasts for survival. Since
kinetoplast are only required for cycling through the insect
vector, the kinetoplast has generally been disregarded as the
critical target. It must be noted, however, that the
kinetoplast, though not essential for energy metabolism in the
blood stream form of the parasite, may still play a critical role
in the regulation of cell replication. Such a role would be
consistent with parasite success since it would mean that
parasites would not replicate nuclei and divide without first
duplicating the kinetoplast . . . thus assuring infectivity in
the insect vector. If antitrypanosomal drugs can work by
interrupting such a regulatory loop, it may be that other, more

*direct means of interrupting this regulatory mechanism may be
devised.*

*Comparison of normal and dyskinetoplastic strains of T. brucei
for sensitivity to ethidium bromide and photoaffinity labeling
with ethidium monoazide* - *A rigorous test of the essential role
of the kinetoplast in the anti-trypanosomal effect of ethidium is
the measure of the antitrypanosomal drug effect in a
dyskinetoplastic organism. There are anecdotal comments in the
literature ascribing both sensitivity and resistance of
dyskinetoplastic strains to the drug. Interpretation is made
difficult because of the persistence in many such strains of some
remnants of kinetoplast structure and perhaps function. We were
fortunate that Dr. Kenneth Stuart of The University of Washington
provided a dyskinetoplastic derivative of EATRO-164 in which
there was no evidence of kinetoplast DNA by DNA hybridization
(38). (Interestingly, this strain grows more aggressively in the
mouse than the parent 164 strain.) This strain was studied for
its drug sensitivity in comparison with the parent strain 164,
and with EATRO-110. Groups of mice were infected with 10^3
organisms administered intraperitoneally. Without treatment all
mice developed heavy parasitemia and died in less than 10 days
when infected with either the parent or the dyskinetoplastic
strain of T. brucei. The appearance of the parasites was
monitored in peripheral blood every 4 hours until the level
reached 3-5 per high power field. At this point mice were
injected intraperitoneally with 0.1 ml. of saline alone, or with
a solution of 10 -3M ethidium chloride dissolved in saline.
Injections were continued every 6 hours until the parasites were
cleared. The total amount of drug required to clear the
parasites from the blood for 2 weeks was designated the curative
dose. The parental strain 164 was 'cured' with an ethidium dose
of 8 +/- .3 mg., while it was not possible to clear the
dyskinetoplastic strain of parasites from the bloodstream with
ethidium treatment (39).*

*Parasites were then tested for in vitro sensitivity to
photoaffinity labeling with ethidium monoazide. Suspensions of
both 164 and dys 164 were photolabeled with a series of
concentrations of the photosensitive drug and injected into mice
to test for subsequent viability of the parasite. Mice injected
with strain 164 which had been photolabeled with 1×10^{-6} M drug
showed complete survival at 30 days indicating complete parasite
kill. In contrast, the dys 164 parasites were completely
eliminated only when photoaffinity labeled with drug
concentrations at or above 1×10^{-5} M (39). The absence of the
kinetoplast, therefore, apparently rendered the organism
insensitive to ethidium bromide administered to the infected
animal, and decreased by an order of magnitude its lethal
sensitivity to photoaffinity labeling in vitro. Our earlier
experiments had shown that photoaffinity labeling at these higher
concentrations resulted in heavy labeling of both nuclei and
kinetoplast.*

These findings are consistent with the hypothesis that the kinetoplast is a critical target for trypanocidal activity by ethidium bromide. The sensitivity of dys 164 to photoaffinity labeling at high ligand concentrations may reflect the toxic effect of the binding known to occur at other sites (probably nuclear) at the higher concentration. The mechanism by which the kinetoplast is critical to survival of the blood stream form of the parasite could provide a new understanding of the biology of the parasite which could lead to new strategies for anti-parasite therapy. Alternatively, the possibility must be considered that the kinetoplast is essential to the action of ethidium bromide because it provides metabolic activation of the drug so that it can act at other sites within the cell. This model is important to consider because of the fact that mitochondria in yeast cells activate ethidium to produce RD mutations (6). If such activation plays a role in drug action here, obviously the identity of such activated species would suggest new drug forms which might be useful in treating trypanosomiasis.

Identification of the ethidium adduct resulting from photoaffinity labeling - Both in vivo and in vitro studies with ethidium and with ethidium photoaffinity labeling have suggested DNA as the putative high affinity target. It has been determined by model studies that ethidium exhibits a preference for C_pG (or at least pyrimidine p-purine) sites in DNA (40). However, previous studies have been limited by the sensitivity in detection of the adduct. Recently we have utilized DNA post-labeling technology (41,42) as a means of detecting adducts at very low adduct/DNA ratios. This powerful technique can measure aromatic adducts, for example, at levels of the order of $1/10^8$ nucleolides (43). We have studied Calf Thymus DNA, synthetic polynucleolides, and intact Trypanosomes and have identified only guanine adducts following photoaffinity labeling with ethidium monoazide. Experiments were done at a wide range of drug/nucleic acid ratios in vitro, and at limiting trypanocidal drug concentrations in vivo. On the basis of these experiments no adducts other than those to guanine could be detected (41,42). These studies, therefore, support the conclusion that initial high affinity binding of ethidium to DNA in vitro or in vivo involves guanine residues.

THE BIOLOGICAL ACTIONS OF ACRIDINES AND THE IDENTIFICATION OF PHOTOAFFINITY PROBES

Acridines (figure 2) exhibit a variety of biological activities including anti-trypanosomal, antibacterial, frameshift mutagenesis, RD induction, and inhibition of cancer cells. In contrast to ethidium, for which all biological actions appear to depend on the same structure, the biological effects of acridines are dependent on a range of structural features. Thus, structure-function studies can distinguish between frameshift mutagenic, RD mutagenic, and antiparasitic activities. Furthermore, the opportunities for derivatization, both photostable and photolabile, are extensive.

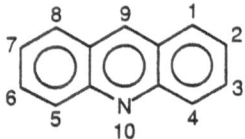

Figure 2: Structure of acridine

Frameshift mutagenesis - Acridines differ from ethidium as frameshift mutagens because they do not require metabolic activation, and they revert Ames strain TA 1537 (44) while ethidium reverts strain TA 1538 either with metabolic activation or by photoaffinity labeling (10,33,34). Substituents on the middle ring (9 and 10 substituents) dominate the activity for frameshift mutagenesis of the acridines. In TA 1537 9-aminoacridine is the standard frameshift compound, and addition of an amino substituent in the 1 position or a methyl group in the 10 position enhanced this frameshift potency; but the presence of other substituents on the side rings interfered strongly with mutagenicity (32). In contrast, acridines lacking a 9 amino group were, in general, either non-mutagenic, or as in the case of 1, or 3-aminoacridine, considerably less mutagenic. The 10-methyl substituent did not interfere with the effectiveness of the 9-amino compound, but G 10-methyl substitution alone was non-mutagenic. Since metabolic activation is not required for mutagenicity, it may be presumed that covalent attachment of the acridine is not required. This conclusion is substantiated by the additional finding that DNA repair proficient strains of *Salmonella* did not show reduced response to 9-aminoacridine (32). The situation with some of the substituted acridines was somewhat different, however. First, there was the appearance of low levels of mutagenicity toward Ames strains TA 1538 in association with persistence of low levels of mutagenicity toward TA 1537 in the 3-nitro derivative of 9-aminoacridine (32). Second, the mutagenic activity associated with this compound was decreased in DNA repair proficient strains, suggesting the possibility of covalent attachment even in the absence of extrinsic metabolic activation of the compounds. Unquestionably, the overwhelming mutagenic activities of the acridine compounds were associated with 9, 10 substitutions, and were not associated with evidence of covalent attachment; but interesting variations can be identified within the larger group of substituted compounds. The acridines, therefore, still represent a most important group of compounds for studies of the mechanisms of frameshift mutagenesis.

Attempts to identify a useful photoaffinity probe corresponding to frameshift mutagenic activity of acridines has involved the synthesis and study of 20 different azido analogs (listed in Table 2) (45). Of these, 3 compounds were of interest. First, 9-azidoacridine was inactive as a frameshift mutagen despite the intense activity observed for 9-aminoacridine. The azido compound, however, did display base pair mutageneic activity toward TA 1535 (46). This activity did not require photoactivation.

Both 9-amino-2-azidoacridine, and 9-amino-2-azido-10-methyl-acridinium chloride proved to have activity as frameshift mutagens. Cells were subjected to brief (10 minute) exposure to the test compound followed by photoactivation with visible light, using time and conditions demonstrated to produce complete photolysis. These conditions emphasized the formation of covalent adducts, since they were insufficient to produce typical mutagenesis by 9-aminoacridine which requires the presence of the drug during replication. It was found that 9-amino-2-azido-acridine produced revertants in strain TA 1537 under these conditions corresponding to the strain specificity of the parent 9-amino-compound. 9-amino-2-azido-10-methylacridinium chloride also produced revertants in TA 1537, but in addition was somewhat mutagenic toward strain TA 1538. The role of covalent attachment was evaluated by parallel testing in the DNA repair proficient strain, TA 1577. Prolonged exposure, in the dark, to either the parent 9-aminoacridine or the azido derivatives resulted in mutations in both TA 1537 and TA 1577, demonstrating the lack of importance of covalent adduct formation in the absence of photoaffinity labeling. However, photolytic activation after only brief exposure to the azido- compounds showed mutagenesis only in the repair deficient strain (32,47). Thus, it appeared that covalent attachment of the drug occurred with photoactivation of the azido- derivatives, in keeping with the attempt to photoaffinity label the mutagenic site(s). These compounds, therefore, may prove useful in studying the molecular bases for frameshift mutagenesis of each strain specific type, i.e., in TA 1537 and TA 1538.

Respiratory Deficient Mutant ('Petite') Induction by Acridines - Petite induction by acriflavine was first reported in 1950 (15). This property of acridines has been studied extensively over the years, and appears to be attributed to the selective loss of mitochondrial DNA from otherwise normal cells. Although acriflavine is equally as effective as ethidium in petite induction, it differs from ethidium in its effectiveness by its requirement for growing cells. It is interesting that the effectiveness of ethidium in resting cells has been associated with metabolic transformation of the drug in the mitochondria to cause its covalent attachment. Covalent attachment of the ethidium moiety is also required for frameshift mutagenesis in Salmonella, in contrast to the acridines which are effective without metabolic activation or covalent attachment in dividing cells. It is of further interest that the degradation of mitochondrial DNA which accompanies petite induction appears to differ qualitatively for the two classes of compounds (48). There would be considerable value, therefore, in contrasting the molecular basis for the two different drug actions. Accordingly, it has been our plan to develop suitable photoaffinity analogs of acridine to assist in these mechanistic studies.

Structure function studies for the induction of petites by acridine compounds revealed interesting contrasts from the induction of frameshift mutagenesis. Without exception, the compounds most active in frameshift mutagenesis were inactive for petite induction. In contrast to the importance of the middle ring (especially the 9 position), the side rings were found to play a dominant role in promoting petite mutagenesis (49). For example, 3,6-diamino-10-methylacridinium chloride, which was the most potent for inducing petites, was inactive as a frameshifter. Similarly, 9-aminoacridine, the classic frameshift mutagen in Salmonella, was completely inactive for petite induction. Thus, it may be concluded that different structural features are responsible for the two biological activities. Based on these differences, therefore, attempts were made to identify an effective photosensitive analog of an active petite inducer. These compounds were tested in resting cells rather than growing cells to avoid any effects corresponding to the parent compounds on resting cells. This was based on the assumption that any covalent change caused by photoaffinity labeling would not require cell division, but would be expressed directly in the progeny cells even in the absence of the compound. These experiments identified two promising candidates, 3-azido-10-methylacridinium chloride, and 3-amino-6-azido-10-methylacridinium chloride (acriflavine monoazide). The former compound appears to be much more favorable because it is associated with much less cell toxicity (49,50). The experimental approach is also complicated by the fact that the active compound as well as the parent compound (acriflavine) on prolonged irradiation with visible light exhibits photodynamic activation of petite induction and cell toxicity, apparently through induction of singlet oxygen (51). The approach of photoaffinity labeling, however, appears to hold promise for the comparison of petite induction by ethidium and acridine compounds.

Trypanocidal Effects of Acridines for Trypanosoma Brucei - Acriflavine has been recognized as a potent antrypanosomal agent and is active in inducing the dyskinetoplastic state in the parasite, presumably through its intense interaction with kinetoplast DNA. The acriflavine-exposed parasites are also photosensitized, corresponding to our experience with acriflavine in inducing petite mutants (16,51,52). A brief survey of the structure-function requirements for antitrypanosomal activity in comparison with frameshift mutagenesis and petite induction was particularly interesting. Of the compounds available to us from our studies with Salmonella and Yeast cells, by far the most active compound against trypanosomes was acriflavine. This was tested by exposing parasites _in vitro_, followed by injection of the drug-treated parasites into mice and monitoring for the survival of the mice. Even more interesting was the finding that antitrypanosomal activity could be enhanced by visible light irradiation. Finally, limited irradiation of parasites exposed to acriflavine monoazide _in vitro_ produced complete parasite killing at drug concentrations as low as $1 \times 10^{-7}M$ (35)! It would appear, therefore, that acriflavine monoazide the photoaffinity label developed to study RD mutagens in yeast is

the most likely candidate for photoaffinity probing of the
trypanocidal actions of acridine drugs.

DISCUSSION

This presentation has summarized experience with two classes
of biologically active compounds whose mechanisms of action have
been approached through the use of photoaffinity labeling.
Ethidium, a well known antiparasitic agent, mitochondrial
mutagen, and DNA reagent has been converted to 2 different
photosensitive azido- analogs (8- azido- and 3-azido-) each of
which has displayed the characteristic biologic and binding
properties of the parent compound. When photoactivated in situ,
there was drastic enhancement of biological response, each
produced covalent adducts which could be retrieved for study, and
adduct formation was clearly competitive with the parent compound
for initial binding both in vitro and in vivo. Photoaffinity
labeling has created useful tools for studying the biological
target(s) for this drug, Based on the approach of identifying
drug targets, for example, high affinity binding to the
kinetoplasts of Trypanosoma brucei has been identified which
correlates with biological effectiveness. It also seems likely
that these probes will be useful in studying the mechanisms for
other biological effects of ethidium such as the induction of
mutagenesis, as well as for modeling the consequences for drug
binding to DNA generally. While the aryl azido compounds may
theoretically not be ideal candidates for photoaffinity labeling
because of their selective reactivity with nucleophiles upon
photolysis, this did not seem to be a problem for nucleic acid
labeling, since photolytic efficiency was high, and covalent
addict formation appeared to be at the biologically relevant
site, in view of the clear enhancement of biological activity
which occurred. Ethidium monoazide, therefore, was considered a
prototype for probe generation using other heterocyclic drugs
which bind to nucleic acids.

Acridine compounds represented a most interesting class of
DNA reagents to study by the photoaffinity labeling approach.
First, acridines display a spectrum of important biological
effects whose mechanisms would be important to model. Second,
these various effects have been shown to depend on different
structural requisites in the molecule, providing the opportunity
of differential study of the structure function relationships.
Third, the acridine molecule presented many opportunities for
derivatization to explore these structural requirements both for
biological activity and for photosensitization. Accordingly, 20
different azido- compounds were synthesized to represent these
different possibilities, and photoprobes then selected which
should be useful in studying, differentially, the antiparasitic
(acriflavine monoazide- - - 3-amino-6-azido-10-methylacridine),
frameshift mutagenic (9-amino-2-azido-acridine), and
mitochondrial mutagenic (3-amino-6-azido-10-methylacridine)
mechanisms displayed. The acridines, therefore, have yielded a

wealth of photoaffinity probes for the study of these different
biological actions.

Photoaffinity labeling, generally, represents a powerful
approach to the study of drug action. It may, in fact, be the
only unambiguous way to identify drug targets *in vivo* under
conditions of drug action. In order to verify the validity of
the approach using a specific probe, it is necessary that
specific conditions be met. First, a photoaffinity probe can be
related to a specific drug action only if, in the absence of
photoactivation, it can mimic the biological actions of the
parent drug. Alternatively, in the absence of drug action, a
specific competitive inhibitor of the drug effect may be directed
at the target site, but this cannot be proved as conveniently,
because competitive effects may actually operate through
distinctly different sites. Ideally, therefore, the probe should
act like the drug in every way. Second, all the conditions
favorable to photoaffinity labeling must be met. These include:
convenient, efficient photoactivation to a highly reactive
species; relative non-specificity of the reaction properties of
the activated state so that the reaction is dictated by the
non-covalent interactions; short lifetime for the activated state
so that reaction occurs before dissociation of the complex; the
activated species which is free in solution must react quickly
with and thus be quenched by bulk solvent; and a readily
identifiable adduct must result from the reaction. While it may
be difficult to satisfy all these requirements in every instance,
these are the standards to be sought.

Operationally, there is great merit in first testing the
biological or other functional acceptability of selected probes.
In the event that they meet the criteria, considerable effort is
then warranted in pursuing in great depth the chemistry of the
target compounds and alternative derivatives.

REFERENCES

1. Daugherty, J. P., Hixon, S. C. & Yielding, K. L. *Biochim. Biophys. acta* 565:13-21. (1979)

2. Woolfe, G. *Ann. Trop. Med.Parasitol.* 46:285-288. (1952)

3. Woolfe, G. *Brit. J. Pharmacol.* 11:330-333. (1956)

4. Slonimski, P.O., Perrodin, G. & Croft, J.H., *Biochim. Biophys. Res. Commun.* 30:232-239. (1968)

5. Nagley, P. and Linnane, A. W. *J. Mol.. Biol.* 66:181-193. (1972)

6. Mahler H.R. and Bastos, R.N. *FEBS. Lett.* 39: 27-34. (1974)

7. Bastos, R. N. and Mahler, H. R., *J. Biol. Chem.* 249: 6617-6627. (1974)

8.   Hollenberg, C. P. & Borst, P. *Biochem Biophys. Res. Comm* 45:
     1250-1254. (1971)

9.   Fukuhara, H. E., Moustacchi, E., and Wesolowski, M. *Mol.Gen.*
     Genet. 162:191-201. (1978)

10. McCann, J., Choi, E., Yamasaki, E. and Ames, B.N. *Proc. Nat*
     Acad. Sci. US. 72:5135-5139. (1975)

11. DeMars, R. I. *Nature* 172:964-965. (1953)

12. Orgel, A. and Brenner, S. *J. Mol. Biol.* 3:762-768.(1961)

13. Lerman, L. S. *J. Mol. Biol.* 3:18. (1961)

14. Crick, F. H.C., Barnett, L. Brenner, S., Watts-Tobin, R. J.
     Nature 192:1227-1232. (1961)

15. Ephrussi, B., and Hottinguer, R. *Nature* 166: 956-958. (1950)

16. Steinert, M and VanAssel, S. *Arch Int. Physiol. Biochem.* 75:
     184-185. (1967)

17. Stuart, K. D. *J. Cell. Biol.* 49:189-195. (1971)

18. Hixon, S., White, W. E., Jr. & Yielding, K. L. *J. Mol. Biol.*
     92:319-329. (1975)

19. Graves, D. E., Yielding, L. W., Watkins, C. L. & Yielding,
     K. L. *Biochim. Biophys Acta* 479:98-104 (1977)

20. Yielding, L. W. and Firth, W. J. *Mutation Res.* 71:161-168.
     (1980)

21. Graves, D. E., Watkins, C. L., & Yielding, L. W.
     Biochemistry 20:1887-1892. (1981)

22. Garland, F., Graves, D. E., Yielding, L. W., & Cheung, H. C.
     Biochemistry 19:3221-3226. (1980)

23. Coffman, G. L., Yielding, L. W. and Yielding, K. L.
     Biopolymers 23:1067-1084. (1984)

24. Cantrell, C. E. & Yielding, K. L. *Photochem. and Photobiol.*
     25:189-191. (1977)

25. Yielding, L. W., Yielding, K. L., Donoghue, J. E.
     Biopolymers 23:83-110. (1984)

26. Yielding, L. W., Graves, D. E. Brown, B. R., & Yielding, K.
     L. *Biochem. Biophys. Res. Comm* 87:424-432. (1979)

27. Morita, T. & Yielding, K. L. *Mutation Res.* 54:27-32. (1978)

28. Morita, T. & Yielding, K. L. *Mutation Res.* 56:21-30. (1977)

29. Hixon, S. C., White, W.E., Jr., and Yielding, K. L. *Biochem. Biophys. Res. Comm.* 66:31-35. (1975)

30. Fukunaga, M., & Yielding, K. L. *Mutation Res.* 80:91-97. (1981)

31. Yielding, L. W., White, W. E., and Yielding, K. L. *Mutation Res.* 34:351-358. (1976)

32. Brown, B. R., Firth, W. J. III, and Yielding, L. W. *Mutation Res.* 72:373-388. (1980)

33. Lumsden, W. H. R., Cunningham, M.P., Webber, W.A.F., and Van Hoeve, K. *Experimental Parasitol.* 14:269-279. (1963)

34. Cox, B. A., Firth, W. J. III, Hickman, S., Klotz, F.B., Yielding, L. W. and Yielding, K. L. *Parasitol.* 67:410-416. (1981)

35. Firth, W. J. III, Messa, A. Reid, R., Wang, R. C., Watkins, C. L. and Yielding, L. W. *J. Med. Chem.* 27:865-870. (1984)

36. Cox, B.A., Yielding, L.W., and Yielding, K.L. *J. Parasitol.* 70:694-702. (1984)

37. Cox, B.A., Prine, L. Byrd, S., Yielding, L.W., and Yielding, K. L. submitted, *J. Parasitol.*

38. Stuart, K. and Gelvin, S. R. *Amer. J. Trop. Med. Hyg.* 29:1075-1081. (1980)

39. Cox, B.A., Prine, L., Omholt, P. & Yielding, K. L. submitted

40. Jain, S. C., Tsai, C. and Sobel, H.M. *J. Mol. Biol.* 114:317-331. (1977)

41. Byrd, S. "Characterization of Ethidium Binding Sites in Nucleic Acids by Photoaffinity Labeling", a dissertation. University of South Alabama College of Medicine, Mobile Alabama. 1987

42. Byrd, Suzanne, Woodley, S., Yielding, L. W. & Yielding, K. L., submitted

43. Gupta, R. C. *Cancer Research* 45:5656-5662. (1985)

44. Ames, B. N., McCann, J. & Yamasaki, E. *Mutation Res.* 31:347-364. (1979)

45. Firth, W. J., III, Rock, S. G., Brown, B. R., & Yielding, L. W. *Mutation Res.* 81: 295-309. (1981)

46. Brown, B. R., Firth, W. J. III, & Yielding, L. W. *Biochem. & Biophys. Res. Comm.* 80:1139-1145. (1979)

47. Brown, B. R. "The Identification of a Photoaffinity Probe for the Frameshift Mutagen, 9-aminoacridine, a dissertation. *University of Alabama in Birmingham, Birmingham, Alabama, USA.* (1982)

48. Faugeron-Fonty, G., Culard, G., Baldacci, G., Goursot, R., Prunell, A., and Bernardi, G. *J. Mol. Biol.* 134:493-537. (1979)

49. Fukunaga, M., Yielding, L. W., Firth, W. J. III, & Yielding, K. L. *Mutation Res.* 82:87-93 (1981)

50. Firth, W. J., III. "The Design & Synthesis of Photoaffinity Probes for Acridine Biological Actions", a dissertation. *University of South Alabama College of Medicine, Mobile, Alabama.* 1984

51. Iwamoto, Y., Mifuchi, I., Yielding, L. W., Firth, W. J. III and Yielding, K. L. *Mutation Res.* 125:213-220 (1984)

52. Simpson, L. *J. Cell. Biol.* 37:660- (1968)

53. Scatchard, G. *Ann. N. Y> Acad Sci.* 51:660-670 (1949)

Quantitative photoaffinity labeling of *Escherichia coli* RNA polymerase transcription complexes by nascent RNA.#

Thomas M. Stackhouse and Claude F. Meares,*

Chemistry Department, University of California, Davis, California 95616

#Supported by research grant GM 25909 from the National Institute of General Medical Sciences, NIH.

Summary

To elucidate the molecular interactions during transcription, we have developed a method to quantify the photoaffinity labeling produced by an aryl azide photoprobe positioned at the leading (5') end of the nascent RNA [Stackhouse, T. M. & Meares, C. F. (1988) *Biochemistry* **27**, 3038-3045]. This identifies those macromolecules adjacent to the probe at each transcript length, and permits ready comparison of the effects of different DNA sequences on the behavior of enzyme subunits during transcription. Here we compare transcription complexes containing the two bacteriophage templates λ P_R and T7 A1. The quantitative analysis provides the percent yield of photoaffinity labeling by each length of RNA in the transcription complex. Significant yields (up to 4%) for the σ subunit are observed the template containing the λ P_R promoter. The σ subunit from this complex is labeled maximally at a transcript length of 11 nucleotides. When transcription is initiated from the template containing the T7 A1 promoter, labeling yields for the σ subunit are less than 0.7% for all transcript lengths observed. Labeling of the β/β' subunits (analyzed together) occurs similarly on both templates, with yields reaching a maximum of approximately 15% for transcripts ≥ 13 nucleotides in length. The photoaffinity labeling of σ in poly[d(A-T)] transcription complexes differs from the results observed with DNA containing either the lambda P_R or the T7 A1 promoter, providing further evidence that the interaction between the nucleic acids and the σ subunit in the transcription complex depends on the nucleotide sequence. They are consistent with early release of σ from the T7 A1 promoter (at an RNA length < 8 nucleotides) and delayed release of σ from the λ P_R promoter (at an RNA length > 14 nucleotides).

INTRODUCTION

The control of gene expression at the level of transcription has been studied most extensively using *E. coli* RNA polymerase. This oligomeric enzyme (E. C. 2.7.7.6) catalyzes the synthesis of ribonucleic acid from a deoxyribonucleic acid template. RNA polymerase from *E. coli* contains five major subunits, with a total molecular weight of 449,000. The primary structures of all the subunits have been determined: α (M_r 36,512;

261

Ovchinnikov et al.,1977); β (M_r 150,619; Ovchinnikov et al., 1981); β' (M_r 155,162; Ovchinnikov et al., 1982); and σ (M_r 70,263; Burton et al. 1981). The core enzyme contains four subunits ($\alpha_2\beta\beta'$) and is capable of elongating, but not efficiently initiating, RNA transcripts from promoter sites on DNA. Efficient initiation of a transcript at promoter sites on DNA requires the holoenzyme, which contains the core RNA polymerase and the σ subunit. The presence of another subunit (ω) has also been observed (Gentry and Burgess, 1986, and references therein); the function of ω is not yet established.

It was discovered by Travers et al. (1969) that σ can be released shortly after the initiation of transcription, and subsequently σ can bind to another core enzyme to initiate another transcript. This is referred to as the σ cycle (Lewin,1983). A detailed understanding of the mechanism of the σ cycle and its relation to the initiation reaction has interested many investigators (Chamberlin, 1974; Hansen and McClure, 1980; Shimamoto et al., 1986). It is believed that release of the σ subunit is accompanied by the formation of a stable elongation complex containing core enzyme, DNA template, and nascent RNA. The elongation complex could subsequently bind other factors involved in control of transcription and translation (McClure, 1985; Greenblatt et al., 1987). For example, it has been shown that nusA protein can bind to the elongation complex only after σ release (Greenblatt and Li, 1981).

In order to determine the RNA length at which σ is released, Hansen and McClure (1980) performed an elegant experiment measuring the amount of σ present in a mixture of *E. coli* RNA polymerase/poly[d(A-T)] transcription complexes containing various transcript lengths. By careful comparison of the amount of σ released with the lengths of RNA present, these authors concluded that σ is released quantitatively from the complex by the time a transcript of 8 or 9 nucleotides has been produced on poly[d(A-T)].

Other investigators have examined the release of σ from transcription complexes that contain the A1 promoter of bacteriophage T7 (Shimamoto et al.,1986). This group analyzed the release of σ as a function of time, and proposed a two-step model that involves a fast triggering step (< 1 sec) followed by a slower dissociation (mean lifetime ≈ 5 sec).

Recently, our laboratory has used an aryl azide photoaffinity probe at the leading (5') end of the RNA (Figure 1) to qualitatively follow the path of the transcript as it moves through an *E. coli* transcription complex containing the T7 A1 and λ P_R DNA templates. The photoaffinity probe is a dinucleotide, allowing specific initiation of transcription from promoters whose transcripts begin with A-U. The aryl azide is converted to a highly reactive nitrene by irradiation with UV light (wavelength λ > 300 nm). This nitrene can insert nonselectively into many types of chemical bonds, thus providing information about which molecules are within its immediate vicinity (Bayley and Staros, 1984; Bayley & Knowles, 1977; Knowles, 1971). The probe also contains a sulfur-phosphorus bond that is cleaved by mercurials. Therefore the enzyme subunits, once crosslinked to the nascent RNA, can be isolated and the RNA cleaved from the protein to determine the RNA chains whose 5' ends were adjacent to each subunit.

The information obtained from the A1 promoter of bacteriophage T7 ΔD111 and ΔD123 DNA (Hanna and Meares, 1983ab) indicates the following: when the RNA is 3-12 bases long, the DNA template is heavily photoaffinity labeled; when the trinucleotide transcript is made in large quantities, the σ and β subunits are just detectably labeled, but not β' or α; for RNA lengths of 4-12 nucleotides, β and β' are labeled but not σ or α. For RNA lengths 12 or longer (and as long as 116 nucleotides on T7 ΔD123), the β and β' subunits are heavily labeled, but the other macromolecules in the transcription complex are not labeled significantly. Thus, σ appears to be in contact with the 5' end of the RNA trinucleotide made from the T7 A1 template, but not with longer RNAs.

Further studies by Bernhard and Meares (1986a) have shown that the photoaffinity labeling of the σ subunit depends on the sequences of the nucleic acids involved. These authors compared the T7 A1 and λ P_R transcription complexes, with the finding that on the λ P_R template, the 5' ends of RNAs with lengths of 9-13 nucleotides heavily labeled σ. Photoaffinity labeling results for β, β', and α were similar on both templates, suggesting that contacts between RNA and the subunits of the core enzyme are not strongly dependent on the nature of the DNA.

Probes with lengths 9-13 label β, β', and DNA on both templates and thus appear to be free to label any molecule in their immediate surroundings. Therefore, these results strongly suggest that the reason σ is not labeled by these lengths on T7 A1 is that σ dissociates from the T7 A1 transcription complex much earlier than from λ P_R. The interaction responsible for σ dissociation on the T7 A1 template evidently occurs before the transcript is 9 bases long, while the 5' end of the RNA is still hybridized to DNA (Hanna and Meares, 1983b).

Using poly[d(A-T)] as the template, quantitative photoaffinity experiments have shown the unexpected result that σ is retained in at least some transcription complexes when the transcript is 10, 11, and 12 nucleotides long (Stackhouse & Meares, 1988). At least 15% of the transcription complexes containing poly[d(A-T)] and RNA chains 10 nucleotides long still contain σ. From the experiments of Hansen and McClure (1980) regarding σ release from poly[d(A-T)], we would expect photoaffinity labeling of σ not to occur for RNA lengths longer than 8 or 9 nucleotides. The suggestion by Bernhard and Meares (1986a), that the early release of σ from the T7 A1 transcription complex could be mainly controlled by the bridging hydrogen bond donor/acceptor pattern within the major groove of the RNA/DNA duplex, is not supported by the results presented here. Considering only the hydrogen-bond donors and acceptors that bridge the two nucleic acid strands, the poly[d(A-T)] transcript sequence has the same property of alternating hydrogen bond donors and acceptors as T7 A1. However, the labeling pattern for poly[d(A-T)] more closely resembles that observed for the λ P_R complex than T7 A1.

Here we discuss the initial results of our endeavor to determine more precisely the nucleic acid sequence dependence of photoaffinity labeling with this 5' RNA probe in *E. coli* transcription complexes. In this context, it is important to measure the per cent yields of photoaffinity labeling on the subunits in transcription complexes as a function of RNA chain length. This adds significantly to the information obtained from these experiments, because it corrects for inequalities in amounts of different RNA chain lengths present, for various losses during sample treatment, and for the limited dynamic range of film. This approach, previously used to analyze poly[d(A-T)] transcription complexes, is used here to more precisely determine the photoaffinity labeling of transcription complexes containing the λ P_R and T7 A1 templates.

EXPERIMENTAL PROCEDURES
Materials

All reagents and solvents were the purest available and used without further purification unless otherwise noted. Nanopure water (Barnstead) was used throughout. *E. coli* MRE 600 cells were purchased from Grain Processing Corp. Bactotryptone and yeast extract were from Difco. HPLC purified ribonucleoside triphosphates were purchased from ICN. Ultra-pure urea and acrylamide were purchased from Schwartz-Mann. Soluble RNA, phenylmercuric acetate, dithiothreitol, tetramethylethylenediamine, bis(acrylamide), cordycepin triphosphate (3'-deoxy-ATP, a chain terminating ATP analogue), agar, TRIS (tris(hydroxymethylaminomethane)), ethidium bromide, and agarose were purchased from Sigma. p-Azidophenacyl bromide was from Pierce, adenosine 5'-O-thiophosphate was from Boehringer-Mannheim, and 3'-O-methyl-UTP, 3'-O-methyl-CTP, and 3'-O-methyl GTP (RNA chain terminator) were from P-L Biochemicals. $[\alpha\text{-}^{32}P]GTP$ and $[\alpha\text{-}^{32}P]CTP$ were purchased from Amersham. Kodak XAR films were used without intensifying screens for autoradiography.

Buffers. These were as follows: *Buffer A* contained 80 mM TRIS-HCl, pH 7.9, 5 mM 2-mercaptoethanol, 0.1 mM Na$_2$EDTA, and 50% (v/v) glycerol; *buffer B* contained 10 mM TRIS-HCl, pH 8.0, and 1 mM Na$_2$EDTA; *buffer C* contained 50 mM TRIS-HCl, pH 7.9, 5 mM 2-mercaptoethanol, 10 mM NaCl, 10 mM MgCl$_2$, and 5% (v/v) glycerol; *buffer D* contained 0.5 mL of buffer F, 2.9 mL of 9-10 M urea , 0.5 g of sucrose, 12 mg of NaDodSO$_4$, 0.1 mL of 0.1% (w/v) bromphenol blue, and 10 mM dithiothreitol (added just prior to use); *buffer E* contained, in 1 L, 4.4 mL of ethanolamine, 4.5 g of glycine, and 1.0 g of NaDodSO$_4$ (pH 9.7); *buffer F* contained 18.6 mL of triethanolamine, 8 mL of concentrated HCl, and 96 g of urea in 200 mL total volume, pH 7.5; *buffer G* contained 89 mM TRIS-borate (pH 8.3), 1 mM Na$_2$EDTA, 7 M urea; *buffer H* contained 15 mM TRIS - HCl, pH 8.5, 10 mM glycine, 0.06% NaDodSO$_4$.

Solvents. *Solvent H* contained 2-propanol/concentrated ammonium hydroxide/water (6:3:1 v/v/v); *solvent I* contained 1% (w/v) ammonium acetate, pH 7.0; *solvent J* contained1% (w/v) ammonium acetate, pH 5.0; *solvent K* contained0.8% (w/v) ammonium acetate, pH 7.0, and 20% (v/v) acetonitrile.

Cleavage solution. First, 0.1% (w/v) NaDodSO$_4$ is saturated with phenylmercuric acetate by shaking with an excess of the pure solid for a minimum of 4 h at room temperature in the dark. The mixture is briefly centrifuged to pellet the excess solid, and the precipitate is discarded. Soluble RNA (1 mg/mL) is added to the cleavage solution as a carrier. The solution is used within one day after preparation.

Methods

RNA polymerase. The enzyme was purified from *E. coli* MRE 600 cells according to the method of Burgess and Jendrisak (1975) as modified by Lowe et al. (1979). It was dialyzed against buffer A and stored at -20° C. Protein concentration was determined by the method of Burgess (1976).

DNA templates. The DNA containing the λ P$_R$ template was obtained from plasmid pGW7, prepared by Geoff Wilson and William Konigsberg and kindly provided by Martin

Schmidt and Michael Chamberlin. The T7 A1 template was obtained from plasmid pAR1707 kindly provided by Michael Chamberlin. Each plasmid was transformed into an *E. coli* JM83 host. The growth of the bacteria containing pGW7 was at 28 °C without amplification, whereas the growth of pAR1707 was at 37°C without amplification. Both plasmids were isolated using alkali lysis and a CsCl density gradient (Maniatis et al., 1982). Isolated pGW7 was digested with *Eco*RI and *Hin*DIII, and the 933 base pair fragment containing the λ P_R promoter and early transcribed region was isolated by electrophoresis in 0.8% agarose. Isolated pAR1707 was restricted with *Sal* I, and the linearized plasmid containing the T7 A1 promoter and early transcribed region was also purified by electrophoresis in 0.8% agarose. The DNA concentration was determined by measuring absorbance at 260 nm (one A_{260} unit = 50 μg/mL). The DNA was stored in buffer B, at -20 °C.

5'-(Thiophosphoryl)adenylyl(3'-5')uridine (SpApU) . The abortive initiation reaction catalyzed by RNA polymerase was used to synthesize SpApU. The method was the same as for pApU synthesis (DeReimer & Meares, 1981) except that AMPS was used as initiator instead of AMP (Hanna & Meares, 1983a). The reaction mixture contained the following in 10 mL: 20 mM Tris (pH 7.9), 5 mM $MgCl_2$, 5 mM 2-mercaptoethanol, 5.3 mM AMPS, 4.5 mM UTP, poly[d(A-T)] (20 μM phosphate), 2 mg of *E. coli* RNA polymerase, 2.8% (v/v) glycerol. The reaction mixture was incubated at 37 °C for 36 h. The reaction mixture was purified by high-pressure liquid chromatography on a Waters semipreparative C_{18} μBondapak column with solvent I. With a solvent flow of 5 mL/min, the product had an 8 min retention time. Ammonium acetate was removed by lyophilization. The product had an R_f value of 0.34 on TLC in solvent H. Concentrations were determined by reading the absorbance at 260 nm, using $\varepsilon_{260} = 2.5 \times 10^4$ M^{-1} cm^{-1}.

5'-[[(4-Azidophenacyl)thiol]phosphoryl]adenyly(3'-5')uridine (N3RSpApU). Synthesis of the dinucleotide photoaffintiy probe was carried out under reduced light. 100 μL of 13 mM SpApU and 10 μL of 0.2 M $NaHCO_3$ were mixed with 30 μL of methanol; 30 μL of 90 mM azidophenacyl bromide in methanol was then added and the reaction was allowed to proceed for 45 min at room temperature. The reaction mixture was extracted three times each with isobutyl alcohol and then with ethyl ether, and ether was removed under pressure at room temperature. The reaction mixture was then applied to the Waters HPLC column and was eluted with a 20-min linear gradient from solvent J to solvent K. With a solvent flow of 2 mL/min, the product had a retention time of 19.5 min. Acetonitrile was removed by lyophilization. The ultraviolet spectrum of each peak from the HPLC column was taken on a Hewlett-Packard Model 8450 UV-vis spectrophotometer. Product concentration was determined by using $\varepsilon_{300} = 2.4 \times 10^4$ M^{-1} cm^{-1} for the azide (Hixson & Hixson, 1975). The product had an R_f value of 0.56 on TLC in solvent H. The product structure was verified by enyzmatic digestions and mercury-facilitated cleavage of the N3RSpApU followed by analysis on TLC in solvent H (Hanna & Meares, 1983a).

Preparation of Transcription Reactions. The experimental procedure is outlined in Figure 1. All transcription reactions were carried out in reduced light. Four separate transcription reactions were performed, one each to terminate the transcript at adenylate, uridylate, guanidylate and cytidylate residues. The main goal was to produce an evenly distributed set of RNA lengths in the transcript mix so that all are represented for photoaffinity labeling.

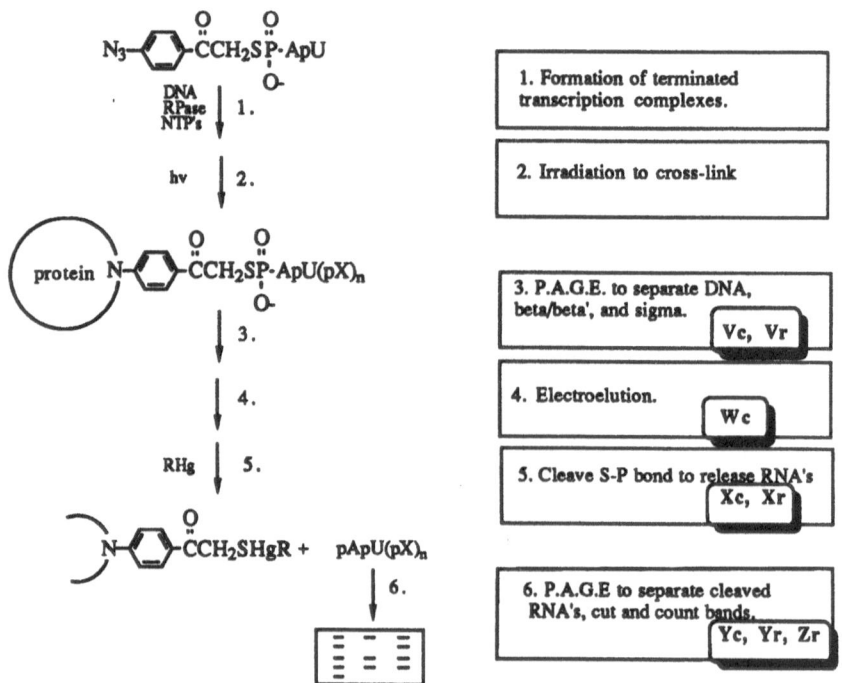

FIGURE 1: Experimental outline, showing the structure of the photoaffinity probe 5'-[[(4-azidophenacyl)thiol]phosphoryl]adenyly(3'-5')uridine at the top. The points where yield factors (Vc,Vr,Wc, etc.) accompanying each process were calculated are indicated. These were necessary for the percent yield determinations, as described under Methods.

First, the *preinitiation mix* was prepared by combining in a total volume of 33.0 μL, 0.61 μM RNA polymerase, 0.15 μM DNA template, and 470 μM N₃RSpApU in buffer C. This was incubated at 37 °C for 10 min to allow the formation of a ternary complex. A 7 μL aliquot from the preinitiation mix was added to each of four elongation mixes A, U, C, and G, forming a complete reaction mixtures of 15 μL. For transcription from λ P$_R$ each elongation mix contained 18 μM ATP, 18 μM UTP, 18 μM CTP and 6 μM [α-^{32}P]GTP (>500 Ci/mmole) in buffer C. The concentration of the RNA chain terminators were as follows: elongation mix A, 1.0 mM cordycepin triphosphate; elongation mix U, 833 μM 3'-O-methyl-UTP; elongation mix C, 833 μM 3'-O-methyl-CTP; and for elongation mix G, 333 μM 3'-O-methyl-GTP. For transcription from T7 A1 each elongation mix contained 18 μM ATP, 18 μM UTP, 18 μM GTP and 6 μM [α-^{32}P]CTP (>500 Ci/mmole) in buffer C. The concentration of the RNA chain terminators were as follows: elongation mix A, 1.2 mM cordycepin triphosphate; elongation mix U, 800 μM 3'-O-methyl-UTP; elongation mix G, 700 μM 3'-O-methyl-GTP; and for elongation mix C, 333 μM 3'-O-methyl-CTP. These complete reaction mixtures were incubated for 10 min at 37 °C. At the end of this elongation period, 1.5 μL aliquots were removed from each reaction mixture and added to 5 µL of RNA sample buffer (5X buffer G containing the tracking dyes bromphenol blue and xylene cyanol [0.25% w/v]) and 13.5 µL of cleavage solution. These unirradiated RNA samples were incubated in the dark for 12 h to cleave the probe. Subsequently, 2 µL of 1 M dithiothreitol was added to these RNA samples, which were incubated for 1 h in the dark at room temperature to reduce the azide and run on an RNA sequencing gel to confirm the lengths of RNA produced.

Unirradiated control protein samples were prepared by removing 5 µL from each final transcript mix and adding it to an equal volume of 2X buffer D. These protein samples were allowed to incubate at room temperature for 1 h in the dark before electrophoresis on a protein gel.

After the control samples had been taken, the terminated reactions were transferred to a borosilicate tube and irradiated for 30 sec in a Rayonet photochemical reactor ($\lambda > 300$ nm). Following irradiation, samples were again removed for both RNA sequencing and protein analysis, as was done for the unirradiated controls above.

Protein Gels. The protein gels used in these experiments employed the $NaDodSO_4$-urea system of Wu and Bruening (1971). The running gel was 6% acrylamide with a length of 20 cm. Except for β and β', which moved as a single broad band, this was found to be sufficient to separate the polymerase subunits from each other, the DNA and the small transcripts. Gels were run for 1 h at 80 V, then at 30 mA until the bromphenol blue tracking dye had run off the bottom of the gel (approximately 3 h). After electrophoresis was complete, the gels were wrapped in plastic wrap and autoradiographed at room temperature for 1-2 h with Kodak XAR film.

Electroelution and Cleavage. The photoaffinity labeled subunits and DNA, which were located by autoradiography and comparison to stained marker lanes, were excised from the protein gel. The individual gel pieces were transferred to 1.5 mL polypropylene Eppendorf centrifuge vials and counted (Cerenkov radiation) in a Beckman LS 6800 scintillation counter prior to electroelution.

Electroelution was carried out by placing the individual gel pieces plus 200 µL of buffer H into dialysis tubing (molecular weight cutoff 12,000-14,000). The dialysis bags were placed into a Bio-Rad Trans Blot apparatus eluted at 10 °C and 75 V (1 mA) for 4-5 h. The eluted protein-RNA conjugates were transferred to 0.5 mL Eppendorf vials. Each dialysis bag was rinsed with 200 µL of freshly prepared cleavage solution. The rinse was pooled with the original eluent; these solutions were counted as were the remaining dialysis bags and gel pieces for each sample. These counts were then used to determine the efficiency of the elution process. Cleavage was allowed to proceed for 18-24 h at room temperature in the dark.

Following the elution and cleavage, the samples were lyophilized to dryness (8 h, 20-50 mtorr) and then resuspended in 10 µL of H_2O and 10 µL of RNA buffer (above) and counted. The samples were then briefly centrifuge to settle the precipitate and the supernatant was removed and counted. These counts were then used to determine the efficiency of the cleavage reaction. 10 µL of each sample was then run on an RNA sequencing gel to determine the lengths of transcript that had crosslinked to the protein or DNA.

RNA Sequencing Gels. Total RNA samples and RNA cleaved from protein or DNA were analyzed on 40 cm X 0.75 mm 25% acrylamide gels (1:29 methylenebis(acrylamide):acrylamide) containing buffer G (Carpousis and Gralla, 1980). Gels were allowed to polymerize for 2 h and were then pre-electrophoresed at 1,000 V for a minimum of 4 h before use. For the actual analysis of the samples, each gel was run at 800-1000 V until the bromphenol blue had run 27.5 cm from the origin. After electrophoresis, the gels were placed between two sheets of plastic wrap and autoradiographed at -80 °C without an intensifying screen.

Quantification of Transcripts. The individual lanes from the RNA sequencing gels were cut out and sliced, starting just below the trinucleotide and working toward the top of the gel; each slice was 3 mm wide. The individual gel slices from each lane were counted (Cerenkov radiation). The gel slices were placed flat on the bottom of the scintillation vials to prevent any differences in counting efficiency due to geometric factors. The gel slices typically extended up the lane to transcripts 20 nucleotides long before resolution became uncertain. Total counting time for a complete gel (12 lanes) was about 72 h. This rather laborious manual procedure may be avoided if automated instruments are available.

The per cent yield of photoaffinity labeling can be calculated by comparing the counts in the lanes that contained RNA released from protein or DNA to the counts in the marker lanes, which were accurate dilutions of an irradiated and cleaved RNA sample from the original transcription mix:

$$\% \text{ YIELD} = \frac{\text{cpm of cleaved transcript}}{\text{cpm of marker transcript}} * F * 100$$

$$\text{where} \quad F = \frac{V_r * X_r * Y_r * Z_r}{V_c * W_c * X_c * Y_c} \quad \text{and}$$

V_c, V_r = volume removed from the original transcription mix for subunit analysis or RNA marker respectively.

W_c = electroelution efficiency.

X_c, X_r = cleavage reaction efficiency for subunit samples and RNA markers respectively.

Y_c, Y_r = fraction of final sample loaded onto the RNA sequencing gel for subunit sample and RNA marker respectively.

Z_r = dilution factor for RNA markers.

The points at which these factors were measured are shown in Figure 1. In this way, the number of moles of a particular transcript in the transcription mix could be directly related to the number of moles of that same transcript photocrosslinked to enzyme subunits or DNA.

The percent photolabeling yield may be analyzed as a ratio of pseudo-first order chemical reaction rate constants. For example, the yield for attachment of the nitrene at the leading end of RNA to the σ subunit is

$$\%\text{Yield}_\sigma = \frac{k_\sigma}{k_\sigma + k_{\beta\beta'} + k_{DNA} + k_{other}}$$

The rate constant k_σ (for example) is proportional to the effective concentration of σ in the vicinity of the nitrene. This effective concentration depends on the rate of collisions between the nitrene and amino-acid residues on σ, and also on the reactivities which those residues exhibit toward the nitrene (Bayley & Staros, 1984). Since the residues involved have not yet been identified, we cannot separate the collision frequencies from the intrinsic reactivities at this time. However, it is clear that the intrinsic reactivities of the competing residues do not play a dominant role in the results, because in most instances the nitrenes on oligonucleotides of a given length (e. g., the decamer) label as many as three different macromolecules (β or β', σ,). Thus it is reasonable to expect that the photoaffinity labeling per cent yields at least qualitatively reflect the frequency of encounters between the nitrene and the labeled macromolecule.

A related question is whether the composite rate constant for all reactions that do not lead to cross-linking, denoted k_{other} in the sum above, is likely to vary with chain length. The experiments of Bernhard and Meares (1986b) on the accessibility of the probe to solvent molecules answer this question affirmatively. Furthermore, the formation of secondary structures (e. g., hairpin loops) in the RNA is likely to vary with chain length; this will cause the self-labeling of RNA to change accordingly. Thus, for probes on RNAs of different lengths in transcription complexes, it is to be expected that the total fraction of nitrene reacting with RNA or solvent molecules (H_2O, TRIS, mercaptoethanol, etc.) can vary widely.

RESULTS

Figures 2 and 3 show the photoaffinity labeling yields from sigma and beta/beta' respectively for each of the templates analyzed, as a function of transcript length. This provides a quantitative analysis of what is seen on an autoradiogram of an RNA sequencing gel. The results of two independent experiments per template are shown in Figures 2 and 3, together with their average. The data in figures 2 and 3 appear somewhat scattered, principally because of the limitations on resolution imposed by the 3 mm slice width when comparing separate gels. The σ subunit is labeled most heavily by transcript lengths of 8 through 14, when transcribing from the λ P_R template with a maximum yield of approximately 4% at a transcript length of 11 nucleotides. When transcription takes place from the T7 A1 template, the yields from sigma for all RNA lengths observed remains below 0.7%. As shown in Figure 3, the yield of photolabeling on the β and β' subunits is similar for both templates, remaining below 1% for transcripts 3-10 nucleotides long, and then dramatically increasing to approximately 15% at a transcript length of 14 nucleotides. The per cent yield on β and β' remains high for the rest of the lengths analyzed.

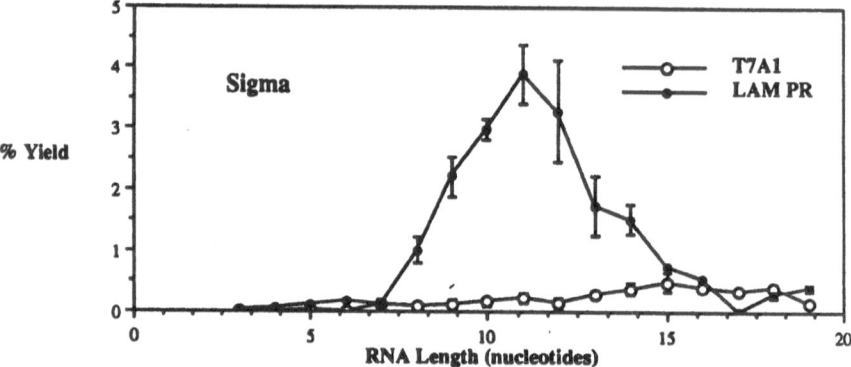

FIGURE 2: Percent photoaffinity labeling of sigma analyzed from transcription complexes containing either T7 A1 (open circles) or Lambda PR (filled circles) as the template. The error bars indicate the range of two independent experiments

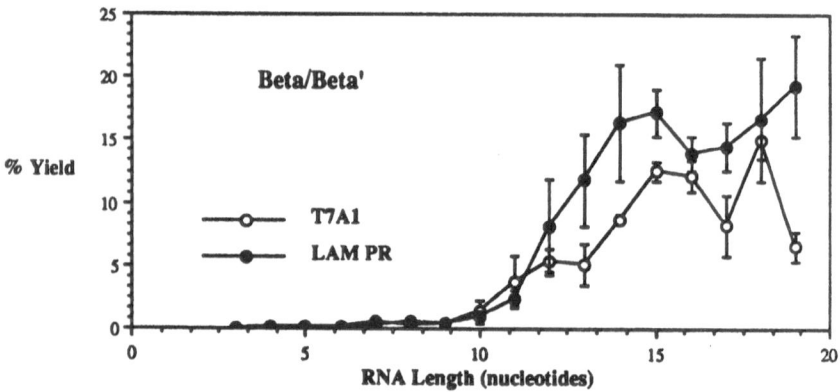

FIGURE 3: Percent photoaffinity labeling of beta/beta' analyzed from transcription complexes containing either T7 A1 (open circles) or Lambda PR (filled circles) as the template. The error bars indicate the range of two independent experiments

DISCUSSION

Methodology. Bayley and Staros (1984) point out that the extent of labeling in different photoaffinity experiments varies from almost 100% to less than 1%. The accessibility of the bound photoprobe to solvent molecules, which can compete with the macromolecule for reaction with the nitrene, may lead to low photoaffinity labeling yields. Also, the covalent bond formed between the photoprobe and the macromolecule may not always be sufficiently stable to permit isolation of a crosslinked product. A third point specifically concerning RNA polymerase is the abortive initiation reaction, which leads to repeated production and dissociation of oligonucleotide transcripts, in some cases up to 12-13 nucleotides in length (Carpousis and Gralla, 1980; Levin et al., 1987). Having been released from the transcription complex, these abortive products react with the solvent; since the percent yield calculation counts all the RNA molecules produced, this contributes to the low percent yields observed for short RNA.

It might be argued that the presence of the photoaffinity probe at the 5' end of the RNA somehow perturbs the system such that some of the σ remains bound. While it is difficult to rule this out entirely, we believe that the high sensitivity of these experiments to subtle variables such as the DNA base sequence supports the notion that the properties of the photoprobe do not interfere with the results.

Cutting and counting polyacrylamide gel bands is a rather crude way to extract the information they contain. The 3 mm width of the gel slices is too large to take full advantage of the resolution afforded by electrophoresis, but narrow enough to produce hundreds of slices to be counted. The slicing procedure appears to be the limiting factor in our current methodology; automated methods would be expected to produce less scatter in the final results (Anderson et al., 1987).

In our determination of % yields, the variation from experiment to experiment usually was larger than the error expected from the statistics of radioactive counting. Only when the yields dropped below 0.5% did the statistics of radioactive counting become important. In Figures 2 and 3, the results of two independent experiments are shown to allow the reader to judge the precision of the data.

A particular complication observed in these experiments is that near the gel band corresponding to a particular terminated transcript, one often finds a band for the unterminated transcript with the same sequence (e. g., one molecule ends with cordycepin, the other with adenosine). Also, the addition of the a residue does not always change the electrophoretic mobility by the same amount. Cross-talk can also occur; each U lane contains small amounts of RNAs mistakenly terminated by O-methyl-C etc. It is difficult to accurately exclude the extra bands present in the lanes. Further, light bands that are clearly evident on the film contain too few counts for brief measurements in a scintillation counter. A relevant discussion of the problems associated with quantitating closely spaced spots in 2-D gels has been presented by Bossinger et al. (1979).

Sigma release. It has been established that the σ subunit of RNA polymerase can be released from the transcription complex after initiation of RNA synthesis (Travers and Burgess, 1969). Hansen and McClure (1980) studied the length of transcript at which the σ subunit is released from a ternary complex using poly[d(A-T)] as the template. Studies of this type are particularly difficult, because RNA polymerase cannot be induced to pause uniquely at each nucleotide position in an RNA chain. Rather, each DNA template possesses some discrete pause sites that can be conveniently studied but are not necessarily located near the point of σ release. After using the chain terminator 3'-deoxy-ATP to produce a mixture of chain lengths, and determining the amounts of the various lengths of

RNA produced and the amounts of dissociated σ, Hansen and McClure developed a mathematical model of the process. This model indicated that all of the σ subunits are released at a transcript length of 8 or 9 nucleotides. However, the photoaffinty experiments using poly[d(A-T)] as template (Stackhouse & Meares, 1988) clearly indicate that σ is still present in at least some transcription complexes when the transcript is 10, 11, or 12 nucleotides long.

The above experiments point out one advantage of photoaffinity experiments versus other techniques such as chromatography for studying RNA polymerase. The reason for the discrepancy as to the exact transcript length at which σ is released from a poly[d(A-T)] transcription complex could be due to the different experimental procedures used. Hansen and McClure performed experiments that involved separation of σ from ternary enzyme/DNA/RNA complexes using gel filtration through Bio-Gel A-1.5m. If the σ-containing transcription complexes were only marginally stable, the added manipulations might lead to the artificial release of σ. Photoaffinity crosslinking is performed in homogeneous solution, with no prior chromatographic separation of components. Secondly, the Hansen and McClure analysis of the RNA products involved quantitation of the transcripts by autoradiography and subsequent scanning of the autoradiogram, after separation of the different lengths by 20% acrylamide/urea PAGE. Recently, it has been observed that 25% gels afford more reliable analysis of short oligonucleotides (Levin et al., 1987). In this investigation, we used the RNA sequencing 25% PAGE of Carpousis and Gralla (1980) to analyze the transcript lengths, and quantitated the transcripts by cutting and counting individual gel slices, compared to a marker of total RNA from the same reaction.

The experiments here indicate that the sigma subunit does not dissociate from the λ P_R transcription complexes before the transcript has reached a length of 14 nucleotides. The results in Figure 2 also make it evident that when λ P_R is used as the template the 5'-azides on transcripts shorter than 7 nucleotides do not label σ. This supports the hypothesis that σ is located near one side of the growing RNA/DNA duplex, and that the labeling only occurs on σ as the 5'-end of the transcript passes across this side. It is also clear that the path of nascent RNA through the transcription complex is not the same for different DNA templates. Photoaffinity labeling of the σ subunit appears to be most sensitive to the nature of the DNA.

For *E. coli* RNA polymerase/T7 A1 transcription complexes, the work of Shimamoto et al. (1986) supports the idea of early σ release. This group analyzed the release of σ as a function of time, and proposed a two-step model that involves a fast triggering step (< 1 sec) followed by a slow dissociation (mean lifetime ≈5 sec). By analyzing the transcription complexes formed under different GTP concentrations and different incubation times, their results indicate no relationship between the transcript length and the slow dissociation step of σ on the T7 A1 promoter. However, the results of Wu et al. (1975) indicate that the fast triggering step does not take place before the formation of the first two phosphodiester bonds. Shimamoto et al. estimate that the triggering step probably occurs before the transcript is 4-6 nucleotides long. These results are consistent with the observation of low photoaffinity labeling on σ when the T7 A1 template is used; the release of sigma from this template should occur before the RNA is 8 nucleotides long.

Sigma is evidently released quantitatively from transcription complexes on some DNAs, such as T7 A1 (Shimamoto et al., 1986) or lac UV5 (Straney and Crothers, 1987). However, on poly[d(A-T)] it appears that σ release can occur over a range of RNA lengths rather than at a single, unique point during transcription, and there may be some transcription complexes that never release σ (Stackhouse & Meares, 1988). Although this was not observed for the T7 A1 or λ P_R transcription complexes, this may apply to other DNA templates. It is possible that the molecular interactions leading to σ release are analogous to those which occur at promoter or terminator sites, in that they have a range of strengths. It is well known that different promoter sites on DNA lead to the initiation of transcription with widely different efficiencies (McClure, 1985). Also, studies of the termination of transcription show that it does not occur with 100% efficiency at most termination sites (Chamberlin et al., 1987).

A point of interest from these studies on T7 A1 and λ P_R becomes evident when one looks at the total yields of photoaffinity labeling in the transcription complex, rather than at the individual subunits. A drastic increase occurs between transcript lengths of 11 nucleotides, at which the overall yield is 2%, and 14 nucleotides, at which the overall yield is 15%. A qualitatively similar increase of photolabeling occurred on poly[d(A-T)]. This indicates a major change in the local environment around the 5'-terminus of the transcript in a majority of the complexes. It is dominated by the movement of the 5'-azide into a region of greater contact with β and β', and less contact with σ. Over a similar range of RNA lengths, Levin et al. (1987) have observed (using natural DNA promoters) that the ternary transcription complex becomes stable and ceases abortive initiation of RNA chain synthesis.

As discussed by Stackhouse & Meares (1988), the interaction between σ and the nucleic acids evidently is not dominated by contacts with the bridging hydrogen-bond donors and acceptors in the major groove of the RNA/DNA hybrid. Therefore, the source of the key interaction remains to be determined. Several possibilities exist: [1] a particular set of contacts with the bridging hydrogen-bond donors and acceptors in the major groove of the RNA/DNA hybrid is necessary but not sufficient, so that the actual requirements are more complex; [2] the release of sigma may depend on a particular set of contacts with the hydrogen-bond donors and acceptors in the major groove within the promoter region of the template; [3] the key interaction involves the non-template DNA strand, rather than the RNA/DNA hybrid (note that this is known to contact σ [Simpson, 1979; Park et al., 1982ab]); [4] other structural parameters such as groove width, or minor groove properties like the presence or absence of the 2-amino group of G, or a homopurine stretch are crucial.

Further quantitative data are being gathered on recombinant constructs, that will allow us to investigate the role of the promoter region vs. the early transcript region on the pattern of σ photolabeling (T. Stackhouse, A. Wieland, M. J. Chamberlin, and C. F. Meares; work in progress). By quantitatively analyzing the photocrosslinking of the various subunits with the different templates, we hope to obtain a more detailed view of the molecular interactions taking place during transcription.

ACKNOWLEDGMENT

We thank Jeffrey Marx and Blaine Bartholomew for helpful comments and discussions.

REFERENCES

Anderson, H. L., Puck, T. T., and Shera, E. B. (1987) *Proc. Natl. Acad. Sci. USA 84*, 4749-4753.

Bayley, H., & Knowles, J. R. (1977) *Methods Enzymol. 46*, 69-114.

Bayley, H., & Staros, J. V. (1984) "Photoaffinity Labeling and Related Techniques," in *Azides and Nitrenes*, pp 433-490, E. F. Scriven, ed., Academic Press, Orlando, FL.

Bernhard, S. L., & Meares, C. F. (1986a) *Biochemistry 25*, 5914-5919.

Bernhard, S. L., & Meares, C. F. (1986b) *Biochemistry 25*, 6397-6404.

Bossinger, J., Miller, M. J., Vo, K.-P., Geiduschek, E. P., and Xuong, N.-H. (1979) *J. Biol. Chem. 254*, 7986-7998.

Burgess, R. R. (1976) "Purification and Physical Properties of *E. coli* RNA Polymerase," in *RNA Polymerase*, pp. 69-100, R. Losick and M. Chamberlin, eds., Cold Spring Harbor Laboratory Cold Spring Harbor, NY.

Burgess, R. R. & Jendrisak, J. J. (1975) *Biochemistry 14*, 4634-4636.

Burgess, R. R., Travers, A. A., Dunn, J. J., & Bautz, E. K. F. (1969) *Nature (London) 221*, 43-46.

Burton, Z., Burgess, R. R., Lin, J., Moore, D., Holder, S., Gross, C. A. (1981) *Nucl. Acids Res. 9*, 2889-2903.

Carpousis, A. J., & Gralla, J. D. (1980) *Biochemistry 19*, 3245-3253.

Chamberlin, M. J. (1974) *Annu. Rev. Biochem. 43*, 721-775.

Chamberlin, M. J., & Berg, P. (1962) *Proc. Natl. Acad. Sci. USA 48*, 81-93.

Chamberlin, M. J., Arndt, K. M., Briat, J. F., Reynolds, R. L., and Schmidt, M. C. (1987) in *RNA Polymerase and the Regulation of Transcription*, pp 347-356, ed.Reznikoff, W. S., Burgess, R. R., Dahlberg, J. E., Gross, C. A., Record, M. T., Jr., & Wickens, M. P., Elsevier, New York.

DeRiemer L. H., & Meares, C. F. (1981) *Biochemistry 20*, 1612-1617.

Gentry, D. R., and Burgess, R. R. (1986) *Gene 48*, 33-40.

Greenblatt, J., Horwitz, R. J., & Li, J., (1987) in *RNA Polymerase and the Regulation of Transcription*, pp 357-366, ed.Reznikoff, W. S., Burgess, R. R., Dahlberg, J. E., Gross, C. A., Record, M. T., Jr., & Wickens, M. P., Elsevier, New York.

Greenblatt, J., & Li, J. (1981) *Cell (Cambridge, Mass.) 24*, 421-428.

Hanna, M. M., & Meares, C. F. (1983a) *Biochemistry 22*, 3546-3551.

Hanna, M. M., & Meares, C. F. (1983b) *Proc. Natl. Acad. Sci. USA 80*, 4238-4242.

Hansen, U. M., & McClure, W. R. (1980) *J. Biol. Chem. 255*, 9564-9570.

Hixson, S. H., & Hixson, S. S. (1975) *Biochemistry 14*, 4251-4254.

Khorana, H. G. (1965) *Fed. Proc. 24*, 1473-1487.

Knowles, J. R. (1971) *Accts. Chem. Res. 5*, 155-160.

Levin, J. R., Krummel, B., and Chamberlin, M. J. (1987) *J. Mol. Biol. 196*, 85-100.

Lewin, B. (1983) *Genes*, Chapter 10, John Wiley & Sons, New York.

Lowe, P. A., Hager, E. A. & Burgess, R. R. (1979) *Biochemistry 18*, 1344-1352.

Maniatis, T., Fritsch, E. F., & Sambrook, J. (1982) *Molecular Cloning*, Cold Spring Harbor Laboratory, Cold Spring Harbor, NY.

McClure, W. R. (1985) *Annu. Rev. Biochem. 54*, 171-204.

Ovchinnikov, Yu. A., Monastyrskaya, G. S. Gubanov, V. V., Guryev, S. O., Salomatina, I. S., Shuvaeva, T. M., Lipkin, V. M., & Sverdlov, E. D.(1982) *Nucl. Acids. Res. 10*, 4935-4044.

Ovchinnikov, Yu. A., Monastyrskaya, G. S. Gubanov, V. V., Guryev, S. O., Chertov, O., Yu., Modyanov, N. N., Grinkevich, V. A., Makarova, I. A., Marchenko, T. V., Polovnikova, I. N., Lipkin, V. M., & Sverdlov, E. D. (1981) *Eur. J. Biochem. 116*, 621-629.

Ovchinnikov, Yu. A., Lipkin, V. M., Modyanov, N. N., Chertov, O., Yu., Smirnov, Yu. V. (1977) *FEBS Lett. 76*, 108-111.

Park, C. S., Hillel, Z., & Wu, C.-W. (1982a) *J. Biol. Chem. 257*, 6944-6949.

Park, C. S., Wu, F. Y.-H., & Wu, C.-W. (1982b) *J. Biol. Chem. 257*, 6950-6956.

Seeman, N. C., Rosenberg, J. M., and Rich, A. (1976) *Proc. Natl. Acad. Sci. USA 73*, 804-808.

Shimamoto, N., Kamigochi, T., & Utiyama, H. (1986) *J. Biol. Chem. 261*, 11859-11865.

Simpson, R. B. (1979) *Cell (Cambridge, Mass.) 18*, 277-285.

Siebenlist, U., Simpson, R. B., & Gilbert, W. (1980) *Cell (Cambridge, Mass.) 20*, 269-281.

Stackhouse, T. M. & Meares, C. F. (1988) *Biochemistry* 27, 3038-3045

Straney, D. C., Crothers, D. M. (1987) *J. Mol. Biol. 193*, 267-278.

Travers, A. A., & Burgess, R. R. (1969) *Nature (London) 222*, 537-540.

von Hippel, P. H. (1979) in Biological Regulation and Development, Vol. 1, pp. 279-347, ed. Goldberger, R. F., Plenum, New York.

von Hippel, P. H., Bear, D. G., Winter, R. B., & Berg, O. G. (1982) in *Promoters: Structure and Function*, pp 3-33, eds. Rodriguez, R. L., and Chamberlin, M. J., Praeger, New York.

Wu, C.-W., Yarbrough, L. R., Hillel, Z., & Wu, F. Y.-H.(1975) *Proc. Natl. Acad. Sci. USA 72*, 3019-3023.

Wu, G. J., & Bruening, G. E. (1979) *Virology 46*, 596-612.

'CAGED' COMPOUNDS TO PROBE THE DYNAMICS OF CELLULAR PROCESSES:
SYNTHESIS AND PROPERTIES OF SOME NOVEL PHOTOSENSITIVE P-2-NITROBENZYL
ESTERS OF NUCLEOTIDES

John F. Wootton
Department of Physiology, College of Veterinary Medicine,
Cornell University, Ithaca, N.Y. 14853

and

David R. Trentham
Division of Physical Biochemistry, National Institute
for Medical Research, Mill Hill, London, NW7 1AA

ABSTRACT.   Photochemical release of substrates or regulatory
molecules is a developing technique that overcomes problems such as
diffusional delays in the kinetic study of biological processes
involving macromolecular assemblies. A wide range of phosphate esters
such as nucleotides and nucleotide analogues can be alkylated on their
phosphate groups with 1-(2-nitrophenyl)diazoethane and 1-(3,4-
dimethoxy-6-nitrophenyl)diazoethane. The phosphate esters so formed
are generally biologically inert and may be photolyzed to starting
material with near-UV (300-360 nm) irradiation. Flash photolysis of
these 'caged' compounds with a pulsed laser or arc lamp releases up to
several millimolar of a biologically active compound.   Subsequent to
photon absorption, dark reactions occur within the caged compounds
before the active compound is released and measurement of the rate of
these processes is necessary to determine the time resolution of the
technique. P-1-(2-Nitrophenyl)ethyl esters of nucleoside 3',5'-cyclic
phosphates photolyze at 5 s^{-1} at 21°C in aqueous solution at pH 7 and
0.11 M ionic strength.   Below pH 7 the photolysis is acid catalyzed.
P-1-(3,4-dimethoxy-6-nitrophenyl)ethyl esters of P_i and ATP photolyze
more slowly than the corresponding P-1-(2-nitrophenyl)ethyl esters.
The P-1-(3,4-dimethoxy-6-nitrophenyl)ethyl esters of cyclic
nucleotides are unstable in water at pH 7.   However the P-1-(3,4-
dimethoxy-6-nitrobenzyl)esters of cyclic nucleotides appear to
photolyze more rapidly than their P-1-(2-nitrobenzyl)ester
counterparts, though with low quantum yield. The implications of
these results for the photolysis mechanisms and the application of 2-
nitrobenzyl phosphate esters in biological research is discussed.

1.   BACKGROUND AND INTRODUCTION

Time resolved analysis of biological processes such as muscular
contraction, active transport and receptor-mediated cellular
regulation is an integral component in understanding their mechanisms.

277

P. E. Nielsen (ed.), Photochemical Probes in Biochemistry, 277–296.
© *1989 by Kluwer Academic Publishers.*

Frequently simple manual addition of effector molecules is inadequate as a means of initiating such processes. For example, with this experimental approach the temporal response of an agonist to a receptor is likely to be limited by diffusion of the agonist or to be complicated by desensitization of the receptor. Over the past decade photochemical approaches have been developed that address this problem. Typically a photosensitive but biologically inert precursor molecule is introduced into the system under investigation. An intense pulse of irradiation causes rapid production of the effector molecule and the biological response is then measured with microsecond to second time resolution depending on the photochemistry and the system under investigation. Recent reviews have addressed the underlying photochemistry and the applications of the approach (1-3).

These photosensitive precursor molecules have been termed 'caged' compounds by Kaplan et al (4). The first compound so named was caged ATP,I, the P^3-1-(2-nitrophenyl)ethyl ester of ATP. Caged ATP and other caged nucleotides have been used to study the mechanism of a range of biological processes or systems that include muscle contraction and its regulation (5-11), ion transport ATPases (12-16), G-protein regulated calcium channels in neurones (17), the cyclic GMP activated cation channel of retinal rods (18) and a second messenger intracellular action in heart cells (19).

Caged ATP has several properties that make it suitable for use in biological research. Like other 2-nitrobenzyl derivatives it is photosensitive in the 300-360 nm region, the excitation being probably associated with an $n \rightarrow \pi^*$ transition (20). Biological systems are frequently optically transparent and their endogenous proteins and nucleotides are not susceptible to photochemical damage in this wavelength range. Caged ATP does not bind significantly to the ATPase active sites of some, but not all, systems in which it has been used (12,21). The quantum yield of caged ATP on 347 nm irradiation is 0.63 and ATP is formed at 118 s^{-1} in a physiological medium at $20^\circ C$, so that biological responses can be monitored with millisecond time resolution (22,23).

The first synthesis of caged ATP was by reaction of 1-(2-nitrophenyl)ethyl phosphate with ADP morpholidate (4). A more convenient synthesis involves treating nucleotides with derivatives of 2-nitrobenzyldiazomethane. Where the nucleotide contains a weakly acidic group, that has always been the site of esterification (22). If no such group is present, as in the case of the nucleoside 3',5'-cyclic phosphates, then esterification has been on the strongly acidic phosphate group (19,24) P[3]-1-(2-Nitrophenyl)ethyl-1,N[6]-etheno-2-aza-adenosine 5'-triphosphate, II (caged ε-azaATP) and the equatorial isomer of P-1-(2-nitrophenyl)ethylformycin 3'5'-cyclic phosphate, III (eq-caged cFMP), are two compounds described in this paper that illustrate these two classes of caged nucleotides.

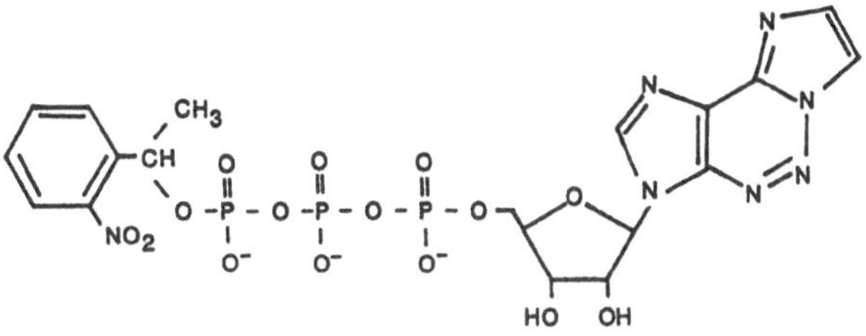

II

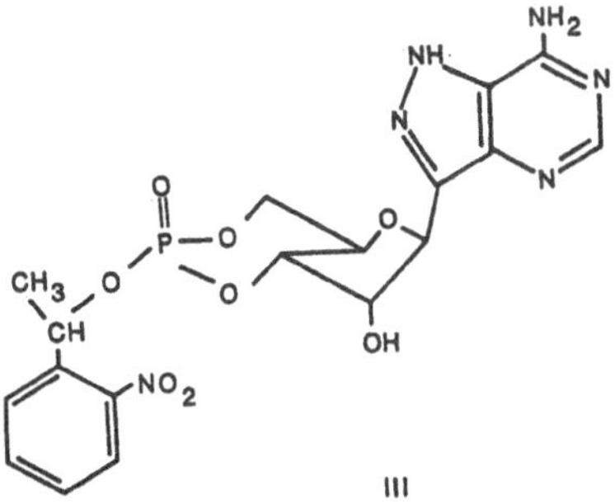

III

It is important when analyzing biological processes through pulse photolysis of caged nucleotides to know how much and how rapidly nucleotide is formed. The mechanism of photolysis of 2-nitrobenzyl caged nucleotides can be conveniently dissected into two steps as in equation 1 where X^- represents the nucleotide:

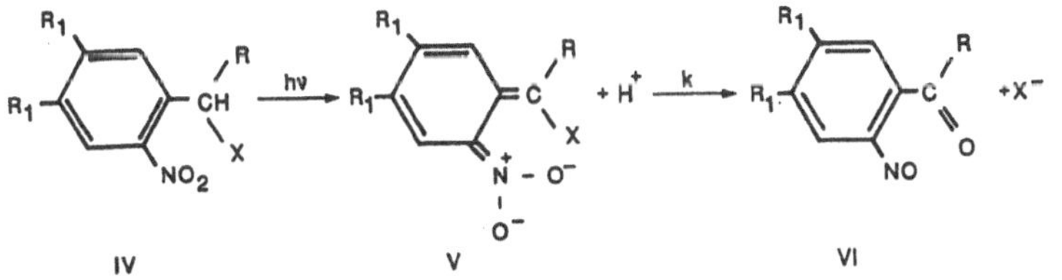

(1)

IV   V   VI

In the first step, primary photochemical excitation occurs and the mode of decay of the excited state determines the extent of photolysis. Following photon absorption, intramolecular hydrogen transfer occurs, probably from a singlet excited state (25), and an aci-nitro compound is formed that is a medium to strong acid and so rapidly loses a proton in neutral aqueous solution yielding the aci-nitro anion, V (3,26,27). V and the released proton can be detected spectroscopically and their rates of formation are greater than 10^5 s^{-1} (21,22).

The second step determines the rate at which the nucleotide is formed and is shown in equation 1 as being controlled by a rate constant k. The step almost certainly contains at least one additional intermediate because otherwise the breakdown of V would be a single-step nucleophilic substitution at an sp^2-carbon atom with the oxygen atom attacking from an unfavorable orientation (22). However in the case of caged ATP, it has been shown that the decay of V occurs concomitant with the rate of formation of ATP (22). Thus in terms of equation 1, other intermediates are either in rapid equilibrium with V or do not accumulate to a significant concentration as V decays. Based on results with caged ATP, it has been concluded that the rate of formation of nucleotides containing a weakly acidic phosphate group equals that of the decay of aci-nitro anion, V (22). In this paper we investigate whether the same conclusion holds for cyclic nucleotides. The approach used is to monitor the fluorescence changes that occur on photolysis of eq-caged cFMP, III, and a model compound, caged ε-azaATP, II. The 2-nitrobenzyl group partially quenches the fluorescence of the nucleotide bases. On photolysis the fluorescence is restored and a comparison is made of the kinetics of decay of V and the rate of nucleotide formation as monitored through its fluorescence.

For caged ATP, k is proportional to H^+ concentration in the range pH 6.5 to 9. This acid catalysis has been postulated as being associated with protonation of the non-bridging oxygen atoms of the γ-phosphate group in caged ATP (22). One might expect therefore that photolysis of caged cyclic nucleotides would not be susceptible to acid catalysis. Whether or not this is the case was investigated for eq-caged cFMP, III.

Substituents in the 2-nitrobenzyl moiety are expected to influence the extent and rate of photolysis of caged compounds in various ways. Nitrobenzyl groups that are substituted with methoxy groups (R_1 = CH_3O in IV) or a dioxymethylene group absorb relatively strongly (ε ≈ 5000 $M^{-1}cm^{-1}$) at 350 nm (19,28). This may afford advantageous properties in the extent and rate of photolysis of caged compounds relative to substituents in which R_1=H as in I, II and III. (For I, ε = 660 $M^{-1}cm^{-1}$ at 347 nm (21)). Substitution of the α-carbon atom with methyl groups (R = CH_3 in IV) has some potential advantages over the parent compound (R = H). For example with methyl substituted compounds the photolysis quantum yield may be greater (29), the photolysis kinetics more rapid (30), and the nitrosoketone by-product, VI, less reactive (and hence less prone to inactivate or modify proteins) (4,31). To investigate this we developed a novel synthesis for caged nucleotides containing both methoxy and methyl groups (R_1 = CH_3O and R = CH_3 in IV) by using 1-(3,4-dimethoxy-6-

VII

nitrophenyl)diazoethane, VII.   The photochemical properties of these dimethoxy substituted caged compounds were analyzed and compared with other caged compounds.

2.   MATERIALS AND METHODS

Formycin   3',5'-cyclic   phosphate   (cFMP)   was   purchased   from Calbiochem, and P-3,4-dimethoxy-6-nitrobenzyl adenosine 3',5'-cyclic phosphate and P-3,4-dimethoxy-6-nitrobenzyl guanosine 3'5'-cyclic phosphate from Molecular Probes. MnO_2, activated for oxidations, was from BDH (Merck 805958).   Other reagents were analytical reagent grade or better.

2.1   Preparation and Characterization of Compounds.

2.1.1.   Hydrazone of 3,4-dimethoxy-6-nitroacetophenone.   3,4-Dimethoxy-6-nitroacetophenone was prepared by nitration at 15-20°C of 3,4-dimethoxyacetophenone with nitric acid (s.g. 1.4) (32).   It was recrystallized from absolute ethanol and characterized by ^{1}H-NMR ($CDCl_3$) δ 7.61 (1H, aromatic), 6.76 (1H, aromatic), 3.98 (6H, methoxy), 2.50 (3H, methyl).   2.3 g (10 mmol) of nitroketone was dissolved in 20 mL of ethanol and treated with 1.12 g (22 mmol) of hydrazine monohydrate and 0.64 mL (11 mmol) of glacial acetic acid. The solution was protected from light and heated under gentle reflux for 3 h.   The   reaction   mixture   was   concentrated   by   rotary evaporation of the solvent and then partitioned between chloroform and water to remove hydrazine acetate.   The chloroform phase was dried over $MgSO_4$.   The product, in which there was no characteristic IR (infrared) carbonyl band, was obtained in 90% yield and was crystallized from $CHCl_3$ solution by addition of light petroleum ether (40-60°C) as rosettes of yellow needles, m.p. (uncorr.) 120-123°C, UV (ethanol): ϵ = 1.79 x 10^4 $M^{-1}cm^{-1}$ at 243 nm, 6.1 x 10^3 $M^{-1}cm^{-1}$ at 290 nm and 6.4 x 10^3 $M^{-1}cm^{-1}$ at 339 nm; ^{1}H-NMR ($CDCl_3$) indicated a mixture of two compounds in the ratio of 2.6:1.   These two compounds represent   the   E (anti) and Z (syn) isomers about the C=N bond. δ 7.60 and 7.76 (2.6:1; total 1H, aromatic), 6.87 and 6.66 (2.6:1; total 1H, aromatic), 5.41 and 4.89 (2.6:1; total 2H, amine (H exchangeable into D_2O in spectrum measured in $CDCl_3$)), 3.96 (br s, 6H, methoxy), 2.04 and 2.19 (2.6:1; s, 3H, methyl).

2.1.2.   1-(3,4-Dimethoxy-6-nitrophenyl)diazoethane VII.   84 µmol of 3,4-dimethoxy-6-nitroacetophenone in 10 mL chloroform was stirred vigorously with 120 mg MnO_2 for 15 min at 22°C in the dark and filtered.   The pseudo first order rate constant for formation of the diazoethane under these conditions was 0.3 min^{-1}, while the first order rate constant for decomposition in the absence of MnO_2 and light was 0.36 day^{-1} at 22°C.   These rate constants were calculated from the single exponential rates of increase and decrease, respectively, of absorbance at 438 nm.   The diazo compound which generally was used directly in $CHCl_3$ solution crystallized from chloroform solution with petroleum ether (80-100°C), m.p. 95-95.5°C (decomp).   UV(chloroform):

$\varepsilon \stackrel{+}{=} 1.9 \times 10^4$ $M^{-1}cm^{-1}$ at 280 nm, 2.9×10^3 $M^{-1}cm^{-1}$ at 438 nm; IR (C=N=N) 2050 cm^{-1}; ^1H NMR (CDCl$_3$) δ 7.62 (s,1H, aromatic), 6.56 (s, 1H, aromatic), 3.96 (s, 3H, methoxy), 3.93 (s, 3H, methoxy), 2.20 (s, 3H, methyl). The diazoethane was highly photosensitive to daylight.

2.1.3.   3,4-Dimethoxy-6-nitrobenzyldiazomethane.   3,4-Dimethoxy-6-nitrobenzyldiazomethane (19) was prepared as VII except that 3,4-dimethoxy-6-nitrobenzaldehyde (32) was used as starting material.   The hydrazone had m.p. 181-183°C and the diazomethane 120-122°C (decomp).

2.1.4.   Synthesis of caged 3',5'-cyclic nucleotides.   A standard protocol for synthesis of caged cyclic nucleotides was based upon that of Nerbonne et al. (19) but using a diazo compound prepared from its parent hydrazone as described above.   The free acid of the 3',5'-cyclic nucleotide (10 to 100 µmol) was dispersed by sonication in 0.1 - 1.0 mL of anhydrous dimethylsulfoxide.   A ten-fold excess of freshly prepared 100 mM diazo compound in chloroform was added with stirring.   The chloroform was removed by rotary evaporation, and the remaining solution in dimethylsulfoxide was stirred at room temperature and protected from light.   Progress was followed by analytical thin-layer chromatography in the dark on K6F Whatman silica gel plates with fluorescent indicator and developed with chloroform: methanol (5:1 v/v).   After 24-48 h the reaction mixtures were clear deep red to orange solutions.   The chromatographic patterns were consistent with completion of the reaction.   They showed only a trace of free cyclic nucleotide at the origin, a pair of major spots with $R_f \approx 0.4$ corresponding to the equatorial and axial isomers of the caged nucleotides (24) and fast moving compounds at the solvent front.

The reaction mixture was applied to a PK6F Whatman preparative silica gel plate with fluorescent indicator, developed with chloroform, and air-dried.   The caged nucleotide remained at the origin while the other reaction products of lower polarity moved ahead.   The plate was redeveloped with chloroform: methanol (5:1 v/v) which separated the caged nucleotide from any unreacted nucleotide.   The band of silica gel with bound caged nucleotide was scraped from the air-dried plate, dispersed in chloroform, poured as a slurry into a small chromatographic column and washed with chloroform until the UV absorption of the eluate was negligible.   Caged nucleotide was then eluted as a sharp band with chloroform: methanol (4:1 v/v) and dried by rotary evaporation.

In the case of P-1-(2-nitrophenyl)ethyl esters of cFMP (caged cFMP) the isomers were purified by HPLC rather than preparative TLC. 20 µmol caged cFMP was purified on two C-18 reverse phase semi-preparative columns in series (2 x 61 cm x 0.78 cm) with CH$_3$CN/H$_2$O (1:3 v/v) as eluting solvent.   Caged cFMP eluted as three peaks A, B and C.   These were pooled separately, concentrated and characterized by ^{31}P- and ^1H-NMR. The NMR spectra measured in DMSO-d$_6$ showed that each compound was >95% pure caged cFMP.   A and B were the two diastereoisomers of the axial isomer (arising because of the chiral center at the benzylic carbon), and C was a mixture of the two diastereoisomers of the equatorial isomer, III. In all cases the ^{31}P signals were upfield of the signals from 85% H_3PO_4 external reference. For A and B isomers: δ -5.38, $J_{5"-P}$ 22 Hz, $J_{benzylic\ proton-P}$ 7 Hz, $J_{5'-P}$ was manifested as line broadening ($J_{5'-P}$ 1.7 Hz in cAMP (33)). For C isomers: δ -3.15 and -3.45 (intensity ratio 1.2:1). The ^{1}H-NMR spectra were fully consistent with the proposed structure. In every case a 7 Hz coupling constant was seen in the ^{1}H multiplet for the benzylic proton due to coupling with the P nucleus.

Concentrations of caged cFMP were based on ϵ = 12,300 $M^{-1}cm^{-1}$ at 295 nm.

2.1.5. 1-(3,4,-Dimethoxy-6-nitrophenyl)ethyl phosphate (DM-caged P_i).
1-(3,4-Dimethoxy-6-nitrophenyl)ethyl phosphate was synthesized in 20% yield as its barium salt from its parent alcohol (prepared by borohydride reduction of 3,4-dimethoxy-6-nitroacetophenone) using 1,2-phenylene phosphorochloridate as phosphorylating agent and Br_2 to remove the catechol group (34). The ester was solubilized by using Dowex 50(H^+ form) to remove the barium ions, and characterized by standard procedures and by its photochemical properties. The ester concentration was quantified from its pH titration curves (pK_a 6.9 at 0.01 M ionic strength). From these concentrations the UV spectrum of DM-caged P_i gave ϵ = 6.1 x 10^3 $M^{-1}cm^{-1}$ at 243 nm and 3.8 x 10^3 $M^{-1}cm^{-1}$ at 350 nm. These estimates of ϵ would be low if contaminating P_i or other weak acid was present (cf. the values of ϵ at 350 nm for P-1-(3,4-dimethoxy-6-nitrophenyl)ethyl esters of nucleotides).

2.1.6. P-1-(3,4-Dimethoxy-6-nitrophenyl)ethyl esters of nucleotides.
Nucleotides containing weakly acid phosphate groups were also converted to caged nucleotides as P-1-(3,4-dimethoxy-6-nitrophenyl)ethyl phosphate esters by use of VII by the method of Walker et al. (22). They were characterized from their ^{31}P NMR spectra (see Table 2 of reference 22, by HPLC and by their susceptibility to photolysis with formation of the parent nucleotide. UV (0.20 M Na_2HPO_4 to pH 7.2 in water): P^3-1-(3,4-dimethoxy-6-nitrophenyl)ethyl ATP (DM-caged ATP), ϵ = 1.72 x 10^4 $M^{-1}cm^{-1}$ at 249 nm, 5.1 x 10^3 $M^{-1}cm^{-1}$ at 350 nm; P^3-1-(3,4-dimethoxy-6-nitrophenyl)ethyl GTP (DM-caged GTP), ϵ = 1.83 x 10^4 $M^{-1}cm^{-1}$ at 247 nm, 4.9 x 10^3 $M^{-1}cm^{-1}$ at 350 nm. DM-Caged ATP was also prepared from (3,4-dimethoxy-6-nitrophenyl)ethyl phosphate and ADP morpholidate by the method of Kaplan et al. (4).

2.1.7. Caged ϵ-ATP and caged ϵ-azaATP. II. Caged ϵ-ATP (P^3-1-(2-nitrophenyl)ethyl-1,N^6-ethenoATP was synthesized from caged ATP, I, based on the procedure of Secrist et al. (35). 1 mmol caged ATP was stirred in 50 mL freshly prepared 1-1.6 M chloroacetaldehyde for 11.5 h at 37°C. The pH was readjusted to 4.5 every few hours with aqueous ammonia. The chloroacetaldehyde was prepared from 100 g of its dimethylacetal by dropwise addition of 129 mL H_2SO_4 (s.g. 1.84) keeping the temperature below 50°C, then heating under reflux for 30 min and distilling under reduced pressure into a cooled receiving flask. Caged ε-ATP was purified and desalted by ion exchange chromatography on DEAE-cellulose with a gradient of triethylammonium bicarbonate at pH 7.5 in 80% overall yield. Concentrations of caged ε-ATP were based on ε = 9900 $M^{-1}cm^{-1}$ at 265 nm (λ_{max} at 260 nm).

Caged ε-azaATP, II, was synthesized from caged ε-ATP according to the procedures of Tsou et al. (36) and Miyaka and Asai (37) and purified by DEAE-cellulose chromatography with a gradient of triethylammonium bicarbonate at pH 7.5 in 26% overall yield. Concentrations of caged ε-azaATP were based on ε = 7500 $M^{-1}cm^{-1}$ at 290 nm (shoulder, λ_{max} at 266 nm). Caged ε-ATP and caged ε-azaATP may also be separated from each other and from caged ATP by preparative reverse phase HPLC on a C_{18} column (Waters) with 10 mM KH_2PO_4 at pH 5.5 and methanol (85:15 v/v) as eluting solvent. Caged ε-ATP and caged ε-azaATP were shown to be pure by their clean photolyses to ε-ATP and ε-azaATP respectively and by HPLC.

2.2. Photolysis Equipment, Spectrometers and other Instrumentation.

Photolysis was effected by a frequency-doubled pulsed ruby laser (347 nm), a frequency-doubled Candella pulsed dye laser (320 nm), or a xenon-arc flash lamp (Figure 2). Absorption and fluorescence spectrometers were linked to the pulse photolysis equipment. The apparatus was as described previously (21,22) and more detailed accounts of the laser set-ups may be found in reference 3.

Steady-state ultraviolet and visible absorption measurements were recorded on a Beckman DU-8B or a Cary 118 spectrophotometer. Fluorescence spectra were recorded on a Farrand Mark I spectrofluorimeter.

^{1}H-NMR spectra were recorded on a Jeol FX90Q or a Bruker WM 200 spectrometer except for caged cFMP for which spectra were recorded at 500 MHz (on a Bruker AM 500). ^{31}P-NMR spectra were recorded at 81 MHz (on a Bruker WM 200 equipped with a broad-band proton decoupler).

HPLC was performed with a Waters system and monitored at 254 nm or 260 nm for nucleotides and the 2-nitrophenyl moiety, and at 340 nm for detection of the 3,4-dimethoxy-6-nitrophenyl moiety. Reverse phase separations were carried out on Merck RP8 or Waters μ-Bondapak C_{18} analytical columns, and Whatman Partisil 10 SAX or Waters μ-Bondapak NH_2 columns were employed for anion-exchange chromatography. Isocratic elution was used in all separations.

Concentrations of caged cFMP, caged ε-ATP and caged ε-azaATP were determined from their absorption coefficients. These were calculated from ε of the nucleotide (cFMP, ε-ATP or ε-azaATP) at their λ_{max} plus ε for 1-(2-nitrophenyl)ethyl phosphate at the same wavelength. This method is potentially prone to errors due to interaction between the two aromatic moieties in the molecules but the approach has been found to be reliable in the case of caged ATP (38).

ε values of DM-caged ATP and DM-caged GTP were determined by the method outlined in reference 38. The caged nucleotides were photolyzed in 0.20 M Na_2HPO_4 adjusted to pH 7.2 in water and the nitrosoketone was extracted into chloroform. The extents of photolysis to ATP and GTP (compounds of known ε values) were followed by HPLC.

The relative extents of steady-state and pulse photolysis of the 1-(2-nitrophenyl)ethyl phosphate ester series and the 1-(3,4-dimethoxy-6-nitrophenyl)ethyl series were compared at 347 nm by photolyzing mixtures of caged ADP and DM-caged ATP in the Farrand spectrofluorimeter (10 nm bandwidth) or with the ruby laser. The products ADP and ATP as well as caged ADP and DM-caged ATP were resolved from each other and photolysis by-products when analyzed by HPLC with a SAX column and an eluting solvent of methanol: 0.25 M $(NH_4)_2HPO_4$ adjusted to pH 5.8 with HCl (13:100 v/v). The extents of photolysis were calculated from the areas of the peaks on the HPLC elution records, taking into account the different extinction coefficients of the compounds being analyzed.

The steady-state product quantum yield, Q_p, of the P-3,4-dimethoxy-6-nitrobenzyl ester of 3',5'-cyclic GMP was measured by the method described by Walker et al. (38). Q_p was calculated from a comparison of the extents of photolysis of DM-caged GTP (for which Q_p = 0.07, equal to that of DM-caged ATP, Table 2) to that of the caged cGMP under identical conditions. A mixture of isomers was photolyzed in 160 mM Na_2HPO_4 adjusted to pH 7.3 in water containing 2% dimethyl sulfoxide, by xenon arc-lamp illumination that was transmitted through a heat filter (water) and a 300-360 nm band pass filter (Schott UG 11). The extent of photolysis was monitored by HPLC with an RP8 reverse phase column and CH_3CN/H_2O (1:3 v/v) as eluting solvent. In this system the axial and equatorial isomers were resolved and the photolysis of each was followed. The thermal hydrolysis of the caged cGMP was monitored under the same solvent and temperature conditions but without illumination. The thermal hydrolysis was comparable in extent to that of the photolysis, especially in the case of the equatorial isomer (see Table 3), and was taken into account in calculating Q_p.

Data from exponential traces were analyzed by use of a non-linear least squares fit (39).

There was no evidence in any of our photolysis studies of significant cleavage of the potentially photolabile methoxy group from the benzyl group in 3,4-dimethoxy-6-nitrobenzyl compounds (40,41).

3. RESULTS AND DISCUSSION

3.1 Photolysis of eq-caged cFMP.

The rate of formation of ATP following pulse illumination of caged ATP, I, equals the rate of decay of an aci-nitro intermediate (V in which R_1=H, R=CH$_3$ and X$^-$=ATP) (22). Proof of this rested in large measure on there being a specific kinetic assay for ATP based on the known rate of ATP-induced actomyosin subfragment 1 dissociation. A similar specific kinetic assay for cyclic nucleotides was not available. It was therefore decided to take advantage of the potential for the 1-(2-nitrophenyl)ethyl group to quench the fluorescence of fluorescent nucleotide analogues by such mechanisms as

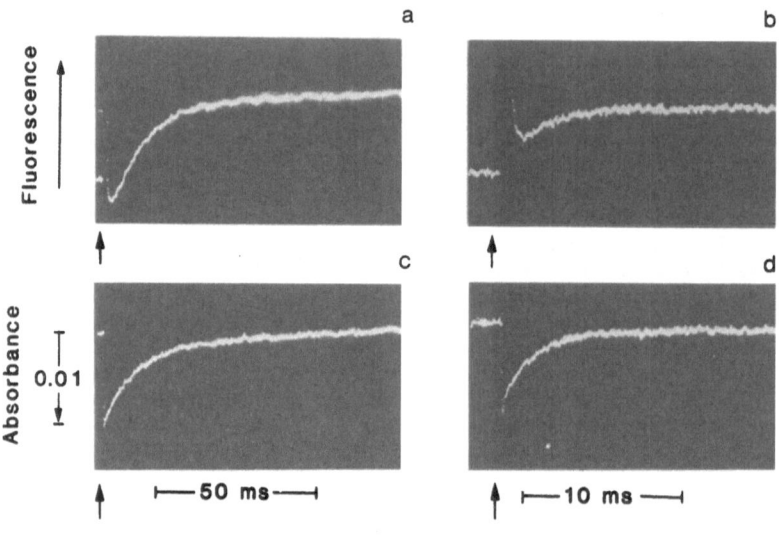

a

b

c

d

├──50 ms──┤   ├──10 ms──┤

Fluorescence

Absorbance

0.01

Time

Figure 1.   Kinetic records of fluorescence and 406 nm absorption changes following photolysis (irradiation 320 nm) of caged ε-azaATP, II, at 22°C.   The aqueous solutions in a square 4-mm observation cell contained 100 μM caged ε-azaATP, 2mM MgCl$_2$, 1 mM dithiothreitol and 100 mM 2-(N-morpholino)ethanesulfonic acid (MES) adjusted in a,c to pH 7.0 and in b,d to pH 6.15 with NaOH.   The fluorescence was excited by light from a xenon arc lamp that was passed through a Schott UG 11 300-360 nm band-pass filter and measured after passing through a 365 nm cut-off filter and a 500 nm interference filter.   Small arrows mark the times of the laser pulse.

Förster energy transfer (42).   Formycin nucleotides were selected because the wavelength of their fluorescence emission (340 nm) (43) overlaps the absorption band of the 1-(2-nitrophenyl)ethyl group as well as the aci-nitro intermediate and possibly other intermediates structurally related to the by-product of the photolysis 2-nitrosoacetophenone (VI in which R$_1$=H and R=CH$_3$) (4,30).

As the caged formycin nucleotide is photolyzed, its fluorescence is expected to increase to that of the free nucleotide.   The rate of the fluorescence increase may be multiphasic with the slowest phase setting a limit on how fast the nucleotide is formed.   Furthermore it is probable that this rate will equal the rate of nucleotide formation because the absorption bands of intermediates associated with 1-(2-nitrophenyl)ethyl photolysis are all expected to overlap to some extent the formycin emission spectrum.   Unfortunately the pulse of laser or arc lamp illumination interfered with the detection of the

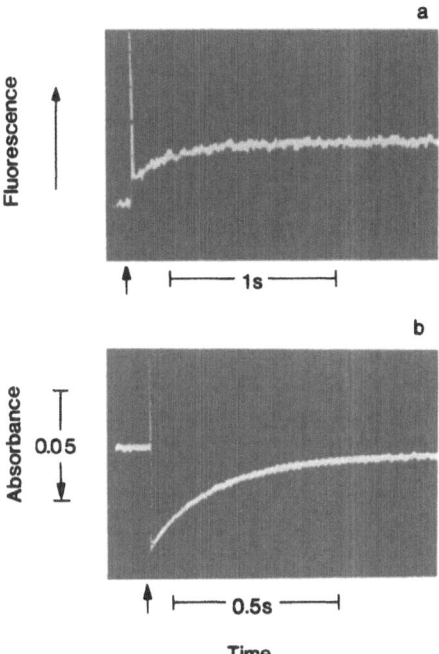

Figure 2. Kinetic records in a of fluorescence and in b of 406 nm absorption changes following pulse photolysis of eq-caged cFMP, III, at 21°C. The aqueous solution in a square 4-mm observation cell contained 100 µM eq-caged cFMP, 0.1 M KCl, 1 mM dithiothreitol and 50 mM N-tris(hydroxymethyl)-2-aminomethanesulfonic acid (TES) adjusted to pH 7.05 with KOH. Pulse photolysis was provided by a xenon arc flash lamp (3,44) equipped with a Schott UG 11 300-360 nm band-pass filter. The exciting light in the fluorescence spectrometer (22) was passed through a 289 nm interference filter and the emitted light through a monochromator set to 340 nm. Small arrows mark the times of the flash lamp pulse.

fluorescence change restricting the time resolution to several milliseconds. This was not a problem with analysis of the photolysis kinetics of eq-caged cFMP, III, but for ATP analogues it was found that caged ε-azaATP (emission 500 nm) was more satisfactory. In the experiments described below the quenching of ε-azaATP and cFMP fluorescence was 3-fold and 16-fold in compounds II and III respectively.

The feasibility of this kinetic analysis was shown in a photolysis study of caged ε-azaATP. Experimental records of absorption and fluorescence changes associated with the photolysis are shown in Figure 1. The absorption changes at 406 nm correspond to the rapid formation of an aci-nitro anion intermediate (V in which

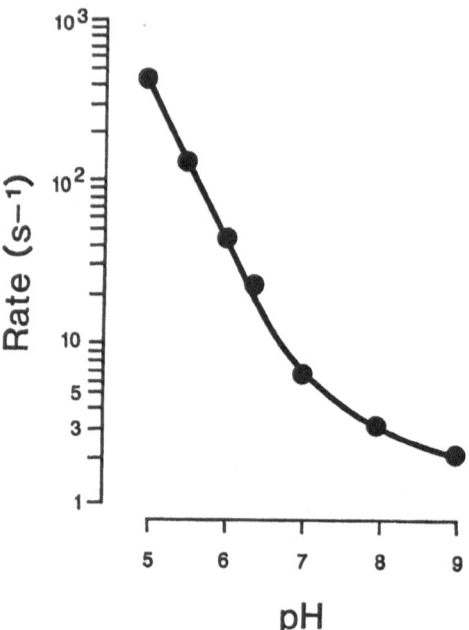

Figure 3. Dependence of the rate constants of decay of aci-nitro intermediates of eq-caged cFMP, III, on pH following photolysis at 21°C. The measurements were carried out as described in Figure 2b in aqueous solution containing 100 μM eq-caged cFMP, 0.1 M KCl, 1 mM dithiothreitol and 20 mM buffer (pH 5, acetate; pH 5.5, 6.0, 6.4, MES; pH 7.0, TES; pH 8.0, tris(hydroxymethyl)aminomethane (Tris); pH 9.0, borate).

R_1-H, R-CH_3 and X^-- ε-azaATP) followed by its decay. Fluorescence increases occur at identical rates to the decay rates of the aci-nitro anion at both pH 7.0 and pH 6.15, which in turn match those of the caged ATP aci-nitro anion under similar conditions (22). Note that interference due to the laser pulse permits only partical observation of the fluorescence increase at pH 6.15. These records establish that monitoring nucleotide fluorescence is a feasible way to measure the rate of release of nucleotide from its caged precursor.

Figure 2 shows that the rate of the fluorescence increase on photolysis of eq-caged cFMP equals that of the absorption decrease associated with the decay of its aci-nitro intermediate. The step increase in fluorescence following pulse illumination is probably because the absorption of the aci-nitro intermediate from 300-350 nm is less than that of III (30). There was no evidence of any contaminating caged FMP for which the aci-nitro decay rate would be expected to be 200 s^{-1} based on that of caged AMP (22). We conclude that the rate of release of cFMP on photolysis of III equals 4.9 s^{-1} at pH 7.05 and 21°C and that the rate of decay of the aci-nitro intermediate may be used to measure the rate of photolysis of other P-1(2-nitrophenyl)ethyl esters of cyclic nucleotides.

In contrast to what is seen in Figure 2b, the aci-nitro decay was not in general a clean exponential process when a mixture of axial and equatorial isomers of caged cyclic nucleotides was photolyzed. However at pH 7 the decay could be described by two exponential processes differing in rate by no more than a factor of five. The rates of photolysis of unresolved isomers of the P-1-(2-nitrophenyl)ethyl esters of cAMP at 21°C could be described by approximately exponential processes with rate constants of 4.2 at pH 7 and 2.3 s^{-1} at pH 8.

In biological preparations caged cyclic nucleotides are expected to be partitioned, at least partially, in hydrophobic environments where the photolysis kinetics may differ from those in aqueous solvents. The aci-nitro decay rate on photolysis of caged cFMP in isopropanol was 28 s^{-1} at 21°C. In chloroform the decay was more markedly biphasic with a halftime of 140 ms (corresponding to a rate constant of 5 s^{-1} for a single exponential decay).

The decay rate of the aci-nitro intermediate as a function of pH is an important indicator of the photolysis mechanism. The result of such an experiment for eq-caged cFMP is shown in Figure 3. The rate only increases 1.5-fold between pH 9 and 8 but becomes proportional to H^+ concentration below pH 7. The decay of the aci-nitro intermediate of caged ATP is acid catalyzed up to pH 9 but also becomes independent of H^+ concentration at pH 10. In the case of caged ATP acid catalysis has been ascribed to protonation of a non-bridging oxygen atom of the anionic γ-phosphate group. This explanation cannot apply to III. It may be that acid catalysis arises through protonation of an oxygen atom in the (cage)-O-P(-O)< structure (3).

The amount of cAMP formed on flash photolysis of caged cAMP (IV in which R_1=H, R=CH_3 and X^-=cAMP) was compared in aqueous solvents with the ATP formed on photolysis of I from the 406 nm absorption amplitude of the aci-nitro intermediate. A product quantum yield, Q_p, of 0.39 was estimated for caged cAMP compared to 0.63 for I. The estimate assumed that the extinction coefficient of the aci-nitro intermediates is the same for both molecules (22).

Q_p and the rate of photolysis at pH 8 of 1-(2-nitrophenyl)ethyl esters of caged nucleotides are similar to those found by Nerbonne for the 2-nitrobenzyl series (45).

3.2 Photolysis of P-1-(3,4-Dimethoxy-6-nitrophenyl)ethyl Esters.

Besides the possible advantages of the higher extinction coefficient and the longer photolysis wavelength of caged compounds substituted with methoxy groups (IV in which R_1 = CH_3O), these compounds may also photolyze more rapidly than caged compounds IV in which R_1 = H. An indication of this is the structure-reactivity study of Weinstein et al. (46) who showed that a photochromic reaction involving 2-nitrobenzylpyridine compounds was 10- to 100-fold faster with the H-atom in the 4-position substituted by an electron-donating group.

A comparison was attempted between the photolysis rates of caged

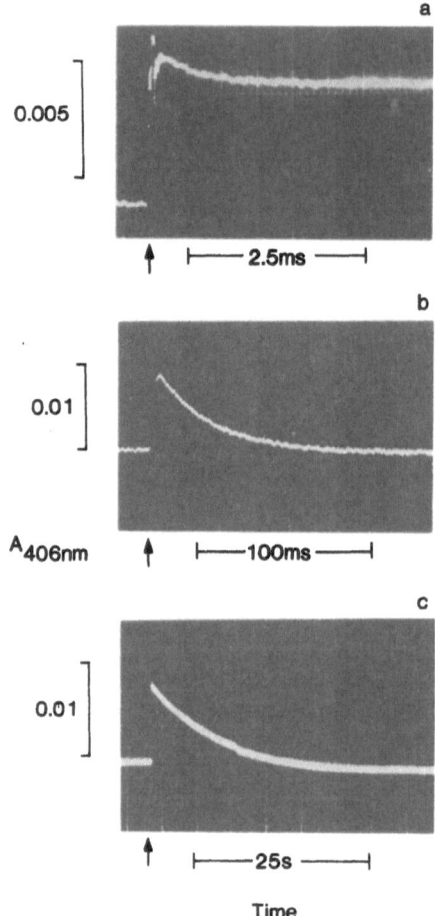

Figure 4. Absorption changes following photolysis (irradiation 347 nm) of the P-1-(3,4-dimethoxy-6-nitrophenyl)ethyl esters of P_i and ATP (DM-caged P_i and DM-caged ATP respectively) at 21°C. The aqueous solutions in a 2.8 mm path-length cell contained: in a 1 mM DM-caged P_i and 100 mM tris adjusted to pH 8 with HCl, in b 1 mM DM-caged P_i, 8mM reduced glutathione and 80 mM triethylamine adjusted to pH 10.7 with HCl, and in c 1 mM DM-caged ATP, 8 mM reduced glutathione and 80 mM triethylamine adjusted to pH 10.7 with HCl. Arrows mark the time of the laser pulse. This experiment was done in collaboration with Dr. J.A. McCray.

P_i, caged ATP and caged cAMP (IV in which R — CH_3 and X^- — P_i, ATP and cAMP with R_1 — either H or CH_3O). However the P-1-(3,4-dimethoxy-6-

TABLE 1

Rate of ATP or P_i formation (s^{-1})

	pH			
	7	8	10	10.7
DM-caged ATP	18 [a]	1.3 [a]	0.14 [a]	0.10 [b]
caged ATP,I [c]	84	9	1.5	
DM-caged P_i	21,000 [d]	2100 [a]	20 [a]	30 [b]
caged P_i [c]	80,000 [d]	8000	84	

[a] Aqueous solution at 20°C contained 0.4 mM DM-caged ATP or DM-caged P_i, 10 mM dithiothreitol and 0.10 M buffer adjusted with KOH or HCl (TES, pH 7; Tris, pH 8; NaHCO$_3$, pH 10). [b] Data from solutions as in Figure 4. [c] Reference 22. [d] Extrapolated from data at pH 8 assuming rate $\propto [H^+]$ (3,22).

nitrophenyl)ethyl ester of cAMP was unstable in aqueous solvents (see below and Table 3), so the comparison was restricted to caged P_i and caged ATP derivatives.

The absorption change at 406 nm associated with the decay of the aci-nitro anion intermediate is relatively small for dimethoxy substituted compounds because of the absorption of the reaction by-product 3,4-dimethoxy-6-nitrosoacetophenone (VI in which R = CH$_3$ and R$_1$ = CH$_3$O). However the nitrosoketone reacts rapidly with thiols to give a relatively transparent product at 406 nm, and the observed absorption decrease of the aci-nitro intermediate shows the photolysis rate. This has been described for DM-caged P_i photolysis (3) and is illustrated in Figure 4. In Figure 4 there is first a rapid increase of absorption followed by a small decrease associated with aci-nitro decay and which indicates the rate of P_i formation. In Figure 4b and c records of DM-caged P_i and DM-caged ATP photolysis show the influence of thiol on the absorption change.

Comparison of the photolysis kinetics of DM-caged ATP, caged ATP (I), DM-caged P_i and caged P_i is shown in Table 1. The photolysis is acid catalyzed below pH 10 for all compounds. It is clear that the introduction of the two methoxy groups offers no kinetic advantage. A more precise interpretation of the data is not meaningful as the photolysis rates are sensitive to ionic strength (3) and possibly the buffer used.

Steady-state comparison of the product quantum yield, Q_p, of P^2-1-(2-nitrophenyl)ethyl adenosine 5'-diphosphate (caged ADP) and DM-

TABLE 2

Quantum yields, Q_p, of caged nucleotides

Cage Group	Nucleotide	Isomer	Q_p
1-(2-Nitro-phenyl)ethyl	ATP [a]		0.63
1-(2-Nitro-phenyl)ethyl	cAMP [b]	mixed	0.39 [c]
1-(3,4-Dimethoxy-6-nitrophenyl) ethyl	ATP [b]		0.07 [c]
3,4-Dimethoxy-6-nitrobenzyl	cGMP [b]	axial equatorial	0.004 [c] 0.003-0.009 [d]

[a] Reference 22. [b] This work. [c] Values of Q_p are estimated to be accurate to one significant figure. [d] The extent of thermal hydrolysis was comparable to that of photolysis, hence the range in estimate of Q_p.

caged ATP at 347 nm showed that Q_p of caged ADP was 9-fold greater than that of DM-caged ATP (Table 2, Q_p values for caged ADP and caged ATP are the same). However laser flash photolysis at 347 nm resulted in only 2-fold less percentage conversion of DM-caged ATP compared to caged ADP in solutions of the same absorbance at 347 nm. This suggests the 50 ns laser pulse was sufficiently long for multiple excitations of DM-caged ATP - a phenomenon discussed in reference 3 and shown by the photosensitive calcium chelator nitr-5 (28).

3.3 Stability and Photochemistry of Caged Cyclic Nucleotides.

Phosphate triesters formed from cyclic nucleotides and derivatives of the 2-nitrobenzyl group are susceptible to hydrolysis (24). The caged cyclic nucleotides were also susceptible to thiolysis and yielded the parent cyclic nucleotide and a substituted nitrobenzyl thioether. The hydrolysis and thiolysis rates, k', of several caged cyclic nucleotides were measured (Table 3). The P-1-(3,4-dimethoxy-6-nitrophenyl)ethyl ester of cAMP hydrolyzed so rapidly that it is probably not practical to use this compound in biological research. Thiolysis rates showed first order dependence on thiol and hydroxide ion concentrations.

We attempted to measure the photolysis rate of the P-3,4-dimethoxy-6-nitrobenzyl ester of cGMP. On laser pulse photolysis in

TABLE 3

Rate of caged cAMP or cGMP hydrolysis and thiolysis at 20°C

Cage group	Nucleotide[a]	Isomer	Medium	$10^3 k'$, min^{-1}
1-(2-Nitro-phenyl)ethyl	cAMP	axial	100 mM P_i, pH 7.0[b]	0.05
		equatorial	100 mM P_i, pH 7.0[b]	0.30
		mixed	20 mM dithiothreitol 50 mM P_i, pH 8.0	1.4
3,4-Dimethoxy-6-nitrobenzyl	cAMP	axial	100 mM P_i, pH 7.0[b]	0.48
		equatorial	100 mM P_i, pH 7.0[b]	2.7
		mixed	20 mM dithiothreitol 50 mM P_i, pH 8.0	10
1-(3,4-Dimethoxy-6-nitrophenyl) ethyl	cAMP	mixed	50 mM P_i, pH 8.0	240
3,4-Dimethoxy-6-nitrobenzyl	cGMP	mixed	100 mM 2-mercapto-ethanol, 50 mM N-tris(hydroxy-methyl)methylglycine (tricine), pH 7.3 [c]	4.5

[a] 0.25 mM [Nucleotide] [b] k' was the same at pH 8.0. [c] $k' \propto [OH^-]$ in pH range 7.3 to 9.0 and $k' \propto$ [2-mercaptoethanol] up to 100 mM thiol.

the absence of thiol a step increase in absorption was observed similar to that seen by Karpen et al. (18). However the signal to noise ratio was sufficiently low that, even if a decay of absorption such as seen in Figure 4 were present, it would not have been detected. It was also impractical to use thiols in order to measure the photolysis rate (as was possible for DM-caged P_i and DM-caged ATP, Figure 4b and c) because the thioether by-product of thiolysis gave rise to a large 406 nm absorption signal following a laser pulse (at a rate of 2 s^{-1} in the presence of 10 mM 2-mercaptoethanol at pH 7 and 20°C). This signal masked any that might have arisen from photolysis of the caged cyclic nucleotide. We have therefore been unable to measure the photolysis rate of the P-3,4-dimethoxy-6-nitrobenzyl ester of cGMP. We know however from the biological assay of Karpen et al. (18) that it photolyzes at probably >3000 s^{-1} at pH 7 and much more rapidly than the rate of 3 s^{-1} of the P-2-nitrobenzyl ester of cGMP (45).

Steady-state measurement of Q_p for the axial and equatorial isomers P-3,4-dimethoxy-6-nitrobenzyl ester of cGMP gave values of

<0.01 (Table 2), much less than the 5% conversion of the caged cGMP on laser pulse photolysis (18). As noted for DM-caged ATP this probably reflects the short life time of the excited state of the caged cGMP compared to the length of the laser pulse.

4. CONCLUSIONS

1. Synthesis of photosensitive phosphate esters has been extended to include those derived from a novel diazoethane, 1-(3,4-dimethoxy-6-nitrophenyl)diazoethane.

2. Selective alkylation of the strongly acidic phosphate groups in 3',5'-cyclic nucleotides may be achieved with 1-(2-nitrophenyl)diazoethane and the equatorial and axial isomers resolved and characterized.

3. The rate of release of fluorescent compounds from their caged precursors can be measured by monitoring the fluorescent changes that accompany photolysis. In the two cases tested, the rates of decay of the aci-nitro anion intermediates are likely to reflect the rates of release of the fluorescent compounds because the slowest phases of the fluorescence changes match the decay rates.

4. The rate of formation of cyclic nucleotides from their P-1-(2-nitrophenyl)ethyl esters following 300-360 nm pulse illumination occurs concomitant with the decay of an aci-nitro anion intermediate. The photolysis is acid catalyzed below pH 7.

5. The rates of photolysis of phosphate mono- and diesters containing the 1-(3,4-dimethoxy-6-nitrophenyl)ethyl group are slower than those containing the 1-(2-nitrophenyl)ethyl group. However the rates of photolysis of phosphate triesters containing the 3,4-dimethoxy-6-nitrobenzyl group appear to be faster than those containing the 2-nitrobenzyl group.

6. Phosphate triesters containing the 3,4-dimethoxy-6-nitrobenzyl and 1-(3,4-dimethoxy-6-nitrophenyl)ethyl groups are susceptible to hydrolysis and thiolysis. This places some limitations on the use of these compounds in biological research. Furthermore thiols are often added in biological experiments with caged compounds because they are effective in limiting inactivation of biological preparations by the 2-nitrosobenzaldehyde or 2-nitrosoacetophenone by-products (4,23).

7. 1-(3,4-Dimethoxy-6-nitrophenyl)ethyl and 3,4-dimethoxy-6-nitrobenzyl phosphate esters have lower product quantum yields than their 1-(2-nitrophenyl)ethyl and 2-nitrobenzyl counterparts. However this is significantly offset in flash photolysis studies because multiple excitations appear to occur in a single flash (> 50 ns) giving rise to relatively high extents of photolysis.

5. ACKNOWLEDGEMENTS

Collaborations with Dr. J.A. McCray on the kinetics of DM-caged P_i and DM-caged ATP photolysis (Figure 4, reference 3), and Mr. G.P. Reid on the synthesis of caged ε-ATP and caged ε-azaATP are gratefully acknowledged. The research is supported by the Medical Research Council, U.K., the National Institute of Health through grant number HL 15835 to the Pennsylvania Muscle Institute, the Muscular Dystrophy Association of America, the EEC Stimulation Action Program, the Cornell University Biotechnology Program and the Wellcome Trust.

6. REFERENCES

1. Gurney, A.M., Lester, H.A. 1987. Physiol. Rev. 67:583-617
2. Kaplan, J.H., Somlyo, A.P. 1989. Trends Neurosci. 12: in press
3. McCray, J.A., Trentham, D.R. 1989. Ann. Rev. Biophys. Biophys. Chem. 18: in press
4. Kaplan, J.H., Forbush, B., Hoffman, J.H. 1978. Biochemistry 17:1929-35
5. Goldman, Y.E., Hibberd, M.G., McCray, J.A., Trentham, D.R. 1982. Nature 300: 701-5
6. Hibberd, M.G., Trentham, D.R. 1986. Ann. Rev. Biophys. Biophys. Chem. 15:119-61
7. Rapp, G., Poole, K.J.V., Maeda, Y., Güth, K., Hendrix, J., Goody, R.S. 1986. Biophys. J. 50:993-98
8. Walker, J.W., Somlyo, A.V., Goldman, Y.E., Somlyo, A.P., Trentham, D.R. 1987. Nature 327:249-52
9. Arner, A., Goody, R.S., Rapp, G., Rüegg, J.C. 1987. J. Mus. Res. Cell Mot. 8:377-85
10. Somlyo, A.V., Goldman, Y.E., Fujimori, T., Bond, M., Trentham, D.R., Somlyo, A. 1988. J. Gen. Physiol. 91:165-92
11. Dantzig, J.A., Walker, J.W., Trentham, D.R., Goldman, Y.E. 1988. Proc. Natl. Acad. Sci. USA 85:6716-20
12. Forbush, B. 1984. Proc. Natl. Acad. Sci. USA 81:5310-14
13. Fendler, K., Grell, E., Haubs, M., Bamberg, E. 1985. EMBO J. 4:3079-86
14. Apell, H.J., Borlinghaus, R., Läuger, P. 1987. Membrane Biol. 97:179-91
15. Hartung, K., Grell, E., Hassebach, W., Bamberg, E. 1987. Biochim. Biophys. Acta 900:209-20
16. Pascolini, D., Herbette, L.G., Skita, V., Asturias, F., Scarpa, A., Blasie, J.K. 1988. Biophys. J. 54: 679-87
17. Dolphin, A.C., Wootton, J.F., Scott, R.H., Trentham, D.R. 1988. Pflügers Arch. 411:628-36
18. Karpen, J.W., Zimmerman, A.L., Stryer, L., Baylor, D.A. 1988. Proc. Natl. Acad. Sci. USA 85:1287-91
19. Nerbonne, J.M., Richard, S., Nargeot, J., Lester, H.A. 1984. Nature 310:74-76
20. Morrison, H.A. 1969. In The Chemistry of the Nitro and Nitroso Groups, Part 1, ed. H. Feuer, pp. 165-213. New York: Wiley. 771 pp.

21. McCray, J.A., Herbette, L., Kihara, T., Trentham, D.R. 1980. Proc. Natl. Acad. Sci. USA 77:7237-41
22. Walker, J.W., Reid, G.P., McCray, J.A., Trentham, D.R. 1988. J. Am. Chem. Soc. 110:7170-77
23. Goldman, Y.E., Hibberd, M.G., Trentham, D.R. 1984. J. Physiol. 354:577-604
24. Engels, J., Schlaeger, E-J. 1977. J. Med. Chem. 20:907-11
25. Yip, R.W., Sharma, D.K., Giasson, R., Gravel, D. 1985. J. Phys. Chem. 89:5328-30
26. McClelland, R.A., Steenken, S. 1987. Can. J. Chem. 65:353-56
27. Schupp, H., Wong, W.K., Schnabel, W. 1987. J. Photochem. 36:85-97
28. Adams, S.R. Kao, J.P.Y., Grynkiewicz, G., Minta, A., Tsien, R.Y. 1988. J. Am. Chem. Soc. 110:3212-20
29. Zhu, Q.Q., Schnabel, W., Schupp, H. 1987. J. Photochem. 36:85-97
30. Walker, J.W., McCray, J.A., Hess, G.P. 1986. Biochemistry 25:1799-1805
31. Barltrop, J.A., Plant, P.J., Schofield, P. 1966. Chem. Commun. 822-23
32. Fetscher, C.A. 1963. Org. Synth. Coll. 4:735-37
33. Lee, C-H., Sharma, R.H. 1976. J. Am. Chem. Soc. 98:3541-48
34. Khwaja, T.A., Reese, C.B., Stewart, J.C.M. 1970. J. Chem. Soc. (C) 2092-2100
35. Secrist, J.A., Barrio, J.R., Leonard, N.J., Weber, G. 1972. Biochemistry 11:3499-3506
36. Tsou, K-C., Yip, K.F., Miller, E.E., Lo, K.W. 1974. Nucl. Acids Res. 1:531-47
37. Miyaka, H., Asai, H. 1982. Biochem. Biophys. Res. Commun. 105:296-302
38. Walker, J.W., Reid, G.P., Trentham, D.R. 1989. Methods Enzymol. 172:288-301
39. Edsall, J.T., Gutfreund, H. 1983. Biothermodynamics: The Study of Biochemical Processes at Equilibrium, pp. 228-36. New York: Wiley. 248 pp.
40. Jelenc, P.C., Cantor, C.R., Simon, S.R. 1978. Proc. Natl. Acad. Sci. USA 75:3564-68
41. Cantos, A., Marquet, J., Moreno-Manas, M., Castello, A. 1988. Tetrahedron 44:2607-18
42. Förster, T. 1959. Disc. Faraday Soc. 27:7-17
43. Ward, D.C., Reich, E., Stryer, L. 1969. J. Biol. Chem. 244:1228-37
44. Rapp, G., Güth, K. 1988. Pflügers Arch. 411:200-3
45. Nerbonne, J.M. 1986. In Optical Methods in Cell Physiology, ed. P. De Weer, B.M. Salzberg, pp. 417-45. New York: Wiley - Interscience. 560 pp.
46. Weinstein, J., Blum, A.L., Souza, J.A. 1966. J. Org. Chem. 31:1983-85

GENERAL COMMENTS ON THE USE OF PHOTOCHEMICAL PROBES
IN BIOCHEMISTRY. ADVANTAGES, PROBLEMS AND PITFALLS

Throughout the course of the workshop most general aspects of the use of photochemical probes in biochemistry were discussed. This chapter is written partly on the basis of this discussion, and is meant both as a help for researchers who wish to enter the "field", providing some hints and describing some of the pitfalls, and as a record of the main conclusions from the workshop including some predictions for future directions.

It was the opinion of the workshop participants that photochemical probes will be of increasing importance for analyzing large biological systems with complex interactions which are not susceptible to physical techniques such as X-ray crystallographic and nuclear magnetic resonans analyses. Examples of such systems were also presented at the workshop, e.g., membranes with embedded proteins (chapters 4 & 5) and protein-nucleic acid complexes: chromatin (chapter 13), ribosomes (chapters 9 & 10), transcription (chapters 13 & 20) and recombination complexes (chapter 12). These systems are highly dynamic, and photochemical probes are ideally suited for studying such systems, ultimately by time resolution experiments (chapter 21).

A large variety of photochemical probes and ligands were discussed at the workshop, each having advantages and drawbacks. It is often warranted to have a photochemical ligand with "global" reactivity, i.e. a ligand which after photoactivation will react equally well with any group (CH_2-NH, -OH, -SH etc.) present proximal to the probe. It must be emphasized that no such ligand has been found to date, irrespective of what is often claimed in the literature. Thus, in interpreting any photolabeling experiment the reactivity of the photoprobe must be taken into consideration and the choice of probe should match the functional groups expected to be present at the site to be analyzed. The micro-environmental lifetime of the active labeling species, and thus the time available for the probe to diffuse ("wander") away from the site of photoactivation is profoundly dependent on the "concentration" of various reactive groups proximal to the site of photoactivation. Although the basic photochemistry of most of the photoactive ligands used in photoprobes today is well described by photochemists, this information is of limited help in understanding the photochemical reactions with biological molecules, so far most often peptides or nucleic acids, in aqueous phase or in hydrophobic pockets surrounded by an aqueous medium containing various buffers etc.

A prerequisite for obtaining any results at all is photoactivation of the probe. This is not as trivial as it may seem. First of all the photoactive ligand must be intact. Secondly, the irradiation wavelength has to correspond to an absorption band of the ligand which is photochemically active and thirdly this radiation has to be absorbed by the ligand without deleterious interference from the rest of the system (other chromophores). The best way to ascertain that these conditions are fulfilled is to monitor the photochemical decompositon of the photoprobe by ultraviolet spectroscopy using the lamp and buffer conditions intended for the actual biological experiment and also measure the absorption of the biological system at the wavelength used for irradiation.

It should also be mentioned that it is often worthwhile before starting elaborate chemical modification to test if any inherent photochemistry of the biologically active compound may be exploited in a photolabeling experiment. This approach has been successful in several cases (e.g chapter 9).

It is generally necessary for any one positive photolabeling experiment to perform numerous control experiments in order to verify that the labeling is actually confined to the binding site of the ligand. These controls usually include: no light, pre-irradiation, competition (with unmodified biologically active ligand), zero concentration extrapolation control, and others, dependent on each particular problem.

Since photolabeling experiments often result in rather poor yields (1% or less of the added probe covalently bound at the target is not uncommon) a successful experiment usually requires high specific (radio)activity and above all (radio)chemically pure probes. Efficient labeling by a small (< 5%) impurity can be disasterous in terms of unspecific labeling and high background, the result being either a false positive or a low signal to noise ratio.

The above-mentioned precautions should, however, not discourage anyone from doing biological photolabeling experiments since the advantages of photochemical probes in our opinion outweigh the disadvantages. The primary advantage is the employment of an external energy source, light, which gives the experimentation full control over the moment and extent of reaction and also allows for low temperature and time resolution experiments.

One should also be aware of the possibility of doing spacially resolved photolabeling experiments, e.g. by selected irradiation (by a focussed laser beam) of specific regions of an organism, a tissue or even organelles in single cells. But these are still experiments of the (hopefully near) future.

Peter E. Nielsen
Ole Buchardt

Sept. 1988

List of Participants

Dr. Jacqueline K. Barton
Dept. of Chemistry
Columbia University
New York, N.Y. 10027
U.S.A.

Dr. Fuat Bayrakçeken
Spectroscopy Lab. Dept of
Engineering Physics
Ankara University
06100 Tandogan
Ankara
Turkey

Dr. Ole Buchardt
Dept. of Chemistry
University of Copenhagen
H.C. Ørsted Institute
Universitetsparken 5
2100 Copenhagen Ø
Denmark

Dr. Barry S. Cooperman
Dept. of Chemistry
University of Pennsylvania
Philadelphia,
Penn. 19104
U.S.A.

Dr. Alex N. Eberle
Laboratory of Endocrino-
logy, Dept. of Research
University Hospital
Basel
Switzerland

Zhan Gao
Max-Planck-Institut für
Biochemie
8033 Martinsried
West Germany

Dr. Maurice Goeldner
Faculté de Pharmacie
Université Louis Pasteur
67048 Strasbourg - Cedex
France

Dr. Niels Harrit
Dept. of Chemistry
University of Copenhagen
H.C. Ørsted Institute
Universitetsparken 5
2100 Copenhagen Ø
Denmark

Dr. John Hearst *
Dept. of Chemistry
University of California
Berkeley, California 94720
U.S.A.

Dr. Claude Helene *
Laboratoire de Biophysique
Inserm U.201, CNRS ERA 951
Museum National d'Histoire
Na., 43 Rue Cuvier
75231 Paris - Cedex 05
France

Dr. Ulla Henriksen
Dept. of Chemistry
University of Copenhagen
H.C. Ørsted Institute
Universitetsparken 5
2100 Copenhagen Ø
Denmark

Dr. Christian Hirth
Faculté de Pharmacie
Université Louis Pasteur
67048 Strasbourg - Cedex
France

* Organizers

Dr. Claus Jeppesen
Dept. of Biochemistry B
University of Copenhagen
The Panum Institute
Blegdamsvej 3c
2200 Copenhagen N
Denmark

Troels Koch
Dept. of Chemistry
University of Copenhagen
H.C. Ørsted Institute
Universitetsparken 5
2100 Copenhagen Ø
Denmark

Dr. Ernst Kuechler
Institut für Biochemie
Universität Wien
Währinger Strasse 17
1090 Wien
Austria

Dr. Vagn Leick
Dept. of Biochemistry B
University of Copenhagen
The Panum Institute
Blegdamsvej 3c
2200 Copenhagen N
Denmark

Dr. Jorge Marquet
Departamento de Quimica
Universidad Autonoma de
Barcelona, 08193 Bella-
terra, Barcelona
Spain

Dr. Claude Meares
University of California
Dept. of Chemistry
Davis, California 05616
U.S.A.

Dr. Cesare Montecucco
Institute of General Pa-
thology, University of
Padova, Via Loredan 16
35131 Padova
Italy

Dr. Marcial Moreno-Manas
Universidad Autonoma de
Barcelona, Departamento
de Quimica, Quimica Organica
Spain

Dr. Peter E. Nielsen *
Dept. of Biochemistry B
University of Copenhagen
The Panum Institute
Blegdamsvej 3c
2200 Copenhagen N
Denmark

Karsten Olsen
Dept. of Chemistry
University of Copenhagen
H.C. Ørsted Institute
Universitetsparken 5
2100 Copenhagen Ø
Denmark

Dr. Francisco de la Rosa
Departamento de Bioquimica
Facultad de Biologia,
APTDO 1095
41080 Sevilla
Spain

Dr. Dietrich Schulte-Frohlinde
Max-Planch-Institut für
Strahlenchemie, Stiftstrasse
34-36, 4330 Mülheim a.d. Ruhr
West Germany

Dr. Gary Schuster *
School of Chemical Sciences
University of Illinois at
Urbana-Champaign, Box 58,
Roger Adams Lab., 1209 W.
California St.
Urbana Illinois 61801
U.S.A

Dr. Martin A. Schwartz
Dept. of Physiol. and Biophys.
Harvard Medical School
25, Shattuck Street
Boston, Mass. 02115
U.S.A.

Dr. José M. Sogo
Institut für Zellbiologie
ETH-Honggerberg
CH-8093 Zürich
Switzerland

Dr. David R. Trentham
National Institute for Medi-
cal Research, The Ridgeway
Mill Hill, London NW7 1AA
England

Dr. K. Lemone Yielding
Dept. of Anatomy
University of South Alabama
College of Medicine
Mobile, Alabama 36688
U.S.A.

Wei-ping Zhen
Dept. of Biochemistry B
University of Copenhagen
The Panum Institute
Blegdamsvej 3c
2200 Copenhagen N
Denmark

Author index

Keyword Index